NEW COMPLETE GI

Charles Hayes

Chapters 30 and 31: Richard Galvin

Gill and Macmillan

Published in Ireland by
Gill and Macmillan Ltd
Goldenbridge
Dublin 8
with associated companies throughout the world

Design and artwork: Design Image
Cartoons: John Byrne
Editorial Consultant: Roberta Reeners
Print origination in Ireland by
Design & Art Facilities and Irish Photo Ltd, Dublin

0 7171 1605 0

By the same author:
Our World and Its People
Complete Geography

CONTENTS

Preface

I am convinced that most post-primary teachers will find the new geography syllabus refreshing, rewarding and easily manageable.

There is nothing for teachers to fear in it. Behind the sometimes esoteric terminology of the syllabus document there is little which we as practising teachers have not encountered already and taken in our stride.

Where the new syllabus differs from the old is in its discouragement of rote learning. This does not mean that the importance of content is being diminished. The contrary, in fact, is the case. But the content of the new syllabus will no longer be learned merely for its own sake. Instead it will be primarily the medium through which students come to grips with geographical concepts, acquire practical skills and develop desirable attitudes.

To teach the many skills which are now at the core of the syllabus, a text must be visual, varied, activity-centred and choice-oriented. Such a text is unlikely to be slender. But it will be absorbing, fruitful and examination-effective.

This two-part work is specifically designed to meet the demands of the new syllabus. The numerous activities which punctuate the text provide a 'bank' from which you can choose the activities best suited to the individual circumstances of your classes.

Part One of this work covers a carefully-chosen selection of course units, so that it offers variety as well as continuity of material. It provides work for up to one and a half years. Part Two covers the remainder of the course.

Charles Hayes, M.A., M.Ed., HDE
St Mary's High School
Midleton
Co. Cork

Acknowledgments

The author expresses special appreciation of the following, whose active assistance has been invaluable to the completion of this work: Tony Dunne, Holy Child Community School, Sallynoggin, Co. Dublin; Richard Galvin, St Angela's College, St Patrick's Hill, Cork; Patrick O'Dwyer, St Joseph's Convent of Mercy, Doon, Co. Limerick.

The author wishes to thank all those who assisted in the preparation of this volume. Special thanks are due to the staff at Gill and Macmillan, to Roberta Reeners and to the staff at Design Image. Special thanks also to Alex Corcoran for his frequent advice and encouragement.

The assistance of the following is also appreciated greatly: ACOT, *Africa* magazine, *African Studies,* Analog Devices, An Bord Bainne, Bord Gais Eireann, Bord Iascaigh Mhara, Bord na Mona, Central Statistics Office, Comhlamh, Comhlucht Siúcra Eireann, Confederation of Irish Industry, *Cork Examiner,* Department of Energy, Department of Fisheries and Forestry (John Molloy), Department of the Marine, Earthwatch, Embassy of the USSR, Rosaleen and Denis Fitzpatrick, Geography Department at UCC (Barry Brunt and Kevin Hourihan), Industrial Development Authority, International Defence and Aid Fund (London), Irish Congress of Trade Unions, Irish Distillers Group plc, *Irish Times,* Irish Transport and General Workers Union, Marathon Oil Company, Merrell Dow, Ordnance Survey Office, Pfizer Chemical Products, *Sunday Tribune,* Trocaire (Anne Kinsella), UCC Library (Pat O'Connell), Womanagh Valley Protection Association.

For permission to reproduce photographs, acknowledgment is made to the following: Barnaby's Picture Library, The J. Allan Cash Photolibrary, Topham Picture Library, Rod Tuach, *Irish Times*, Bord na Mona, Bord Gais Eireann/Finbarr O'Connell, Swiss National Tourist Office, Bord Failte, Office of Public Works, Electricity Supply Board, Irish Distillers Group plc, University of Cambridge Committee for Aerial Photography, Industrial Development Authority, Pfizer Chemical Products, City of Sheffield/Sheffield Made Products.

1 THE RESTLESS EARTH

Inside the earth

Figure 1.1 shows that the inside of the earth consists of several ***different layers***.

Figure 1.1 The layers inside the earth's crust ▷

The ***outer crust*** of the earth consists of solid rock. But this outer crust is extremely thin (about 65km).

A ***transition zone*** lies immediately inside the crust. It consists of hot, molten rock and semi-solid materials.

The ***mantle*** consists of heavy rocks. They are generally solid, but can flow at times because of the great heat (up to 4000°C).

The ***core*** is made up largely of nickel and iron. Temperatures are highest at the core.

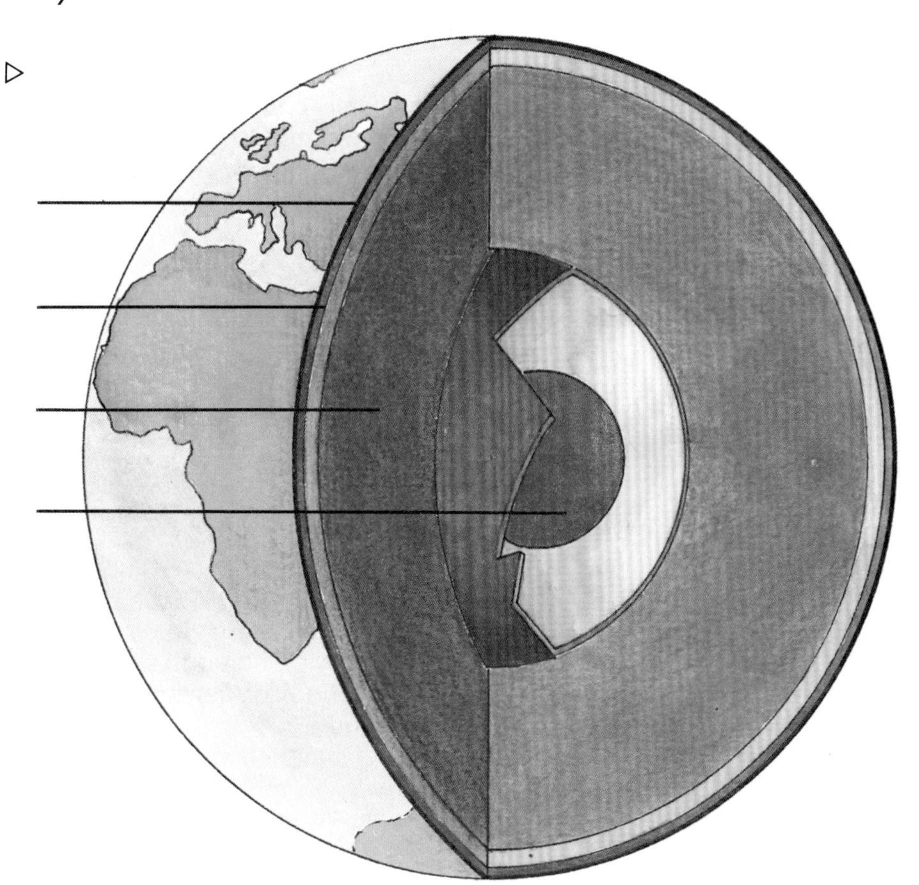

The plates of the earth's crust

The outer crust of the earth can be compared to the hard shell of a nut. But it is like a shell which has been cracked in many places. The broken pieces are known as ***plates*** (figure 1.2).

Figure 1.2 The earth's plates. ▷

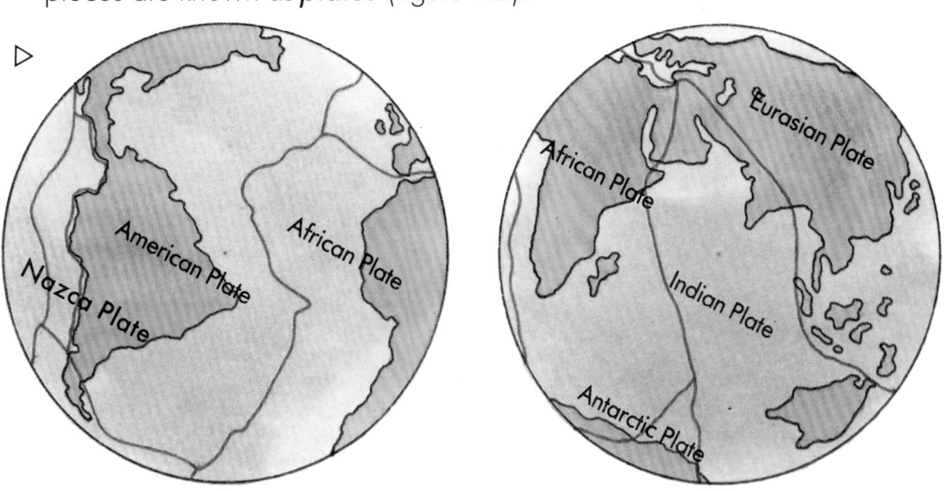

These huge plates of rock:

- ☐ ***float*** on heavier, semi-molten rock beneath them
- ☐ are pushed along by currents of molten rock beneath them. So the plates ***move around*** slowly and carry our continents with them as 'passengers' (see 'Continental Drift')
- ☐ constantly ***collide*** with and ***separate*** from each other. These movements cause ***folding, volcanic activity*** and ***earthquakes***.

Why plates separate and collide

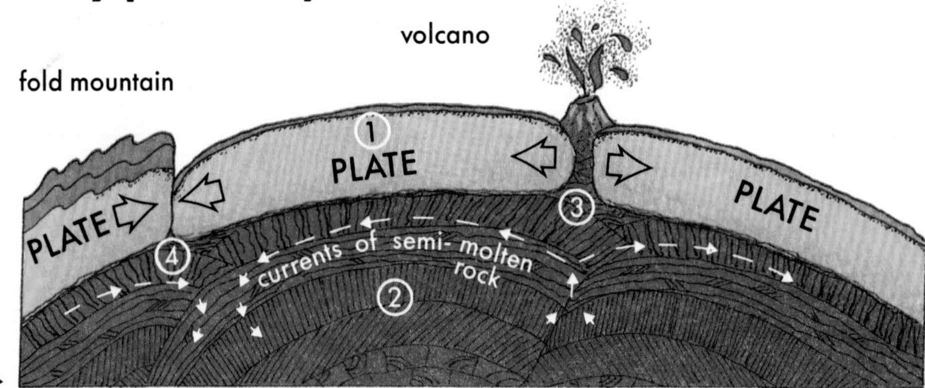

Figure 1.3 ▷

1. The plates of the earth's crust ***float*** on heavier, semi-molten rock (figure 1.3).
2. The semi-molten rock contains ***currents*** which move about very slowly but constantly.
3. The currents sometimes ***drag the plates apart*** and cause them to separate from each other.
4. The currents sometimes ***push the plates together*** and cause them to collide with each other.

☐ *Continental Drift – our moving continents*

Figure 1.4 Continental drift
▽

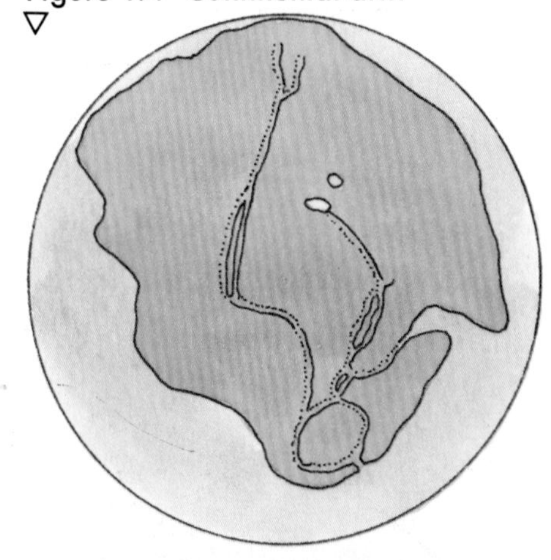

1. About 200 million years ago, all the world's landmasses were joined together in a single landmass.

2. But this huge landmass or continent broke up as its various pieces drifted apart because of plate movements.

3. Continental drift still continues today.

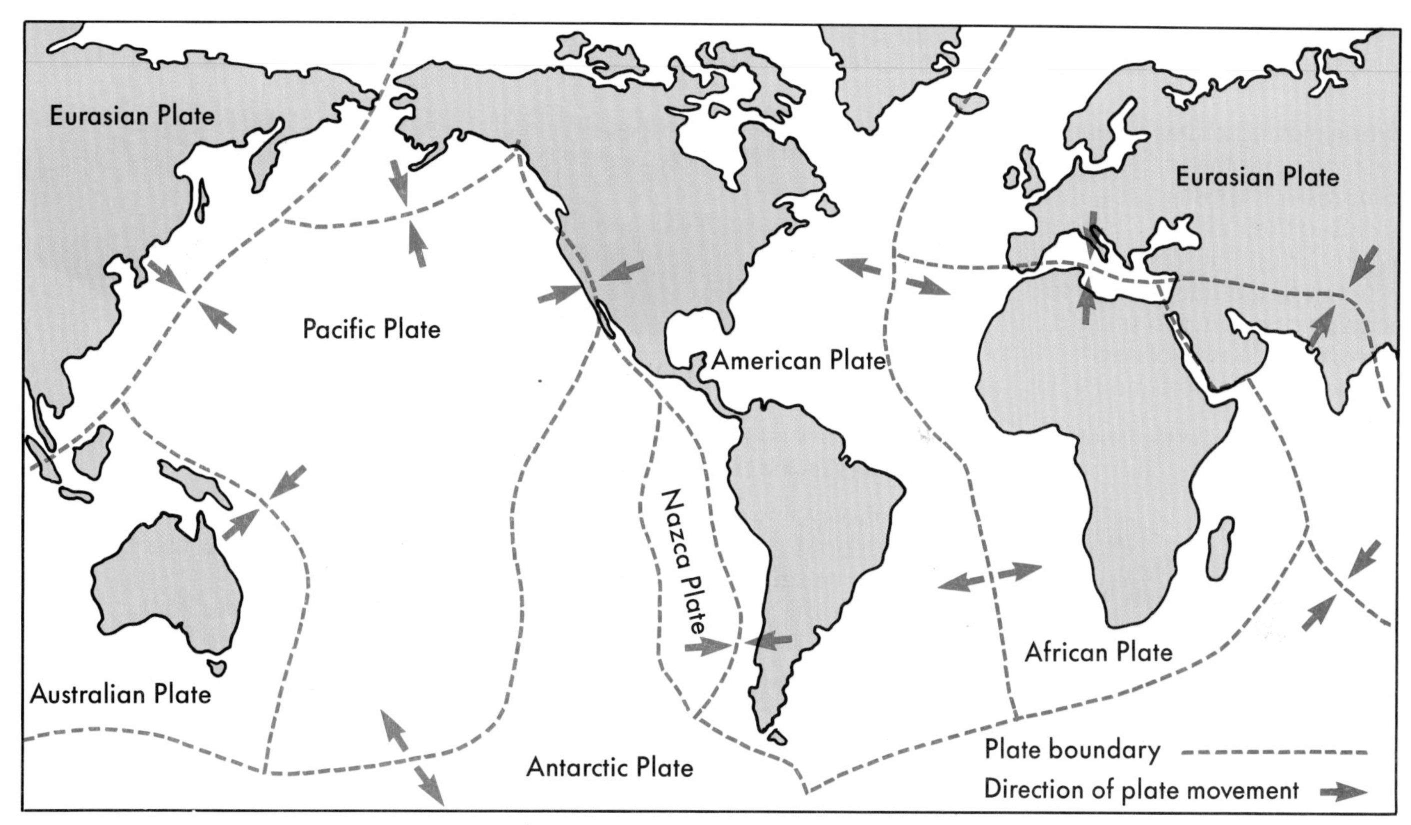

△ **Figure 1.5** The crustal plates, continents and oceans as they are today.

Figure 1.5 shows that, as the plates move slowly, the edges of some places ***separate*** from each other, while the edges of other plates ***collide*** with (or slide against) each other.

When plates separate and collide, many important ***natural activities*** occur. These activities result in the making of some of the biggest ***landforms (physical features)*** on the earth's crust (see box).

Study figure 1.5 and the areas listed below. Mark the appropriate blanks to show the areas in which plates are separating and the areas in which plates are colliding.

	Area of separation	*Area of collision*
California (Western USA)	______	______
Mid Atlantic Ocean	______	______
Himalaya Mountains	______	______
West of the Rocky Mountains	______	______
Mediterranean region	______	______

Your atlas will help you find these places.

Where?	*Activities caused by plate movements*	*Some resulting landforms*
where plates collide	***folding***	fold mountains
where plates separate or collide	***volcanic activity***	volcanic mountains, volcanic plateaux, mid ocean ridges
where plates separate or collide	***earthquakes***	

Folding

When moving plates collide with each other, tremendous ***compression*** (pushing together) occurs at their zones of contact (figure 1.6a). This compression may cause the earth's crust to become ***buckled and arched upwards*** at these zones of contact, forming fold mountains (figure 1.6b).

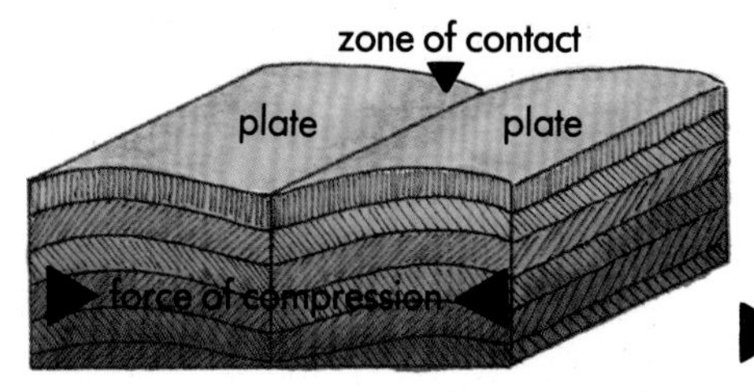

△ **Figure 1.6a The plates collide**

△ **Figure 1.6b Rocks at the point of collision become wrinkled and arched**

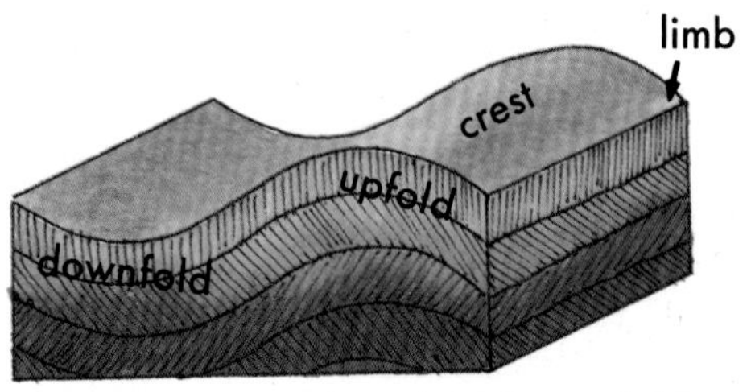

△ **Figure 1.7 A downfold, an upfold, a crest and a limb**

Fold mountains

Fold mountains sometimes occur in the form of ridges called ***upfolds*** and depressions called ***downfolds***. The top of an upfold is called its ***crest***. Each of its sides is called a ***limb*** (figure 1.7).

The world's youngest fold mountains include the Alps, the Rocky Mountains and the Himalayas. These mountain ranges were formed during the ***Alpine foldings*** only about 35 million years ago. They are very high because they have not yet become worn down as severely as other fold mountains.

Ireland's fold mountains were formed hundreds of millions of years ago. These fold mountains are so old that they have been worn down to quite low heights. Our most recently-folded mountains occur throughout Munster. They were formed during the ***Armorican foldings*** about 250 million years ago.

△

Small-scale rock folding. This rock was once laid down in horizontal (flat) layers, but it later became buckled because of folding.

Draw a *simple sketch* of this picture. Show and label an *upfold*, a *downfold* and a *limb*.

△

△

These photographs show Ireland's highest mountain, the Macgillicuddy's Reeks (left) and North America's highest mountain, the Rocky Mountains (right).

- ☐ **Which mountains are the highest? Why?**
- ☐ **Which two plates collided to form the Rocky Mountains?**

Volcanic activity

△ A volcano erupting in Hawaii. Volcanic ash produces fertile soil for farms on the lower slopes of the mountain. The volcano is also an important tourist attraction. But violent eruptions can still endanger the lives and property for those living nearby.

Beneath the earth's crust there is hot, liquid rock called ***magma***. Where plates separate from or collide with each other, the magma can sometimes force its way up through cracks in the crust until it reaches the surface. When the magma reaches the surface, it cools and hardens. It is then called ***lava***.

- ☐ Lava may pour quietly through long cracks in the earth's surface. Where this happens, it may build up high, flat ***lava plateaux*** such as the Antrim and Derry plateau. The lava may also build up ***mid ocean ridges*** such as the Mid Atlantic Ridge.
- ☐ Lava may also force its way violently through a small hole called a ***vent***. When this happens, a ***volcanic mountain*** is formed.

The Mid Atlantic Ridge

Deep below the middle of the Atlantic Ocean there lies a long, narrow chain of mountains called the Mid Atlantic Ridge. This ridge runs roughly in a north-south direction, with some of its peaks rising above the surface of the sea to form volcanic islands.

The Mid Atlantic Ridge is volcanic in origin. It lies along a zone where the American and the African/Eurasian crustal plates are slowly moving away from one another. Figure 1.8 shows how the Mid Atlantic Ridge was formed.

Study a wall map or atlas for signs of the Mid Atlantic Ridge. Try to find some of the islands which form part of the ridge.

Figure 1.8 How the Mid Atlantic Ridge was formed ▽

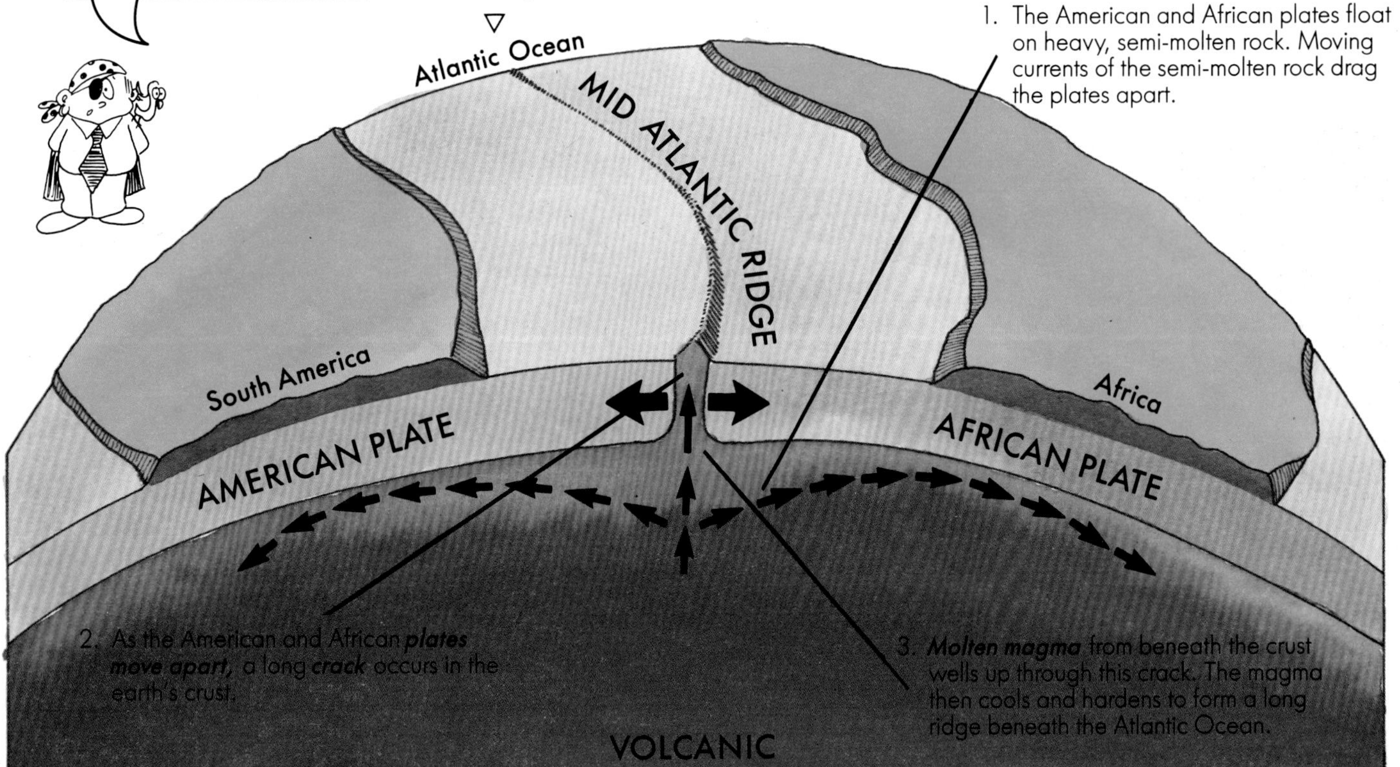

Iceland – Land of Fire and Ice

Did you know . . .?

- ☐ Iceland is not only a place of winter snow and ice. It is also a volcanic island which contains ***volcanic plateaux*** and ***mountains***.
- ☐ Volcanic activity causes ***hot-water springs*** to rise from the ground in parts of Iceland. This water is used to provide central heating for houses in Reykjavik, Iceland's capital city.

△
What evidence in this photograph suggests that Iceland experiences volcanic activity?

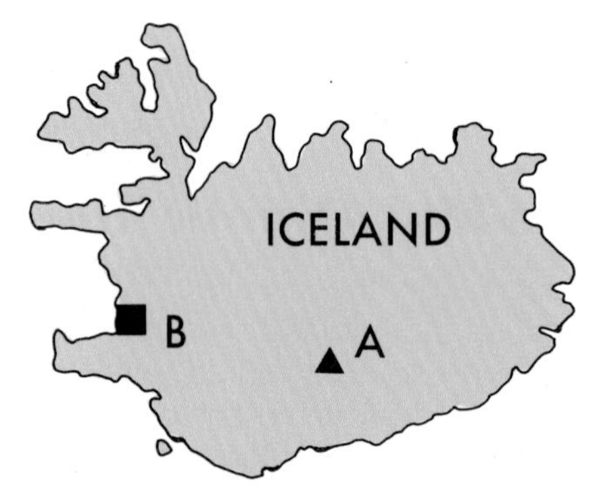

◁ **Figure 1.9**

- ☐ **Explain why Iceland is volcanic. Use your atlas and figure 1.8, if necessary.**
- ☐ **Use your atlas to identify the Icelandic volcanic mountain (A) and the city (B) in figure 1.9.**

Volcanic mountains – their formation

Figure 1.10 The formation of a volcanic mountain
▽

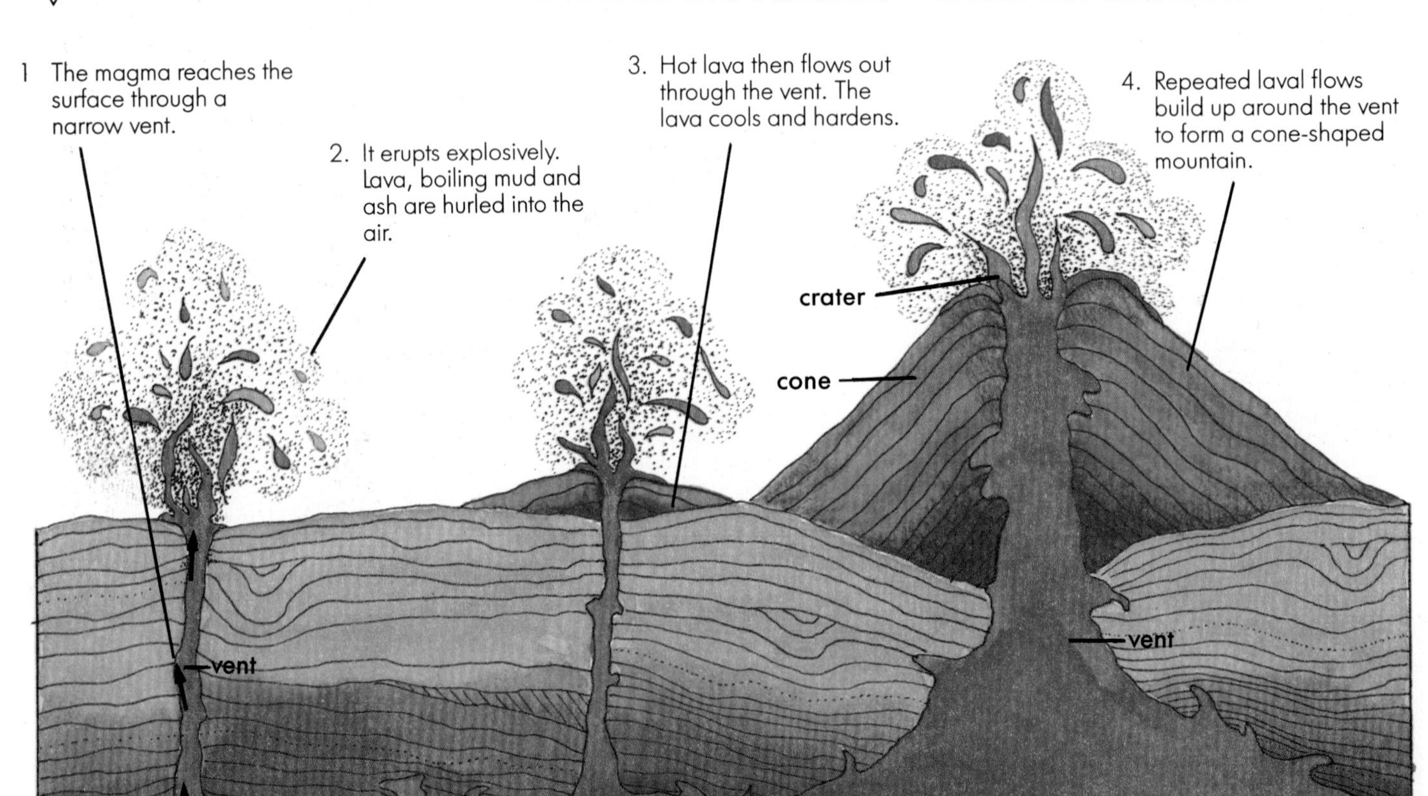

A terrible eruption – Vesuvius destroys Pompeii

The ancient Roman city of Pompeii was completely destroyed in the year AD79 when nearby Mount Vesuvius erupted. This passage, which is based on the account of a survivor of that terrible eruption, describes the destruction of the once-beautiful city. Read the passage and then answer the questions.

August 24 began like any other day, with the citizens of Pompeii going about their usual business. Then, in the early hours of the afternoon, a terrifying roar was heard throughout the city. Vesuvius was erupting.

The eruption was so extreme that one side of the mountain split open suddenly. A black cloud rose from the volcano, showering down hot cinders and fragments of lava. Then a heavy cloud of deadly sulphur fumes came tumbling down from the direction of the mountain. This poisonous gas penetrated every part of the city, suffocating those who sought shelter indoors. A minority, those who tried to escape by fleeing into the open, also suffered heavy casualties as they were hit by fragments of lava and stone.

There then came a sea of volcanic mud and lava. For three days it poured from the open wound in the mountain. Then, after the third day, the sun emerged onto a desolate landscape. Pompeii had disappeared under more than twenty feet (6 metres) of volcanic matter.

Types of volcanoes

Active	still erupt regularly	Etna and Vesuvius (Italy)
Dormant	have not erupted for a long time, but may erupt again	
Extinct	have not erupted in historic times	Slemish, Co. Antrim

◀ 1. How many years ago was the city of Pompeii destroyed?
2. Which active volcano destroyed the city?
3. How did the citizens first realise that the volcano was erupting?
4. What evidence do we have that this eruption was a severe one?
5. How did most of Pompeii's people die?
6. For how long did this eruption last?
7. Why did Pompeii 'disappear'?
8. Vesuvius is an active volcano. It erupts because it is situated in an unstable part of the earth's crust where two crustal plates collide. Name these two plates.
9. Draw a simple map of Italy. Show and name Vesuvius and one other active Italian volcano.

△ The ruins of Pompeii, with Vesuvius in the background.

Earthquakes

Earthquakes are sudden vibrations or ***tremors in the earth's crust***.

An eye-witness account of a strong earthquake

The quake lasted well over a minute – officially 120 seconds – as walls waved violently back and forth like giant fans, buildings wobbled from side to side like jelly, towers and balconies fell into the streets, and plaster and dirt rained everywhere.

Screaming, fainting and running people created a sense of fear and confusion as the electricity went out, and the waves of tremors continued to undulate under the floors, buildings, everywhere, everything. People ran out into the surrounding patios and laid out those who had fainted or who were injured on the ground. The mountain range in the background, the towering Andes, sent up clouds of billowing dust from the landslides taking place at that very moment for hundreds of kilometres along its flanks.

The quake reached 7.7 on the Richter scale in the capital and claimed one hundred and forty-five lives. Nearly one million were left homeless, in a country where the housing situation is already critical. Block after block of what is known as 'Old Santiago', comprised entirely of two- or three-storey houses and buildings built fifty or sixty years ago, simply crumbled into heaps and left a scene of a city bombarded.

(from *Far East* magazine, November 1985)

1. How long did the earthquake last?
2. Describe its effects on people, buildings and mountains.
3. What is meant by the phrase: 'the quake reached 7.7 on the Richter scale'?
4. In which country did this earthquake occur? Which parts of the account help to identify the country?
5. Which two crustal plates collide near the country identified in 4 above?

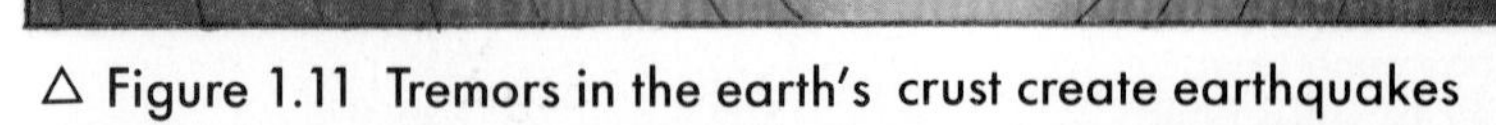

△ Figure 1.11 Tremors in the earth's crust create earthquakes

California in the USA has suffered severe earthquakes because it lies close to where the Pacific and American plates collide. In 1906, for example, a strong earthquake rocked the city of *San Francisco*. Buildings collapsed. Gas pipes were broken, causing fires which destroyed much of the city.

Compression is known to be building up at present in the earth's crust beneath San Francisco. Many fear that the city will experience another severe earthquake. Wide streets and specially-reinforced buildings may lessen the destructive effects of another earthquake.

Locate San Francisco in your atlas.

Earthquakes, volcanoes and Ireland

Figure 1.12 shows that earthquakes and active volcanoes occur near the meeting places of the earth's great crustal plates. The largest earthquake and volcano zone lies along the edges of the Pacific Ocean. This zone is known as the ***Pacific Ring of Fire***.

Ireland is situated ***far away from the meeting places of plates***. It ***suffers little*** from earthquakes and volcanic activity.

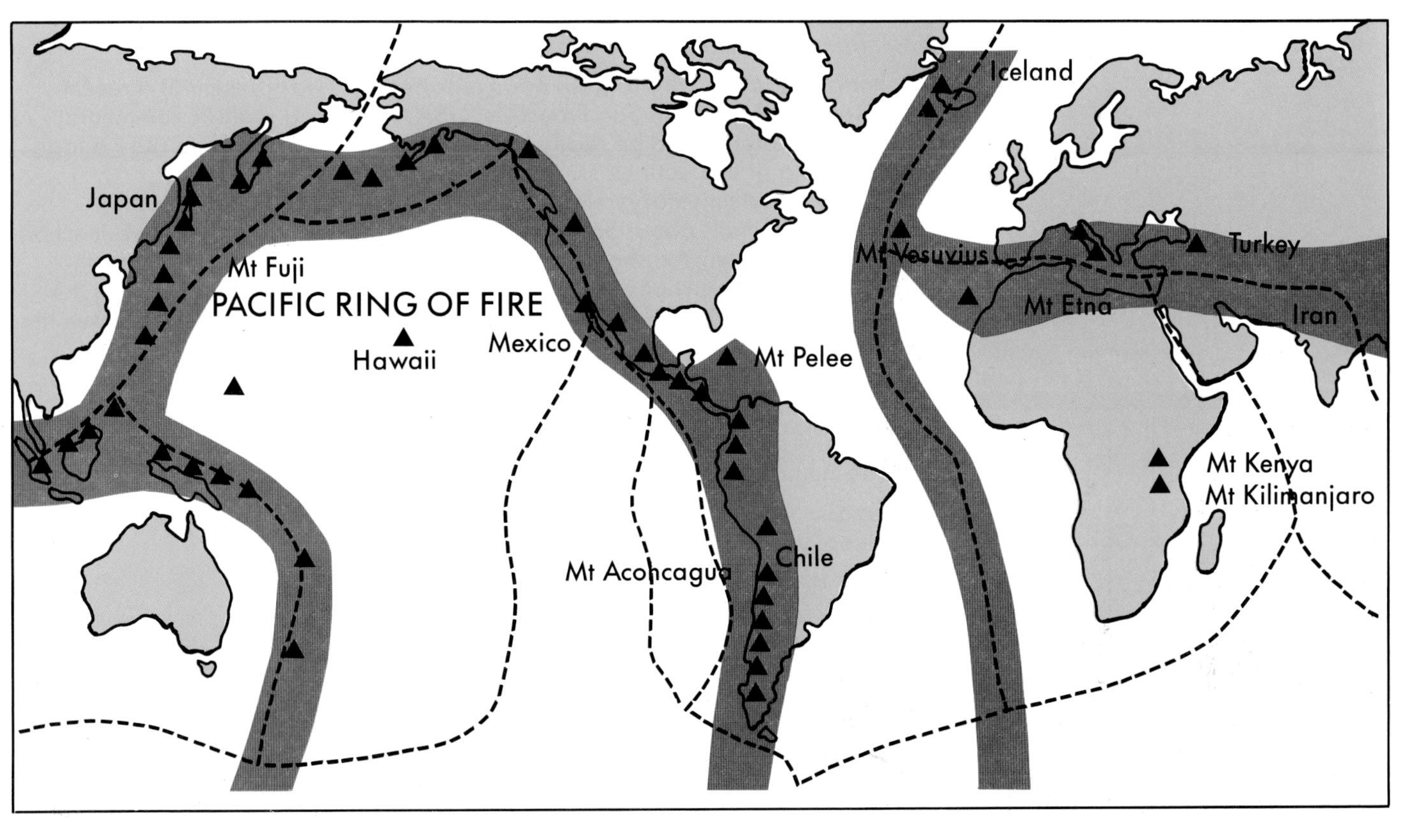

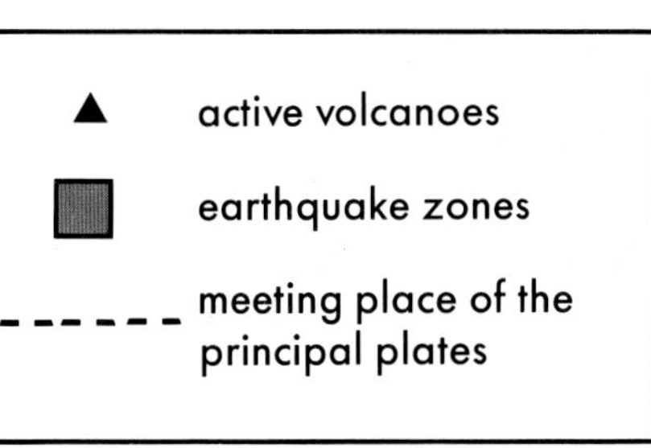

△ Figure 1.12 The location of active volcanoes and earthquake zones

The Pacific Ring of Fire

1. Why do you think this area is called a 'ring of fire'?
2. Why do volcanoes, earthquakes and fold mountains occur near the Pacific Ring of Fire? Answer precisely. Consult figure 1.5 if necessary.
3. Which three major continents are bounded by the area?
4. Name ***ten countries*** on or bounding the Pacific Ring of Fire.

Activities

1. (a) Name one place in the world where crustal plates separate, and one place where crustal plates collide.
   (b) Name one major landform which was formed as a result of plates separating and one formed as a result of plates colliding.
   (c) In the case of each of the landforms you have named: (i) describe its appearance; (ii) describe how it was formed. (You may use diagrams to illustrate your answer.)

2. *Fold Mountain, volcanic island*
   (a) Describe each of these features.
   (b) Give an example of a fold mountain range in Ireland and a volcanic island in the Atlantic.
   (c) With the aid of a diagram, explain how *one* of these features was formed.
   (d) Describe the effects of a very severe earthquake on a densely populated region.

3. *Information retrieval and oral skills – a project*

   Imagine that you are a radio news reporter who was present at a recent severe earthquake in San Francisco, USA. Prepare a realistic commentary of your experience and be ready to read your report aloud or to play a tape recording of it in your class.

   In your commentary, refer to some accurate geographical and/or historical details about San Francisco. (Refer to encyclopedias in your school or local library.) You should also refer to the various effects of the earthquake, to the reactions of the people and to your own feelings during the quake. Conclude your report by stating what needs to be done to lessen the effects of this disaster for both the people and the environment.

4. *A Word Game*

   Down:
   2 The method by which the strength of an earthquake is measured.

   Across:
   1 Sleeping volcano.
   3 Opposite of compression.
   4 Young fold mountains in North America.
   5 A volcanic mountain in Iceland.
   6 In Antrim and Down there is a lava _ _ _ _ _ _ _.
   7 Magma may surface through this narrow hole.
   8 The top of an upfold.
   9 May be caused by an earthquake.
   10 The centre part of the earth.
   11 The top of a volcanic mountain.
   12 The side of an upfold.
   13 An extinct volcano in Ireland.

5. *Mapwork*

   Study the Slieve Mish (fold) mountains on the Tralee Bay Ordnance Survey sheet in the Map Supplement which accompanies this book.
   (a) In which direction do the Slieve Mish mountains run? From which direction(s), therefore, were these mountains compressed?
   (b) During which mountain-folding period were the Slieve Mish mountains formed? Approximately how old are the mountains? How high is their highest peak?

2 ROCKS

The earth's crust is made up of many kinds of *rocks*. Rock types differ from each other in a number of ways.

- ☐ Rocks have different ***physical characteristics***. The ***colour*** of one rock type may be different from another type. Rock types differ in their ***hardness***, in their ***density*** (weight in relation to size) and in their ***texture*** (the way they feel). Some rocks, such as peat, are powdery. Others, such as oil, occur as liquids.
- ☐ Rocks vary according to their ***origins***. Some began their existence as hot lava, while others started out as crushed sea shells or as heaps of dead vegetation.
- ☐ Rock types are ***usually divided into three groups, depending on how they were formed***.

Group	*How Formed*	*Examples*
IGNEOUS	They were formed when hot, molten ***volcanic material cooled down*** and became solid.	granite, basalt
SEDIMENTARY	They were formed from the ***compressed remains*** (sediments) of animals, plants and other rocks.	limestone, coal, sandstone, shale
METAMORPHIC	They were once igneous or sedimentary rocks which were ***changed by great heat or pressure***.	marble, quartzite

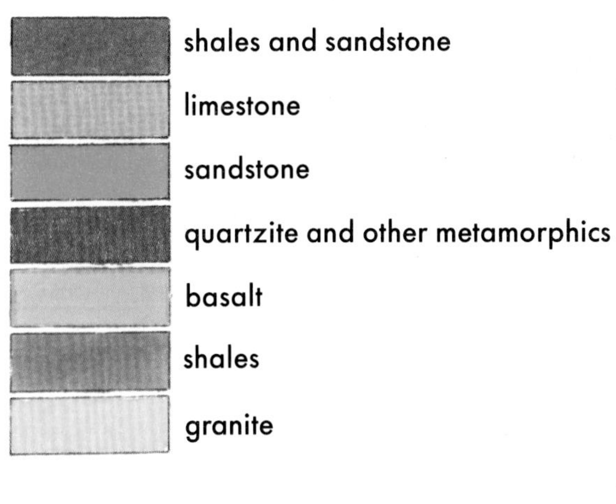

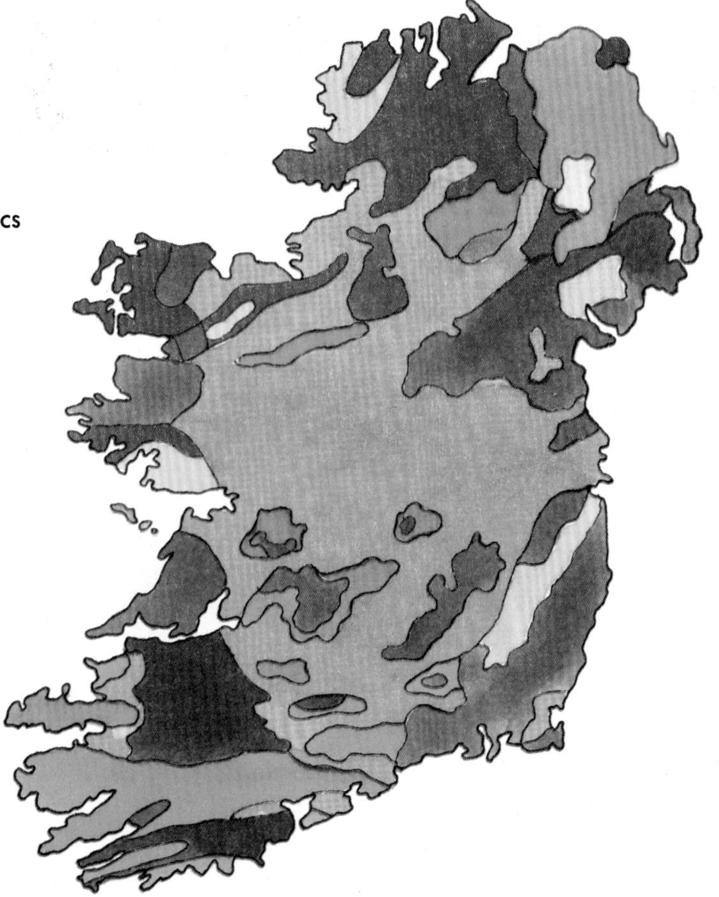

Figure 2.1 This is a geological map ▷ of Ireland. It shows the general distribution of the most common rock types. Which rock is most common: (a) in Ireland; (b) in your own county; (c) in the Antrim Mountains; (d) in the Kerry Mountains; (e) in the Wicklow Mountains; (f) in the Burren (north Clare); (g) in the Central Plain?

Some igneous rocks

Granite

△ Granite

Granite is a ***hard, coarse, multi-coloured rock.*** It contains pink or grey feldspar and crystals of mica and quartz.

It formed when magma cooled deep within the earth's crust. The magma cooled so slowly that large crystals had time to form.

Granite is found in the Wicklow and Mourne mountains.

Basalt

Basalt is a ***heavy, black rock.***

It formed when lava cooled at the earth's surface. The lava cooled too rapidly for large crystals to form.

Basalt is found on the Antrim Plateau and at the Giant's Causeway.

Basalt formations at the Giant's Causeway in Co. Antrim.
This former lava cooled very rapidly on the earth's surface. As it cooled and dried, the lava cracked. The cracks formed a uniform pattern and split the rocks up into thousands of *regular-shaped columns.* ▷

Some sedimentary rocks

There are many types of sedimentary rocks, each of which was formed from the crushed-together remains of different materials. Ireland's main sedimentary rocks are outlined in the box.

Type	*Formed from*	*Characteristics*	*Where found*
LIMESTONE	remains of ***sea creatures***	usually smooth, grey and well jointed	Most of Ireland, especially the Central Plain and the Burren, Co. Clare
COAL	***decayed vegetation***	soft and black	Arigna, Co. Leitrim; Beneath Dublin Bay
SANDSTONE	***grains of sand***	coarse and usually red/brown	The mountains of Co. Cork and Co. Kerry
SHALE	***particles of mud***	smooth and usually dark	Co. Clare and Co. Limerick

Limestone – Ireland's most common rock

Origin

Limestone is made from the ***remains of fish and other sea creatures.*** As generations of these creatures died, their ***skeletons*** were piled up on the beds of shallow seas. The skeletons were crushed by the weight of later deposits and ***cemented*** together by the sea water until they formed slowly into solid rock.

Characteristics

Figure 2.2 The characteristics of limestone

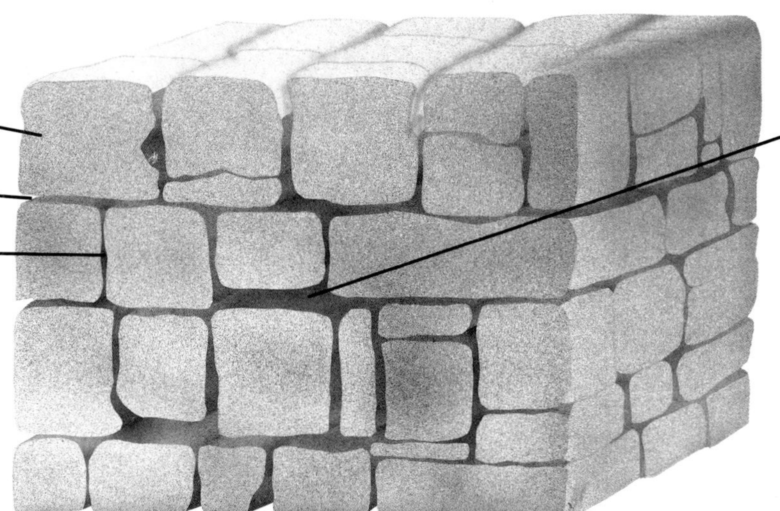

5. Limestone may contain ***fossils.*** A fossil is the preserved remains of a plant or animal.

A fossil.

St. Patrick's Cathedral in Dublin is made mainly of limestone.

Uses of limestone

Limestone is used in many trades.

- ***Manufacturers*** use it to make headstones and cement. It is also used in the production of iron and steel.
- ***Builders*** use blocks of limestone to make public buildings. Limestone chippings are used to surface roads.
- ***Farmers*** use ground-up limestone as a soil conditioner.

Some metamorphic rocks

Igneous or sedimentary rocks can sometimes be changed completely when they come into contact with great ***heat*** (from magma) or with great ***pressure*** (due to folding). These rocks can be changed into hard, metamorphic rocks such as marble or quartzite.

limestone

changes to

marble

A crystalline rock which is used widely as an ornamental stone. It may be white (Rathlin Island), green (Connemara), red (Cork) or black (Kilkenny).

sandstone

changes to

quartzite

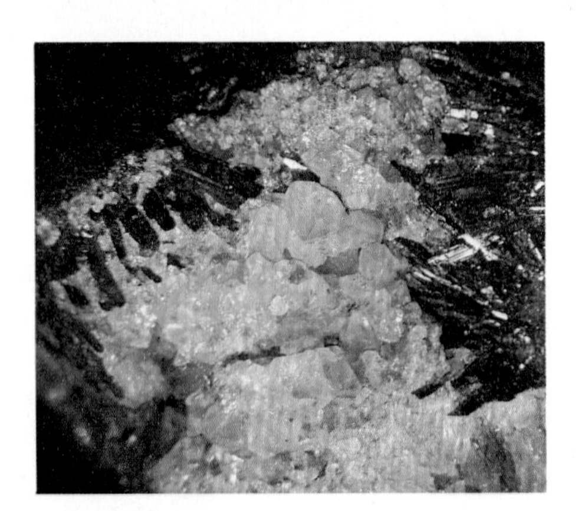

A light-coloured, hard rock often used to surface roads. It occurs at the tops of many hills and mountains such as Errigal (Co. Donegal) and the Hill of Howth (Co. Dublin).

Rocks and people

Valuable natural resources

Rocks provide us with many of the ***natural resources*** (useful natural materials) which we need for our comfort and survival. These resources include:

- ☐ ***Common rocks*** such as limestone.
- ☐ ***Common minerals*** contained in rocks. Such minerals include iron ore (from which iron is made) and quartz (used in making watches and radios).
- ☐ ***Rare minerals*** such as gold and silver.
- ☐ ***Fuels*** such as turf, coal, oil and gas.

Copy the table in figure 2.3 into your geography copybook. Use the information in figure 2.4 to complete the table. Some answers have been filled in to help you get started.

Natural resource	*Example of natural resource*	*Two products of the natural resource*	
A. Common rocks	1. limestone	1. cement	2. ________
	2. ________	1. ________	2. ________
B. Common minerals	1. ________	1. ________	2. ________
	2. ________	1. ________	2. ________
C. Rare minerals	1. ________	1. ________	2. ________
	2. ________	1. ________	2. ________
D. Fuels	1. ________	1. ________	2. ________
	2. ________	1. ________	2. ________

Figure 2.3

RESOURCES

angrite
nori roe
dogl
qutazr
loi
letmiseno
cola
dimdanos

Figure 2.4 The products of natural resources

All mineral resources are ***non renewable***, though some can be ***recycled***.

What do these terms mean?

Extracting the rocks

Removing the rocks and mineral resources from the earth's crust is known as an ***extractive industry***. The removal or extraction of rock and mineral deposits can be carried out in three different ways (figure 2.5).

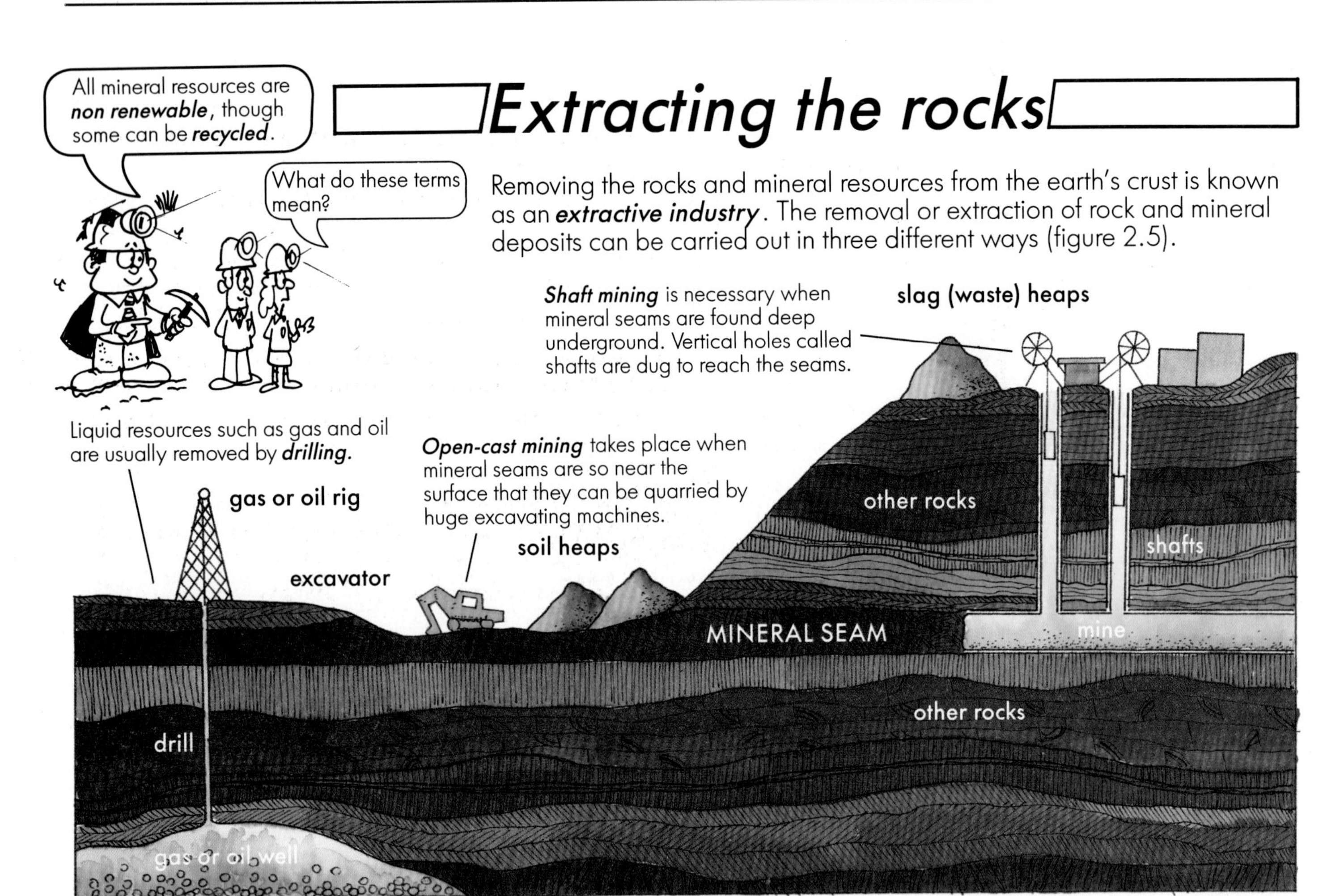

△ Figure 2.5 Methods of mining

Which method of mining is:

- ☐ most dangerous?
- ☐ least expensive?
- ☐ most likely to damage the surface environment? . . . and WHY?

See Activity3(c)–The study of a local quarry, page 18.

Some fuel-providing rocks in Ireland

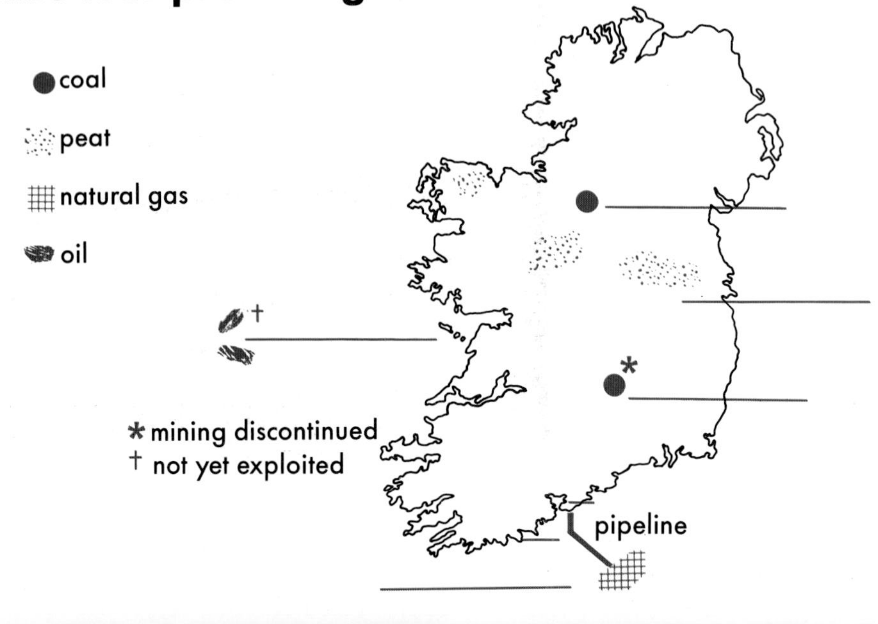

Figure 2.6 The main Irish locations of various energy-giving rocks

On the spaces provided, fill in the following places: Arigna; the Bog of Allen; Ballingarry; Kinsale field; Porcupine bank. What type of fuel is present at each of these locations?

LOCAL STUDIES–See Activity 3, page 18.

Activities

1. Limestone, quartzite, granite, basalt, coal and marble are examples of six types of rock found in Ireland.
   (a) Classify (group) the six rock types as igneous, sedimentary or metamorphic.
   (b) Draw or trace an outline map of Ireland. On the map, show and name *one* location at which *each* of the six rock types may be found.
   (c) Describe the formation of limestone.
   (d) Limestone is a permeable, jointed, layered and easily-eroded rock. Explain the meaning of each of these four terms.
   (e) State how limestone could be used by a farmer, a manufacturer and a builder.

2. *The main rock groups and change*

   Figure 2.7 illustrates change and the main rock groups. Study the diagram and then answer these questions.
   (a) Which group of rocks was formed first?
   (b) Which group or groups of rocks may be formed from either of the other two groups?
   (c) Which rock group owes its formation to heat and pressure?
   (d) In which ways, not shown in figure 2.7, may sedimentary rocks be formed?
   (e) Suggest a suitable title for the arrow marked *A*.

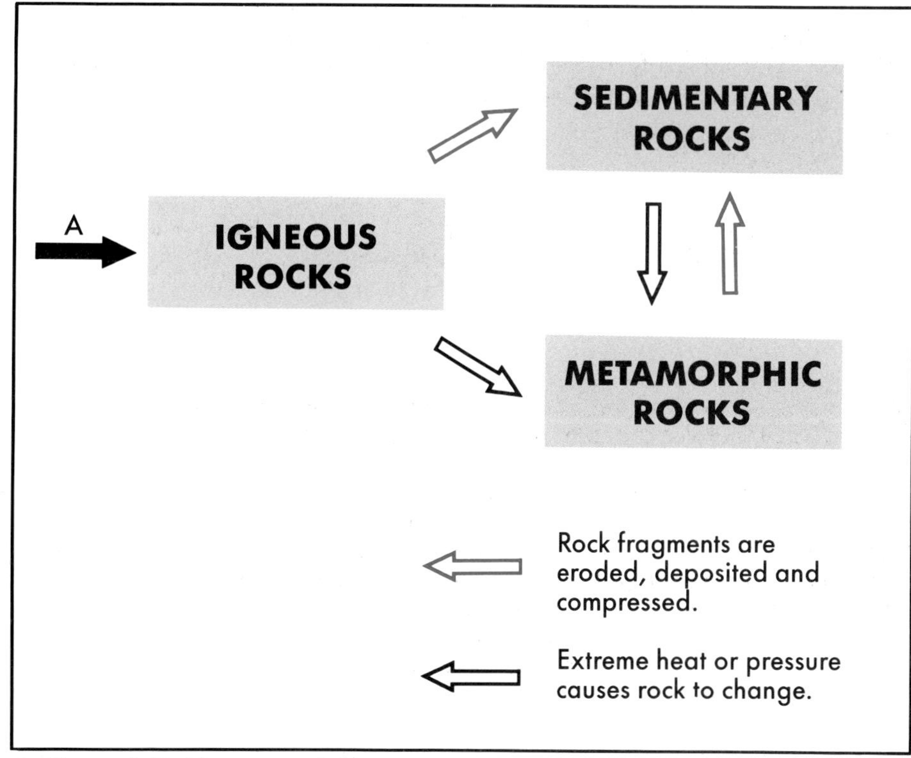

△ Figure 2.7 Change and the main groups of rocks

3. *A selection of field activities*

(a) Collect a sample of the most common type of rock in your area. Find out all you can about its formation, characteristics, uses etc. Bring the sample to school and be prepared to tell your class what you have learned about it.

or

(b) Find out which types of stone are being used or have been used in your area for public buildings, road surfacing, walls, fireplaces, headstones etc. Find out where each type of stone was quarried.

or

(c) *Group field work–The study of a local quarry or sand/gravel pit*
Arrange an accompanied visit to a local quarry or sand/gravel pit which is currently in use. Find out all you can about the following during your visit.

- ☐ The type of rock being quarried: its origins, characteristics, how it was formed.
- ☐ How the rock is quarried: methods used and machinery.
- ☐ Processing: is the rock processed after quarrying? If so, how and why?
- ☐ What is the rock used for? Where is it being used? How is it delivered? What is its selling price?
- ☐ How is the rock suited to its particular uses?
- ☐ Working at the quarry: types of work, any dangers etc.
- ☐ Good and bad effects of the quarry on the locality.

If possible, take photographs of the quarry. Collect small rock samples during your visit.

After your visit, produce a written account of your findings. Illustrate your report with drawings, photographs, diagrams, rock samples and a labelled map showing the location of the quarry. Quarrying and processing methods could be presented in the form of a flow chart.

WARNING! Be sure you are accompanied by an adult on your visit to the quarry. While at the site, obey the directions of your guide. Watch out for dangers such as cliff edges, moving machinery and loose rocks.

△ At work in a quarry.

4. Imagine that plans have been announced to open a large open-cast or shaft mine in your area. Write a short article for a local newspaper explaining why you are either in favour of or opposed to this type of development. Your article should be carefully and clearly written. You may use a sketch map to help make your point.

3 WEATHERING AND EROSION: AN INTRODUCTION

Rocks at the earth's surface are constantly being ***worn away*** or ***denuded*** by the forces of ***weathering and erosion***.

Types of weathering

Mechanical weathering breaks up rocks into smaller pieces. Its main agents (causes) are:

- ☐ *frost*
- ☐ *sudden temperature changes*
- ☐ *plants and animals*

Chemical weathering causes rock to dissolve or otherwise ***decompose***. Its main agent is:

- ☐ *rain.*

△ Figure 3.1

Denudation is carried out by . . .

Weathering
This is the simple ***breaking down of rocks*** which lie exposed to the weather. There are two types of weathering: ***mechanical*** and ***chemical*** (figure 3.1).

and . . .

Erosion
This involves the breaking down of rock and the carrying away of rock particles. Erosion is caused by ***moving water*** (seas and rivers), ***moving ice***, and ***moving air*** (wind).

Transport

Deposition
Materials which are carried away by the forces of erosion are eventually ***dropped*** (deposited) in other areas.

◁ This 'young' river is eroding rock from a mountain . . .

. . . and depositing the eroded rock particles on lowlying ground. △

Activity: Explain the main differences between *weathering* and *erosion*.

4 MECHANICAL WEATHERING: THE EFFECTS OF FROST

In Chapter 3, we learned that frost is one of the chief agents of mechanical weathering.

How frost can weather rocks

1. By day . . .

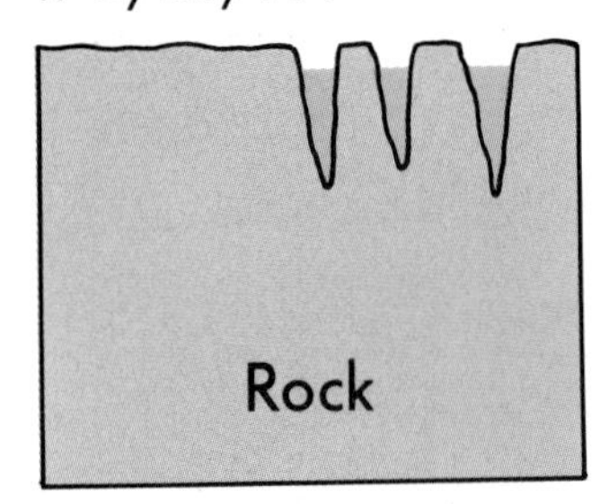

. . . *water collects* in cracks in rocks.

2. By night . . .

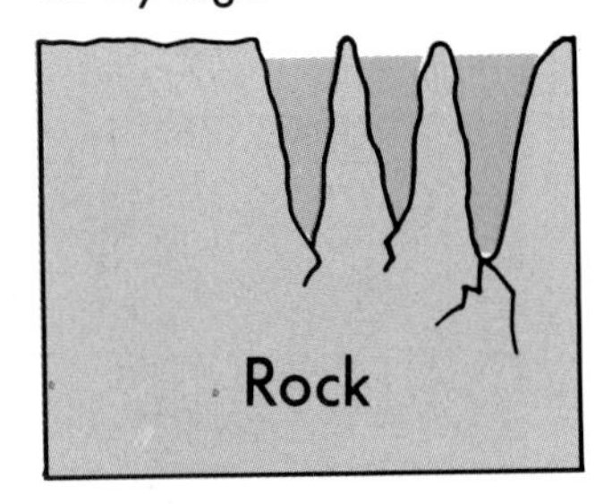

. . . when temperatures drop to 0°C, the *water freezes* and turns to ice. As it does so, it expands by about 10%. This widens the cracks and puts a *strain* on the rock.

3. Eventually . . .

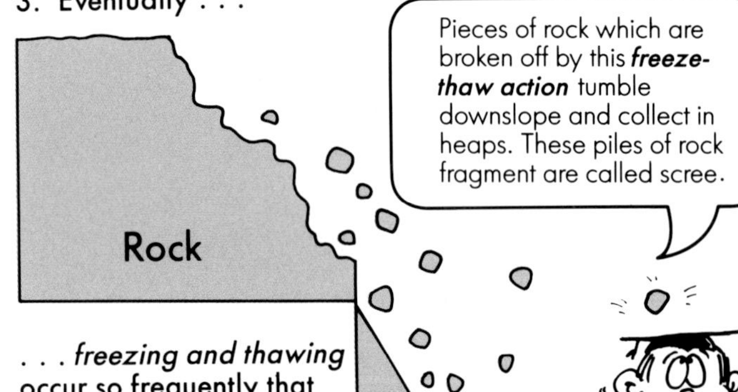

. . . *freezing and thawing* occur so frequently that the rock is gradually split and broken up.

◁ 'The King', a mountain in Norway. The jagged top of this mountain is a result of weathering by frost. Identify the scree and give its location on the mountain.

See Activity 1, page 21.

Where frost weathering occurs

Frost action occurs in areas where:

☐ ***temperatures*** frequently rise above and fall below freezing point

and

☐ ***precipitation*** (rain, snow etc.) is sufficient to provide plenty of water.

Such areas include the mountains of Ireland, where frost action is common in winter. Frost weathering is particularly severe in the great glaciated mountain regions of the world. These regions include the Himalayas, the Alps and the Rocky Mountains.

Activities

1. *Areas where frost action occurs – A word and map game*

(a) Write the names of *8 Irish* and *8 European* mountain areas in the blanks in figure 4.1a. (The first letter of each area is already written.)

(b) Name each of these areas, shown in the maps in figure 4.1b.

Figure 4.1a ▷

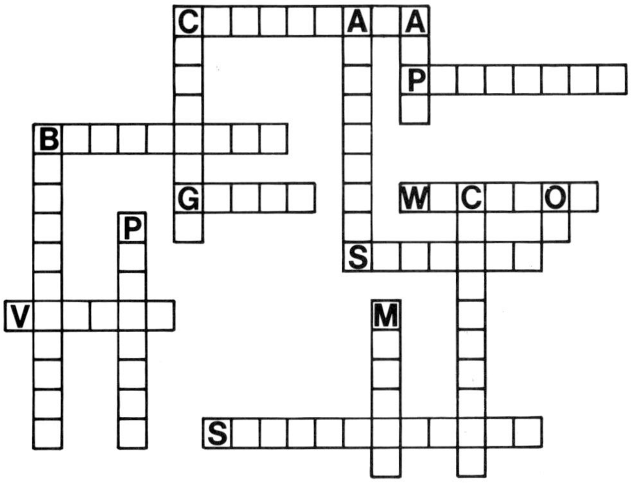
C
A
A
P
B
G
W
C
O
P
S
V
M
S

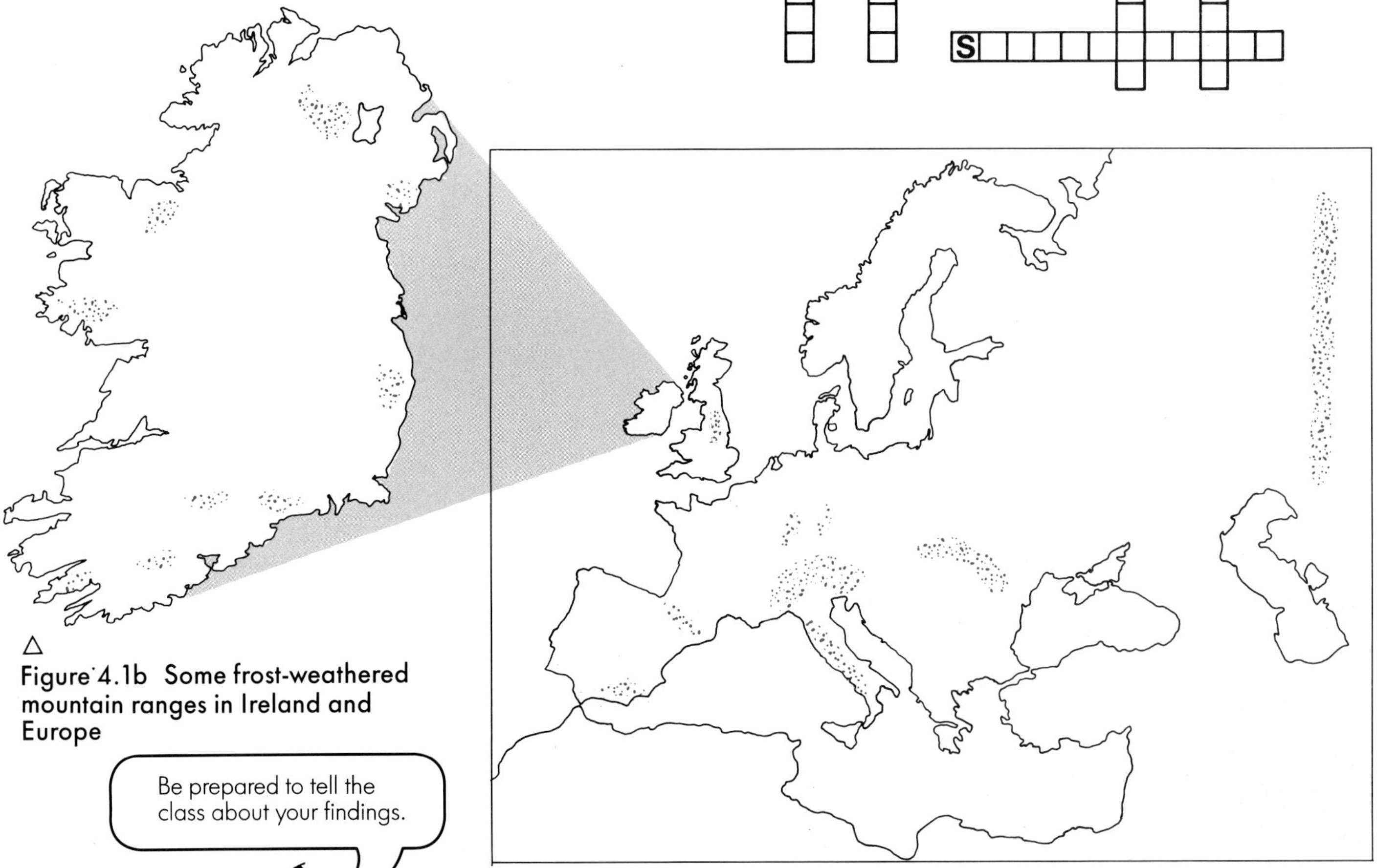

△
Figure 4.1b Some frost-weathered mountain ranges in Ireland and Europe

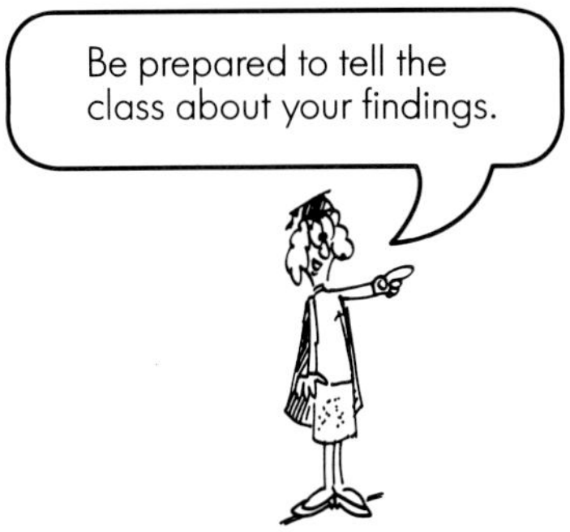
Be prepared to tell the class about your findings.

2. *Find out for yourself!*

(a) Why do people insulate water pipes in frosty weather?

(b) Why do people put anti-freeze in their cars during winter? What might happen if they didn't?

(c) Why do many people dig their gardens at the beginning of winter?

5 CHEMICAL WEATHERING BY RAINWATER: THE BURREN AREA

In the northern part of Co. Clare there is a limestone area known as the Burren (figure 5.1). It is a rare and famous ***karst*** area – a place where the soluble limestone is exposed at the surface.

The Burren contains some very unusual scenery because the limestone is severely weathered by rainwater (see box 'How limestone is weathered by rain water'). In this chapter we shall examine the effects of weathering by rain water both on and beneath the limestone surface.

Figure 5.1 The Burren of Co. Clare

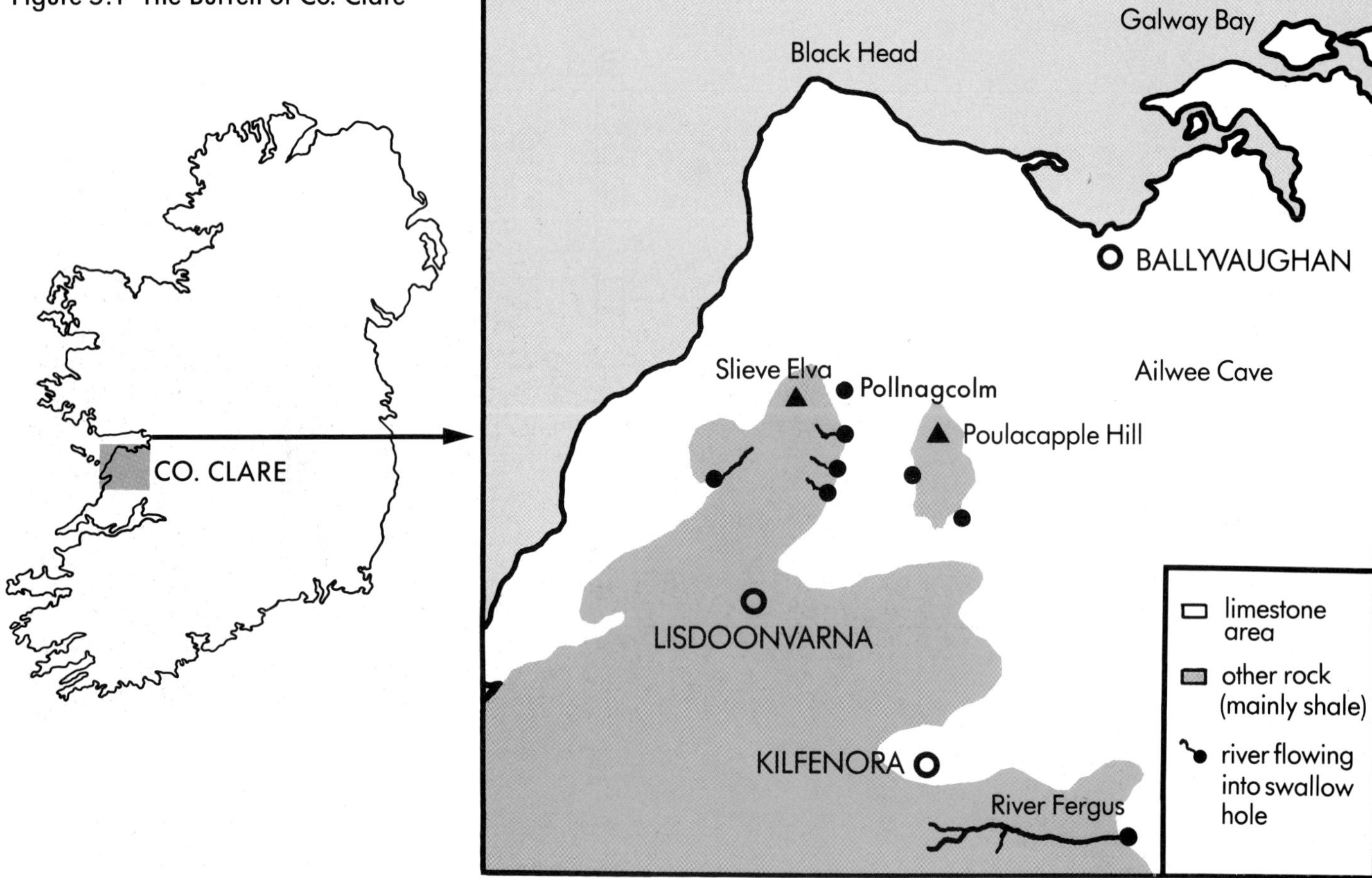

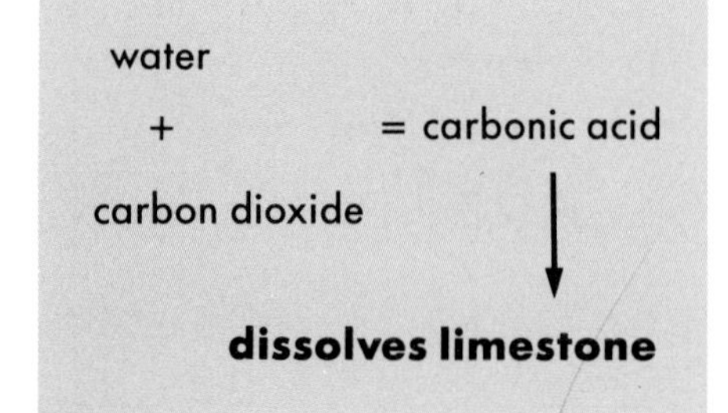

Figure 5.2

How limestone is weathered by rainwater

1. ***Rainwater dissolves soluble limestone.***
   As rainwater passes through the atmosphere, it takes in carbon dioxide and then becomes a weak ***carbonic acid***.
   This acid reacts with calcium carbonate in limestone, causing the limestone to be dissolved slowly (figure 5.2).

2. ***Limestone is permeable.***
   This means that limestone ***allows water to pass through it.*** It is easy for this rainwater to pass through the many vertical ***joints*** and horizontal ***bedding planes*** in limestone (see Chapter 2, 'Rocks').

The Karst Surface

The surface of the Burren consists largely of ***bare rock***. It often takes the form of flat or gently-sloping ***terraces***, separated by occasional ***cliffs***.

The bare ***limestone pavements*** are criss-crossed by deep grooves called ***grikes***. These are vertical joints in the limestone which have been widened by rainwater weathering.

The blocks of limestone which separate the grikes are known as ***clints***.

This is what one of Cromwell's generals had to say about the Burren. What does his description tell us about the Burren's surface? Why is so little water visible in the Burren? Why are there so few trees?

△ **Cliffs and terraces in the Burren.**

Rivers cannot flow for long distances on the karst surface. No sooner do they enter it from areas of other rock than they are likely to ***disappear underground*** through the vertical grikes. When this happens, the grikes may become enlarged into vertical holes called ***swallow holes***.

One Burren swallow hole, called Pollnagcolm, is 6m wide and 16m deep. ***Locate Pollnagcolm on figure 5.1.***

◁ **Limestone pavements.**

Beneath the surface

Underground rivers and seeping rainwater continue to dissolve the limestone beneath the surface of the Burren, gradually forming ***passages*** and ***caves***. These caves contain features such as ***stalactites***, ***stalagmites***, ***pillars***, and ***curtains*** (figure 5.3).

Figure 5.3

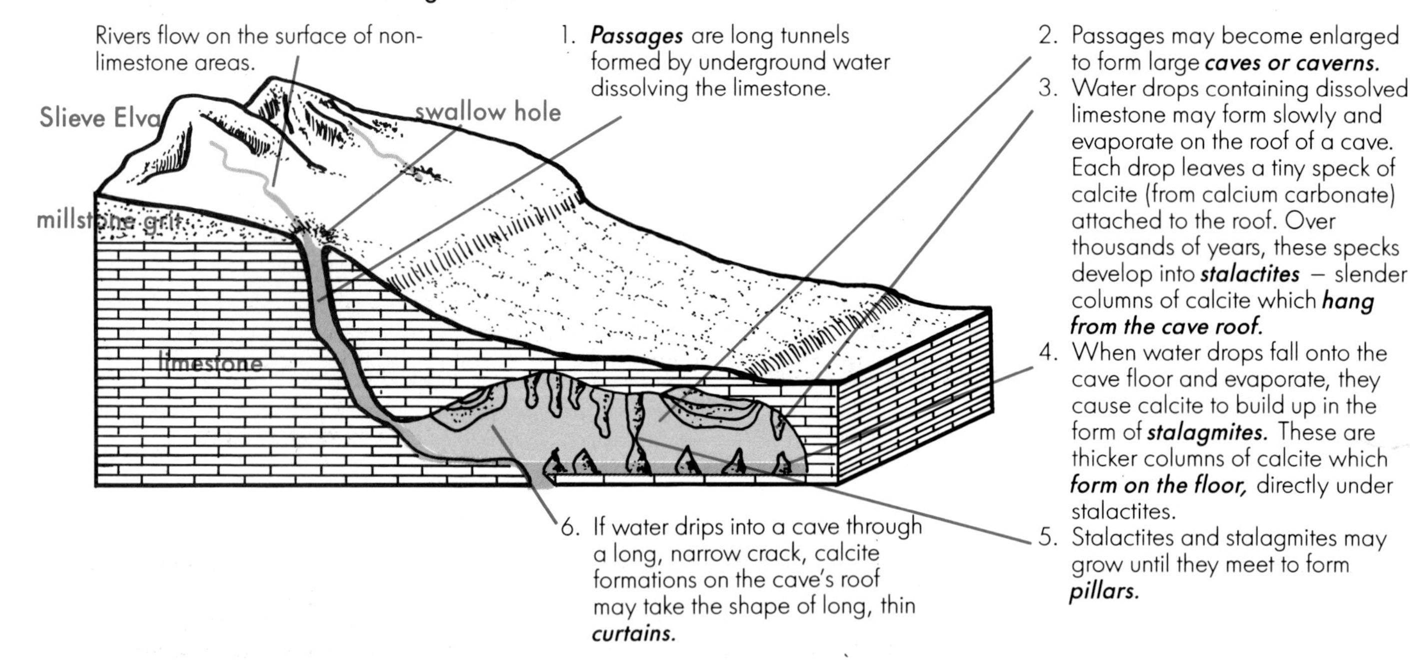

◁ Stalactites in Ailwee Cave, Co. Clare.

Attractions of the Burren

The dry, rocky parts of the Burren's surface are not suitable for farming. Only sheep and goats are reared in the area. But the Burren holds many attractions for a wide variety of people.

- ☐ ***Geographers*** study the limestone features.
- ☐ ***Potholers*** and ***cavers*** explore the underground cave systems.
- ☐ ***Botanists*** and other ***naturalists*** examine the rare plants which grow throughout the Burren, often in the grikes of the limestone pavements.
- ☐ ***Tourists*** admire the unusual ***karst*** scenery.

Class activity

A land-use disagreement in the Burren area

Imagine that an American holiday company, AmVac Incorporated, has applied for planning permission to build a major holiday home centre in the scenic Burren area of Co. Clare.

Fact sheet on the proposed centre

- ☐ The centre would consist of 50 detached holiday bungalows, along with a large central amenities complex. The central complex would contain bars, restaurants, an indoor swimming pool, a large discotheque and a games room with 6 pool tables and 30 gaming machines.
- ☐ The centre would cover 50 hectares and would overlook Galway Bay from a site between Black Head and Ballyvaughan (see figure 5.1). This site is in the scenic area shown in the photograph on page 23.
- ☐ It is estimated that 150 people would be employed in the construction of the centre. When completed, the centre would offer 15 permanent jobs and over 30 part-time summer jobs. Top managerial positions would be filled by Americans.
- ☐ AmVac hopes to rent most of its Burren bungalows on a monthly basis to American tourists.

Some of the arguments for and against the granting of planning permission to AmVac are outlined in the speech bubbles on the next page.

Divide the class into five groups. Each group should consider these arguments carefully and then make a group recommendation about whether planning permission should be granted or refused. A spokesperson for each group should report on his/her group's decision and the reasons for it. Reasons should be listed in order of importance.

When all groups have reported their recommendations, a general class vote should be held on whether or not planning permission should be granted for the proposed centre.

Suggested homework: Draft a carefully-written letter to the president of AmVac Incorporated in New York city. Tell him/her whether planning permission for the Burren development has been granted or refused. Outline the reasons for the decision in order of importance.

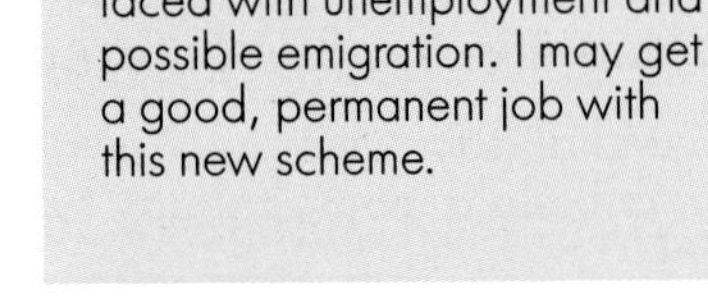

a I'm a local school leaver faced with unemployment and possible emigration. I may get a good, permanent job with this new scheme.

b I'm a former building worker, but now I'm on the dole. The AmVac scheme would give me a year's work—maybe more!

c I own the land on which the new centre will be built. I'll get rich by selling my land to AmVac!

d I'm a local milkman. The new scheme would greatly improve my business. It would also increase business for local publicans, shopkeepers and other suppliers.

e They might let the local people use the swimming pool and other facilities provided at the new complex. Wouldn't that be great!?

f The proposed scheme is far too big! It would destroy the peaceful beauty of this quiet area. It would also look out of place in this unique natural region.

g Traffic would be greatly increased on our quiet—and narrow—country roads. There will be more accidents and more traffic jams.

h I love the traditional music and rich culture of this locality. This scheme would swamp the area with too many tourists and threaten our culture.

i I'm a botanist. Like other naturalists, I love the rare flora and fauna of the Burren. A sudden influx of tourists might threaten the survival of rare plants and animals. Tourists may pick the rare plants or trample them.

j The Burren holds many ancient historical sites. A huge influx of tourists could mean that these priceless monuments may be damaged.

Can you think of any other arguments for or against the proposed scheme?

And beyond the Burren . . .

The Amazon and Zaire river basins suffer heavy chemical weathering. So do large areas of countries such as Indonesia, Venezuela and Panama.

Find these river basins and countries in your atlas.

Chemical weathering in the Tropics

Wet regions within the Tropics suffer from so much chemical weathering that the rocks in these regions can be decomposed to a depth of up to 60m. This happens because these regions have:

- ☐ large amounts of ***rainfall***
- ☐ ***high temperatures*** (which speed up the chemical reactions of water on rock)
- ☐ large amounts of ***rotting vegetation*** (producing acids which help to weather the rock).

Acid rain

A Case Study of how people contribute to chemical weathering

Some rock types can be chemically weathered by natural rainwater. ***Acid rain***, caused by human activities, is an even more powerful weathering agent.

Figure 5.4

△ A statute damaged by the effects of acid rain.

The various effects of Acid Rain

Acid rain damage in West Germany is now affecting over 50% of that country's forests—a dramatic increase from the 8% recorded in 1982. Conifers are particularly vulnerable, suffering needle loss, branch distortion, bark injuries and damage to root fibres. Millions of hectares of trees are now affected in several countries.

Lakes have been affected in almost every European country, as well as in North America, many to the point where they will no longer support fish life. In Scandinavia, liming (to neutralise the acidity) has been carried out on a massive scale, but this is only regarded as a stop-gap measure rather than a real solution.

Crop losses due to acid rain, in combination with high ozone levels, have been estimated at £25 million per year in Scotland and £200 million per year for the whole of the UK. In the US, the figure for losses from reduced yields caused by acid rain is estimated at $3100 million per year for wheat, corn, soya beans and peanuts alone.

Toxic heavy metals are leached out of the soil and out of piping systems by increased acidity, causing contamination of drinking water. In Athens, the Acropolis has aged more in the past twenty years than in the previous 2400, and Cleopatra's Needle more in the past eighty years of London's pollution than in its 3000 years by the Nile. Buildings, pylons, fencing materials, cars, tents—all have their lifespans shortened by acid rain.
(from ***Disarm***, January/February 1985)

1. Describe the effects of acid rain on forests, lakes, crops, drinking water and buildings. Name five countries which have been affected by acid rain.
2. Acid rain is caused by the burning of *fossil fuels* such as coal and oil. List some of the reasons for the burning of coal and oil. Suggest one way of reducing the acid rain problem.

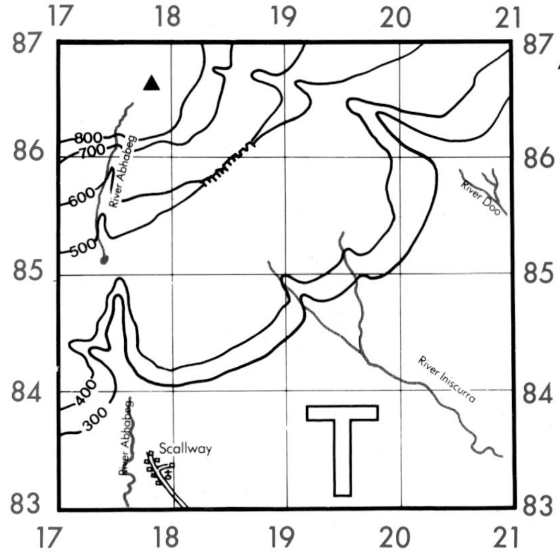

Figure 5.5

Activities

1. Imagine that you have just returned from a visit to the Burren. You want to write to a friend who has never seen a karst area. Describe the surface of the Burren to your friend as fully as you can.

2. Describe each of the following features: passage, cavern, stalactite, stalagmite, pillar, curtain. Use a single large diagram to illustrate your answer.

3. Use encyclopedias or other books in your school or local library to find out about *dry valleys* and *springs*.
   Identify the location of a dry valley and a spring in figure 5.5. Locate a terrace, a cliff and a place where limestone is *not* exposed at the surface. Suggest a reason for the location of Scallway village.

4. *A word game*

   If all the clues across are filled in correctly, one of the lines down will spell the name of a common cave feature.

   Clues across

   1. A cave feature which forms beneath a stalactite.
   2. A flat surface-area.
   3. A cave in Co. Clare.
   4. A hill in the Burren area.
   5. People who study the plant life of the Burren.
   6. Rainwater is a weak _ _ _ _ _ _ _ _ acid.
   7. Stalactites are made of this.
   8. A deep groove in limestone pavement.
   9. This may form on a cave roof.
   10. Home county of the Burren.

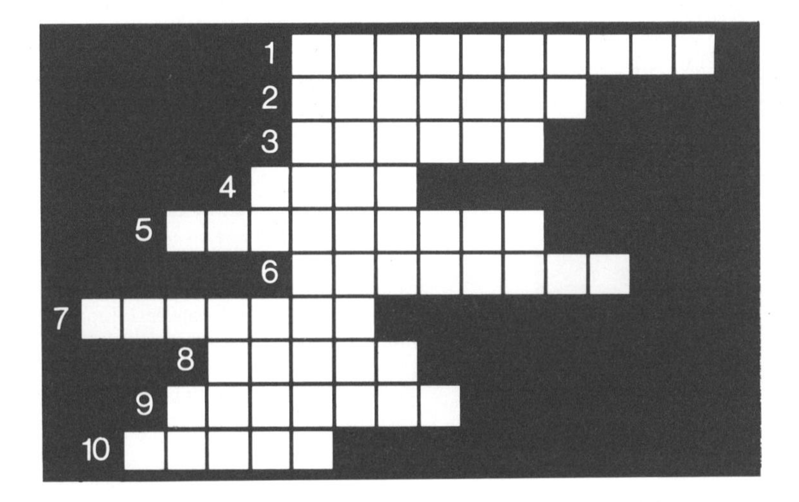

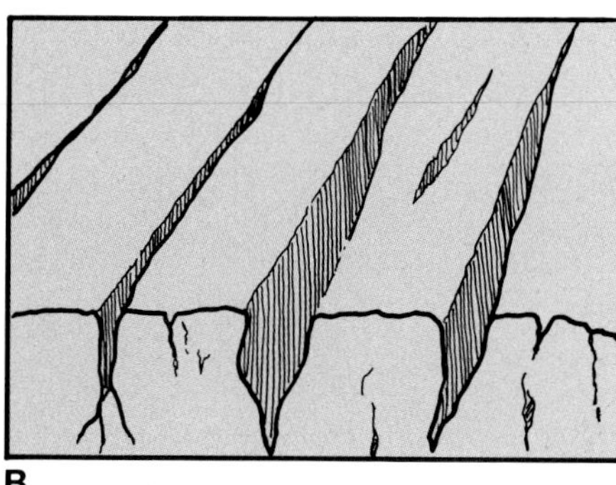

5. (a) *Name* and *describe* the weathering processes which are likely to have affected the rocks shown in illustrations A and B.

   (b) The illustrations show weathering under natural conditions. Describe one way by which people may speed up the weathering of rocks.

6. *Local field activity*

Visit an old graveyard in your locality. Examine the inscriptions on the oldest headstones. Are these inscriptions always easy to read? Explain fully why some inscriptions are so illegible.

Are inscriptions easier to read on limestone or marble headstones of a similar age? Why?

7. *Suggested fieldwork*

Organise a supervised class field trip to the Burren or to limestone caves such as those at Mitchelstown, Co. Cork or Dunmore, Co. Kilkenny. Before your trip, discuss the kind of information which you should gather and make up worksheets to take with you. Upon your return, write a report of your visit using diagrams and photographs to illustrate it.

The best time to visit the Burren is in spring when its unusual flowers are in bloom. It would be worthwhile to contact the Burren Centre at Kilfenora and the Ailwee Caves, both of which may be able to organise guided tours for your group.

Guided tours of the Mitchelstown Caves can be arranged by phoning (052) 67246. For the Dunmore Caves at Ballyfoyle, Co. Kilkenny, ring (056) 67726.

△ How has weathering affected this old headstone? How do you know?

6 MASS MOVEMENT

In chapters three and four, we learned that weathering loosens and breaks up the rocks of the earth's surface. This loose, weathered material is called ***regolith***.

Regolith moves down slopes under the influence of gravity. This action is known as ***mass movement***. Mass movement can occur wherever the land surface is sloping. It is therefore a widespread and important process in shaping the earth's surface.

Mass movement from frost-shattered peaks in the Swiss Alps, where weathered rocks have fallen to the bottom of steep slopes. ▷

(a) Make a sketch of the photo. Show and name: a frost-shattered peak; a steep slope down which regolith falls; scree.

(b) Why do you think there is no vegetation on the scree slope?

Factors affecting mass movement

The amount and type of mass movement which occurs in any place depends on a combination of several factors (figure 6.1).

Figure 6.1 ▷

Water content
Water in the regolith assists some types of mass movement such as earthflows and mudflows.

Slope
The steeper the slope, the faster the mass movement.

Factors affecting the amount and type of mass movement

Vegetation cover
This may hinder mass movement.

Human activities
These may assist or hinder mass movement.

Can you think of some human activities which could assist or hinder mass movement?

Types of mass movement

Some of the main types of mass movement are outlined in the box (figure 6.2). They are distinguished by their ***speed***, which depends largely on the steepness of the slopes on which the movements occur.

Figure 6.2 Some of the main types of ▷ mass movement

Type of mass movement	*Speed*
soil creep	slow
earthflows, mudflows	slow to rapid
landslides	rapid

Over time, this road has been △ completely destroyed by the effects of soil creep.

Soil creep

Soil creep is the ***slowest*** type of mass movement. It can occur even on ***very gentle slopes***. Soil creep is so slow (sometimes only 1mm per year) that it might not be noticed at all, except for the effect that it has on surface objects such as walls and fences (figure 6.3). Soil creep is greatest near the surface and decreases with depth.

Anything which loosens the soil, including freeze-thaw action or the burrowing of animals, assists the movement of soil downslope.

Watch for any signs of mass movement near your home or school. Report your findings to your class.

▽ Figure 6.3 Some damaging effects of soil creep.

Suggest a reason why: (a) poles, fences etc. lean downslope; (b) soil creep decreases with depth (hint: friction between lower soil and bedrock).

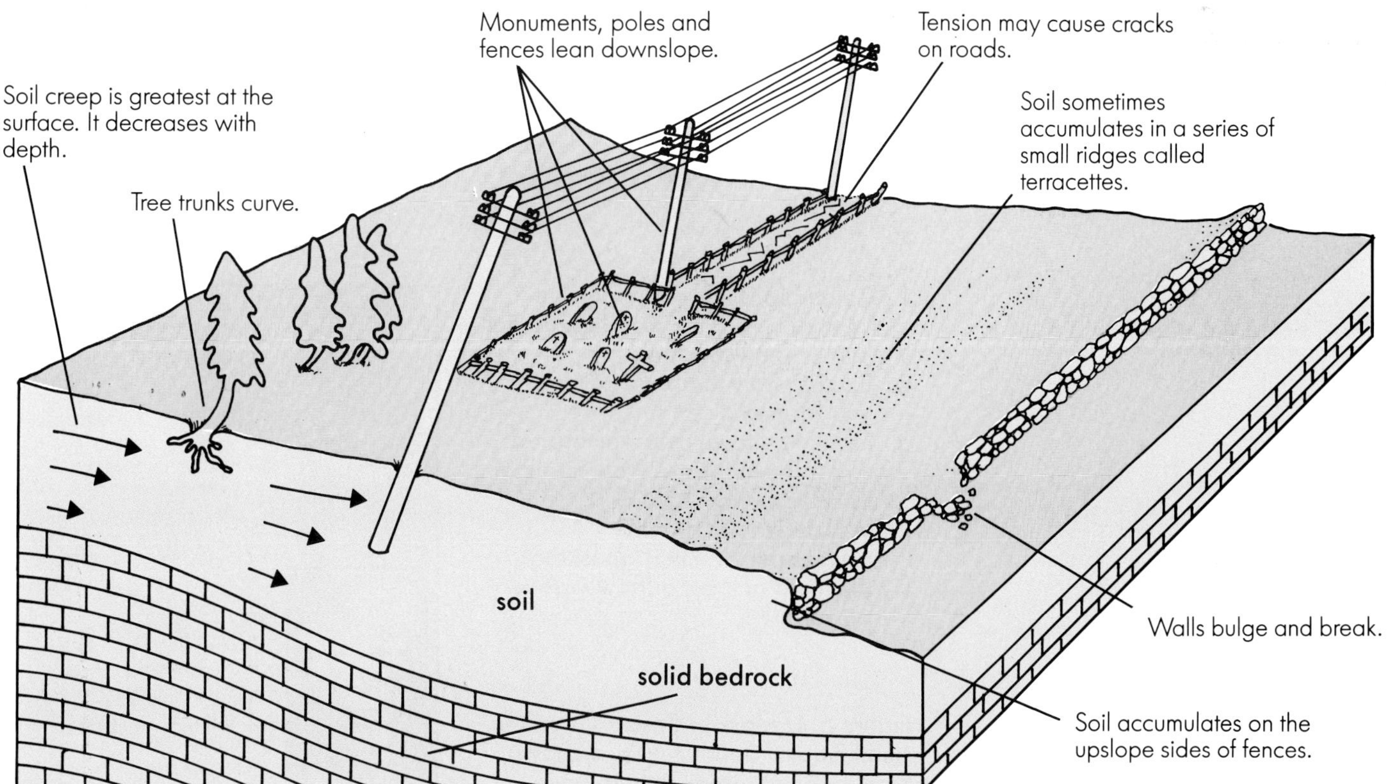

Earthflows and mudflows

Earthflows and mudflows often occur on hillsides after periods of heavy rain. These flows are ***more rapid*** than soil creep, with regolith tending to flow downslope like thick treacle. The regolith tends to flow rather than slip downslope because it is saturated with water. Earthflows and mudflows are relatively rapid because they occur on ***steeper slopes*** than soil creep.

Small mudflows are common in Irish hill areas, along the Antrim coast road and in the Wicklow glens, for example. But not all mudflows are so small or harmless. In 1985, a rapid and terrible mudflow killed 20 000 people in Colombia, South America (see 'Buried alive').

Buried Alive

20 000 feared dead

One of the mightiest volcanic eruptions of the century caused the deaths of an estimated 20 000 people when the resultant mudslide buried a Colombian town beneath a river of mud.

The snowcapped 5000-metre Nevado del Ruiz mountain erupted yesterday, having stood dormant for 140 years. The hot volcanic ash from the eruption mixed with and instantly melted the thick mountain snows. This created a vast river of mud which surged down the mountainside, causing fearsome scenes of devastation in Armero, a town at the foot of the mountain with a population of some 21 000.

Armero police sergeant Jose Victor Otilavaro told reporters that 23 of the 25-man police force in the town had died when their station was washed away by a crashing sea of mud. Sergeant Otilavaro himself had climbed a tree, along with other survivors, to escape the advancing mud.

One terrified Armero resident, Marina de Huez, said: 'More than half the population was buried under the torrent of mud that came with a horrible noise. It dragged away houses, cattle, tree stumps, even gigantic rocks. There was not time for anything. The church was buried, the school, the theatre.'

Another survivor said that he had awakened to find watery mud all around him. 'I heard a sound like a huge locomotive at full steam. Then I felt the mud swirling around my neck,' he said.

(based on a newspaper report of 15 November 1985)

△

1. Colombia is near the equator. Why do you think the Nevado del Ruiz is snowcapped?
2. Explain how the 'river of mud' was formed.
3. Name the town that was buried under the mudflow. How many of its citizens were killed?
4. Which two parts of the newspaper article suggest that the force of the mudflow was very powerful?

Figure 6.4 Mark in Colombia on the small map of South America shown here. Use your atlas to help you. ▷

Landslides

A landslide is a ***very rapid*** slipping of earth or rock down a cliff or ***very steep slope***. Landslides are most common in mountainous areas which contain large quantities of regolith. They also occur along coastal cliffs which have been undermined by sea erosion.

Anything which undercuts the base of a slope—such as sea erosion, road building or quarrying—can trigger a landslide (see figures 6.5 and 6.6). So can frost action or earth tremors, which may loosen the regolith on the slope.

How sea erosion can cause landslides on a coastal cliff

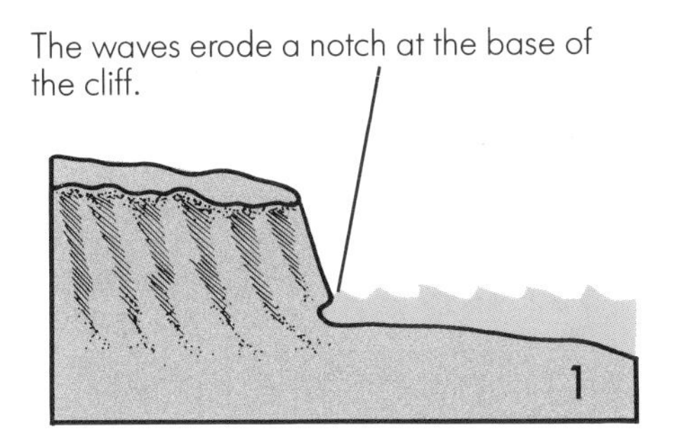

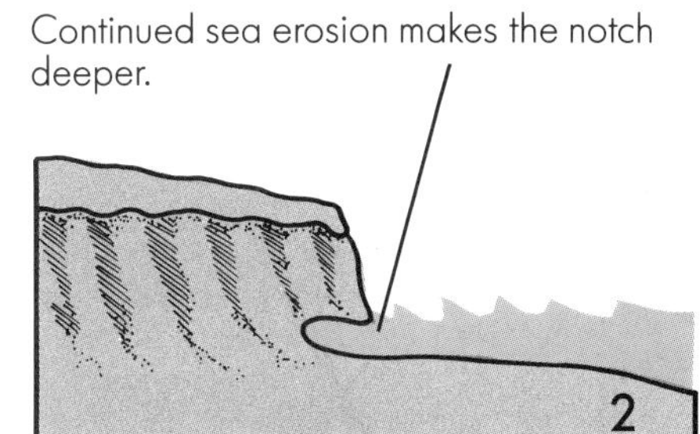

△ Figure 6.5

How road building can cause landslides

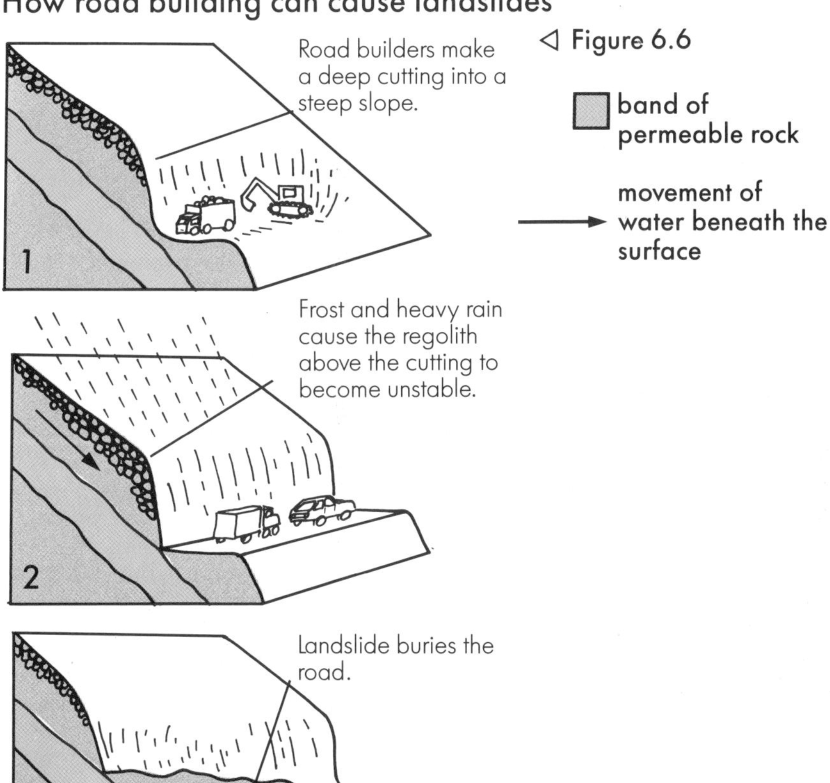

◁ Figure 6.6

△
What danger is being warned of in this sign? How might human activities contribute to this danger?

Major landslides can result in ***death*** and ***the destruction of property***. This happened in 1987 in the city of Medellin, Colombia, not far from the location of the disastrous mudflow which had engulfed Armero two years earlier.

Frantic parents search for lost children in shanty town

Landslide Kills 135 in Colombia

Rescuers using spades and pickaxes yesterday dug through an 11-metre-high mound of earth to find victims of the landslide that had buried a shanty town in Medellin, Colombia's second-largest city.

Hospital spokesmen said that at least 135 had been killed and another 53 seriously injured. Many more were missing, including a group of children taking part in their First Communion service.

The majority of the dead were trapped in their houses. The landslide destroyed everything on a 550-metre front. It was not halted by the pine trees which had been planted a few years ago to stop erosion.

Colombia is a mountainous country which has had many landslide disasters. In 1983, more than 100 workers building a hydroelectric dam were swept away when the site was struck by a huge mountain of rock.

(*adapted from* The Cork Examiner *29 September 1987*)

◁ 1. What is a shanty town?
2. What measure did the local people take to stop erosion and landslides on the mountain slopes?
3. What previous landslide disaster is referred to in the extract?
4. Disastrous landslides can occur in any part of the world, in rich countries as well as poor ones like Colombia. Find out about the tragedy which occurred in the Welsh mining town of Aberfan in 1966. How was the Aberfan landslide different from the Colombian one?

Activities

1. (a) What is meant by mass movement?
   (b) Name one very slow and one very rapid type of mass movement.
   (c) Describe any *one* of the mass movement types named in (b).

2. Make lists of the damaging effects of each of the following types of mass movement: (a) soil creep; (b) landslides.
   Suggest some precautions which might be taken to lessen the damage done by any one of these forms of mass movement.

3. *Mapwork*
   Study the Clew Bay OS map in the Map Supplement which accompanies this book. Identify a place on the map where, in your opinion, landslides are likely to occur. Explain your answer.

4. *Fieldwork*
   In the company of an adult, visit a sloping field or a beach beneath a sea cliff near your home. Look for evidence of any form of mass movement and make notes on what you observe. Illustrate your notes with photographs or sketches. Include a description of the location of the place you have visited in your notes.
   Display your findings as part of an exhibition on 'Local Mass Movement' in your classroom.

The Syllabus requires that TWO of the next three chapters be studied.

7 THE WORK OF RIVERS

Rivers: some common features

The diagram in figure 7.1a shows some common river features. Figure 7.1b gives a definition of each of these features. In the spaces provided in figure 7.1b, write the name of each feature next to the appropriate definition. Then learn these definitions.

Figure 7.1a Common river features

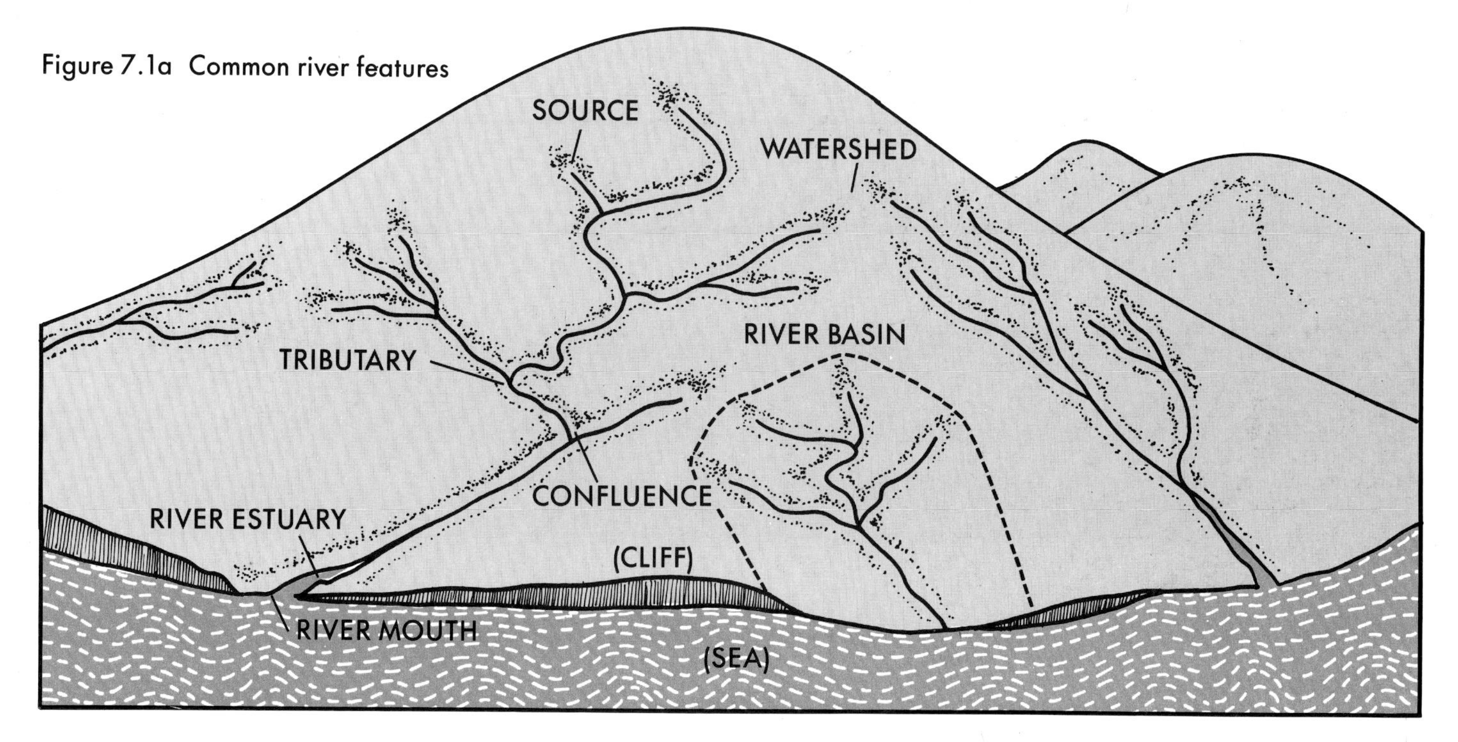

Figure 7.1b Common river features: ▷ definitions

Feature	*Definition*
______________	the place where a river begins
______________	a stream or river which joins a larger river
______________	the place at which rivers join
______________	the place where a river enters the sea, a lake or another river
______________	that part of a river mouth which is tidal
______________	the total area drained by a river and its tributaries
______________	the high ground which separates one river basin from another.

See mapwork activity 3, page 45.

The three stages of a river

A river rises in high ground (its ***source***). It then flows down a slope (its ***course***) until it reaches the sea (its ***mouth***).

Three stages may occur in the course of a river. These are the upper or ***youthful stage***, the middle or ***mature stage***, and the lower or ***old age stage***. The main characteristics of these stages are shown in figure 7.2.

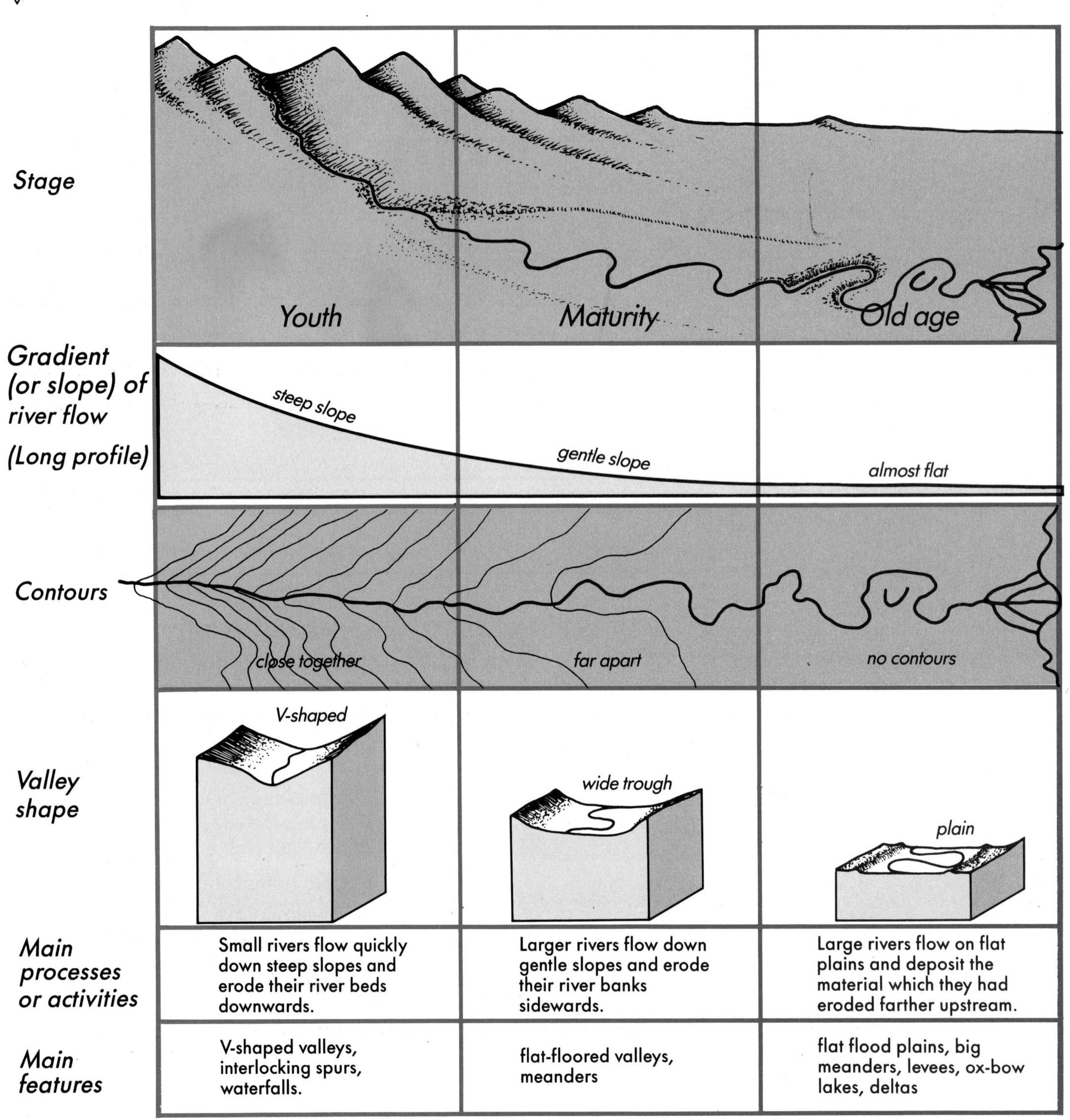

Figure 7.2 The three stages of a river
▽

△
The River Liffey in its youthful stage. How is this area being eroded by the river?

The youthful stage

Youthful rivers are usually small in volume (size). They usually flow quickly down steep slopes. Such fast-flowing rivers generally erode the land over which they flow.

How rivers erode

- ☐ ***The force of the moving water*** breaks off fragments of materials from the bed and the banks of the river.
- ☐ The material is then carried along by the river and is called its ***load***. The load hits against the bed and banks of the river, eroding them rapidly. (The deepening of a river bed is called ***vertical erosion***. The widening of the river banks is called ***lateral erosion***.)
- ☐ ***Chemicals*** contained in the water may help in the erosion of some rocks. Carbonic acid, for instance, helps to erode limestone.

The ***amount of erosion*** depends on several things, including the following.

The resistance of rock	Soft rock is usually more easily eroded than hard rock.
Water volume	Rivers with large amounts of water usually erode more powerfully than rivers with less water.
Speed of flow	Fast-flowing rivers usually erode more powerfully than slow-flowing rivers.

Examples occur in the upper courses of the Liffey and the Barrow.

Features of erosion by young rivers

V-shaped valleys (figure 7.3)

These valleys have ***steep sides*** and very ***narrow floors***. Their cross sections are shaped like the letter V.

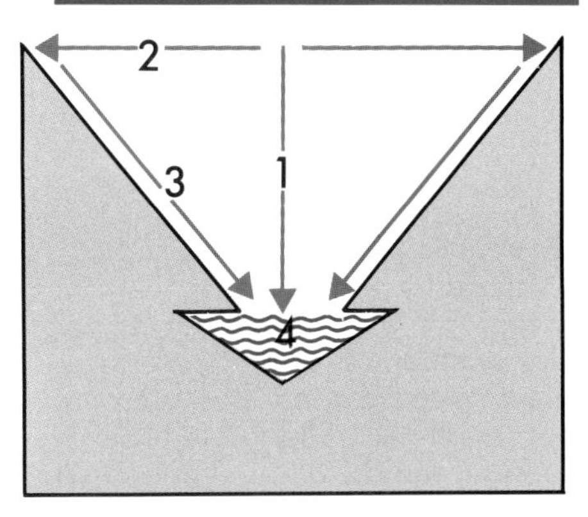

Figure 7.3 How V-shaped valleys are formed

Formation

1. ***Vertical erosion*** by the river deepens the valley floor. Meanwhile . . .
2. . . . ***weathering*** breaks off materials from the valley sides . . .
3. . . . ***gravity*** causes these loose materials to tumble or slide into the river . . .
4. . . . and the ***river*** carries the materials away.

All these activities combine to form V-shaped valleys.

Figure 7.4 A V-shaped valley with interlocking spurs

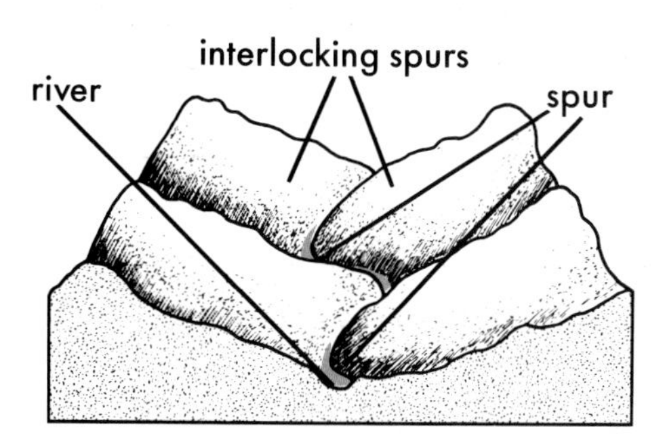

Interlocking spurs (figure 7.4)

These are ***projections*** of high ground which seem to ***'lock into' each other*** from both sides of a young river valley.

> ***Examples*** occur in the upper courses of the Liffey and the Barrow.

Formation

The young river tends to develop a winding course as it is forced to flow around ***obstacles*** of hard rock.

As it develops its valley along this course, the river leaves interlocking spurs of high ground jutting out on both sides of the valley.

Waterfalls (figure 7.5)

These are ***falls of water*** caused by sudden drops in the courses of rivers.

> ***Examples*** include the Powerscourt waterfall on the River Dargle in Co. Wicklow and the Torc Waterfall near Killarney, Co. Kerry.

Formation

Figure 7.5 The formation of a waterfall

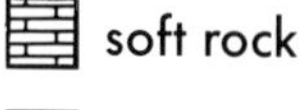

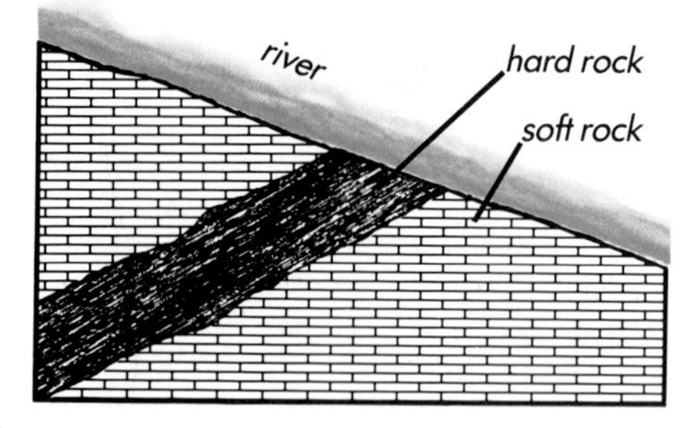

1. ***A band of hard rock*** crosses the river bed. The river cannot erode the hard rock quickly.

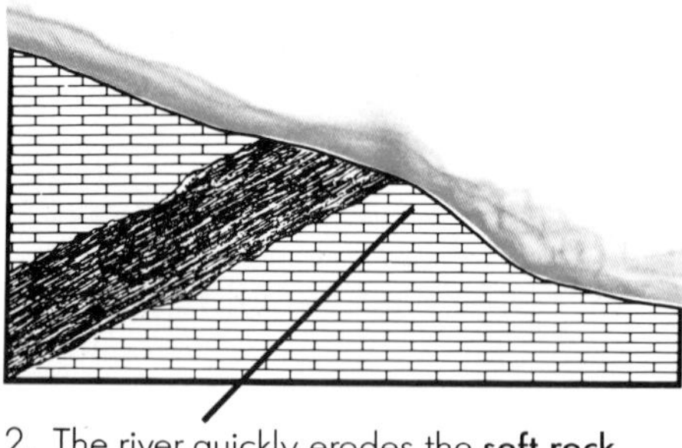

2. The river quickly erodes the **soft rock** downstream from the hard rock.

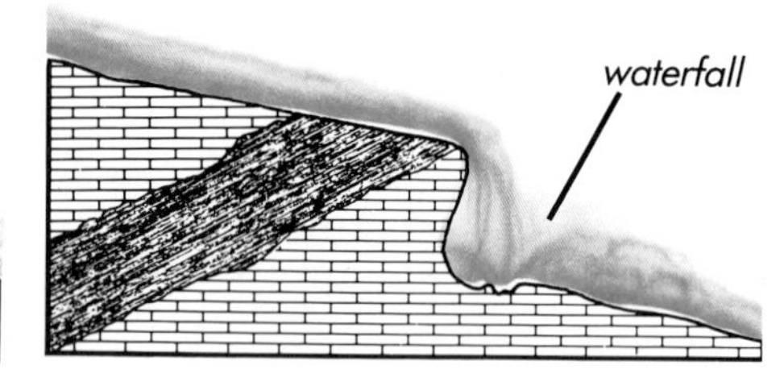

3. So a ***sudden drop*** in the river's course develops just downstream from the hard rock. The river **falls** over this drop.

◁ A river in its youthful stage.
Make a tracing of this photograph. Show and label the following features of a young river valley: steep sides; narrow valley floor; small river; interlocking spurs; waterfall; rapids.

△
Figure 7.6 How materials are transported

The mature stage

By the time they have reached the mature stage, rivers will have received the waters of many ***tributaries***. So they will have increased in volume. As these larger rivers sweep along, they ***transport large loads of materials*** with them.

How rivers transport materials

The lightest particles are suspended in the water and carried along by it.

Heavier particles are 'bounced' along the river bed.

The heaviest pebbles and stones are rolled and dragged along the river bed.

Features of mature river valleys

Valley troughs (figure 7.7)

These are ***wide-floored*** valleys with ***gently-sloping sides***.

> ***Examples*** occur in the middle courses of the rivers Lee and Nore.

Formation

As it moves down towards sea level, a river erodes laterally rather than vertically. So its valley becomes wider. Weathering also continues to wear away the valley sides which become more gently sloping.

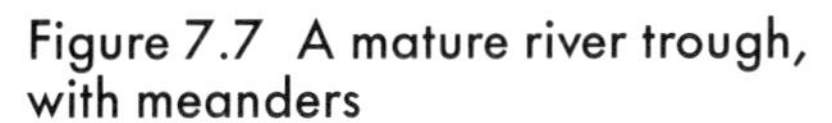

Figure 7.7 A mature river trough, with meanders
▽

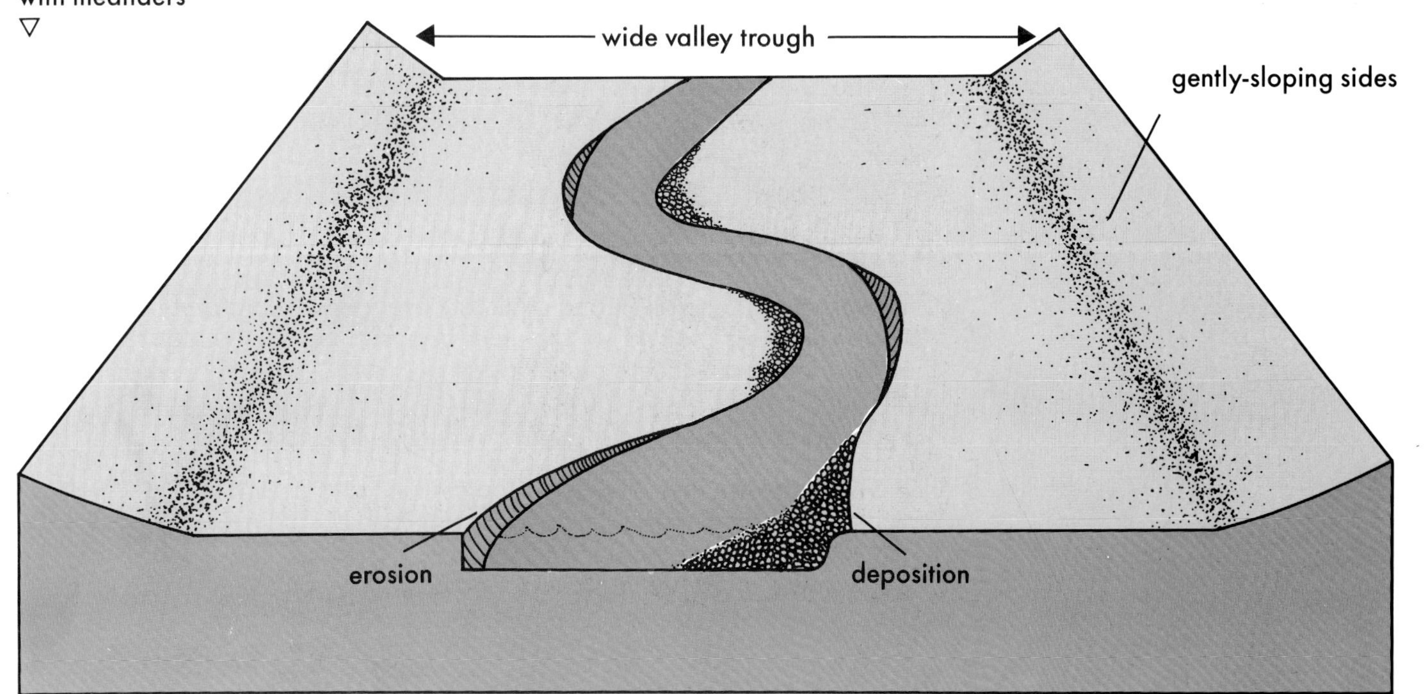

Examples occur in the middle and lower courses of the Shannon and the Boyne.

Meanders (figure 7.8)

These are ***pronounced curves*** in the course of a river.

Formation

Once a river has developed a winding course, the curves in the river become more and more pronounced until they form meanders. Meanders are developed both by ***erosion and deposition***.

Figure 7.8 The formation of meanders ▷

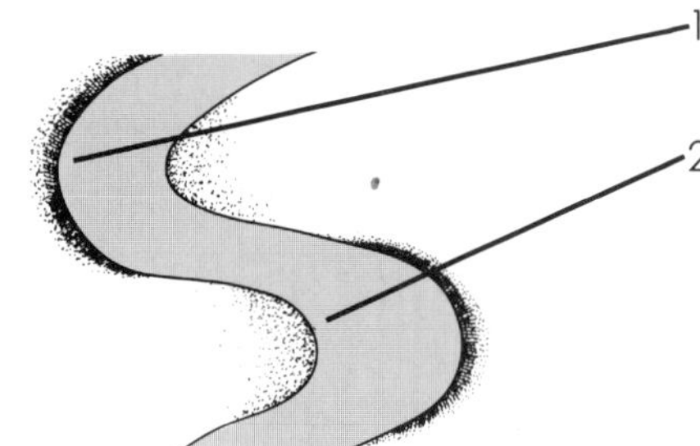

1. Water flows more ***quickly on the outside of each bend*** and so erodes back the river bank there.
2. Water flows ***more slowly on the inside of each bend*** and deposits material there.

Erosion of the outside banks and infilling of the inside banks cause meanders to become more and more pronounced.

A river in its mature stage. ▷
List the differences between the landscape shown here and the landscape with the youthful river on page 37.

A river's old age

Old rivers ***meander slowly*** and aimlessly over almost ***flat plains*** as they near sea level. As they do so, the rivers ***deposit*** much of the loads which they have carried in their mature stages.

Why old rivers deposit their loads

Rivers ***slow down*** when they enter the almost-flat plains of old age. Slow-moving rivers ***cannot carry large loads***, as fast-flowing rivers can. So the old rivers deposit their loads.

The loads of old rivers consist of light particles such as mud and grains of sand. This material is called ***alluvium***. It is deposited along the river bed. When the river overflows its banks, the fertile alluvium is also deposited on the plains which border the river.

Examples occur in the lower stages of the Suir, Barrow and Nore.

Features of deposition by old rivers

Flood plains (figure 7.9)

These are almost ***flat plains*** which border old rivers and which are sometimes flooded by the rivers. They are covered in fertile alluvial soils which are deposited by the river.

Formation

Unable to carry all of its load, the river ***deposits*** some of it on its bed. This ***raises the level of the water*** so that it almost reaches the tops of the river banks.

In times of ***heavy rainfall***, the river water may ***rise still further***, overflowing its banks and ***flooding the nearby plains***.

Figure 7.9 An old river and its features ▷

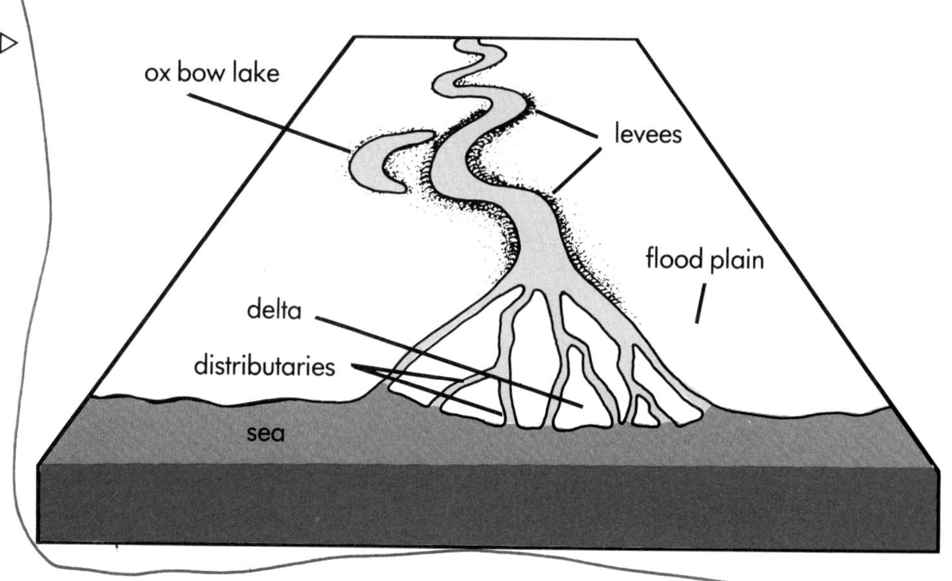

Examples: Levees occur along the Shannon where they have been heightened by people to help prevent the river from flooding.

Levees (figure 7.9)

These are ***narrow ridges of alluvial deposits*** found along the banks of old rivers.

Formation

When a river floods, its ***flood water spreads out***, and loses much of its force. Such water deposits most of its load as soon as it leaves the river banks.

After repeated flooding, these deposits build up to form long, narrow ridges called levees along the river banks.

△
An ox-bow lake, with meanders.

Ox-bow lakes (figure 7.10)

These are ***horseshoe-shaped lakes*** found near old rivers.

Examples occur in the lower stages of the Mississippi in the USA and the Mekong in Vietnam.

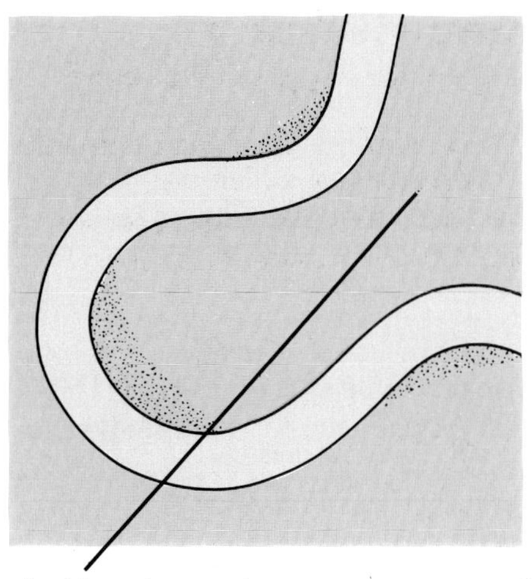

1. Meanders can become so pronounced that only a narrow neck of land separates their outer banks.

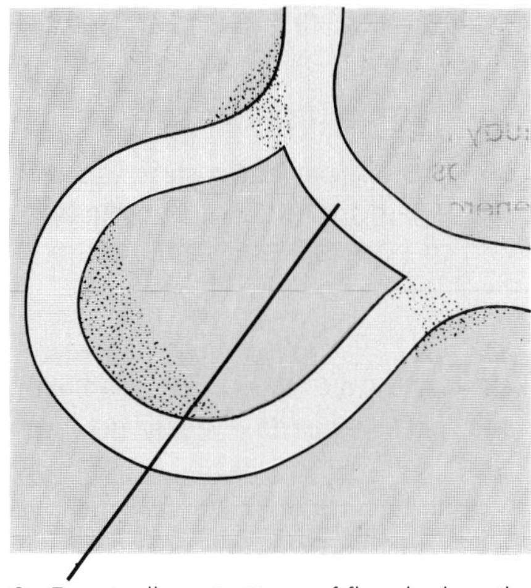

2. Eventually – in times of flood when the river flows more quickly – the water cuts through the neck of land.

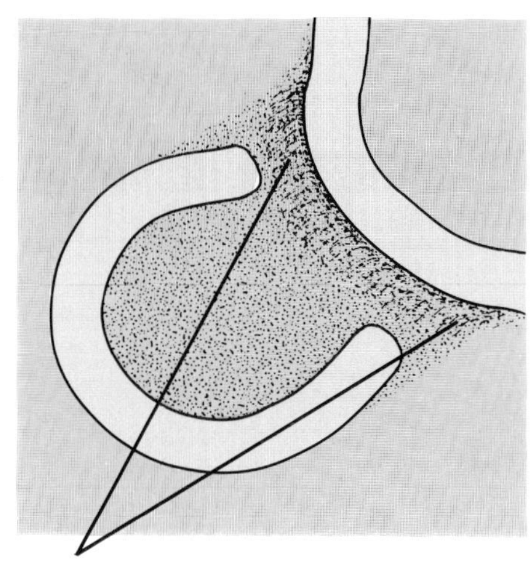

3. Levee deposits later seal off both ends of the abandoned meander, which become an ox-box lake.

△
Figure 7.10 The formation of ox-bow lakes

See Activity 4, page 45.

Examples occur at the mouths of the Nile in Egypt and the Rhone-Saone in France.

Cities flooded as Rhine continues to rise

As the Rhine roared in a brown torrent through leafy streets and gardens, thousands of families moved up to the top floors of their flooded houses in Cologne, Bonn and Koblenz . . . .

The swift current swept away caravans, chalets and trees from low-lying areas, washed away earth and plants from people's gardens and broke through walls and fences.

All shipping has been forbidden for virtually the length of the Rhine; trains have had to be diverted from flooded railways and many roads are under water and impassable.

Several people have already been drowned in accidents caused by the floods.

(from *The Irish Times*, 30 May 1983)

Deltas (figure 7.9)

These are ***triangular-shaped tracts of land*** which form at the mouths of rivers. They are made up of ***alluvium*** (soil deposited by the river).

Formation

When a river enters the sea, its flow is finally checked and it deposits what remains of its load.

If the sea cannot wash away these deposits, they will build up to form a ***delta*** at the river's mouth. Soon, the river channel becomes choked and the river is forced to break up into a number of smaller channels called ***distributaries***.

Flood danger

Read the newspaper report 'Cities flooded as Rhine continues to rise'. Then list the effects of the flood described in the article.

Some ways of preventing flooding

1. ***Dams*** can be used to store water temporarily in times of flood danger.
2. ***Dredging*** can deepen river beds, thus preventing river water from rising over its banks.
3. ***Artificial levees*** can be built to help prevent old rivers from breaking their banks.

Human activities such as dam building interfere with the natural activities of rivers

Study the photograph of Ardnacrusha on the River Shannon where the ESB has built a hydroelectric power station which uses rushing water to generate electricity. As part of this hydroelectric scheme, a large dam was built across the river valley. Behind the dam, trapped water has risen to form a large artificial lake.

The damming of our rivers provides us with many benefits. It also has some disadvantages.

Figure 7.11 Ireland's hydroelectric power stations. What Irish rivers have been harnessed for HEP? Name the power stations on each river.

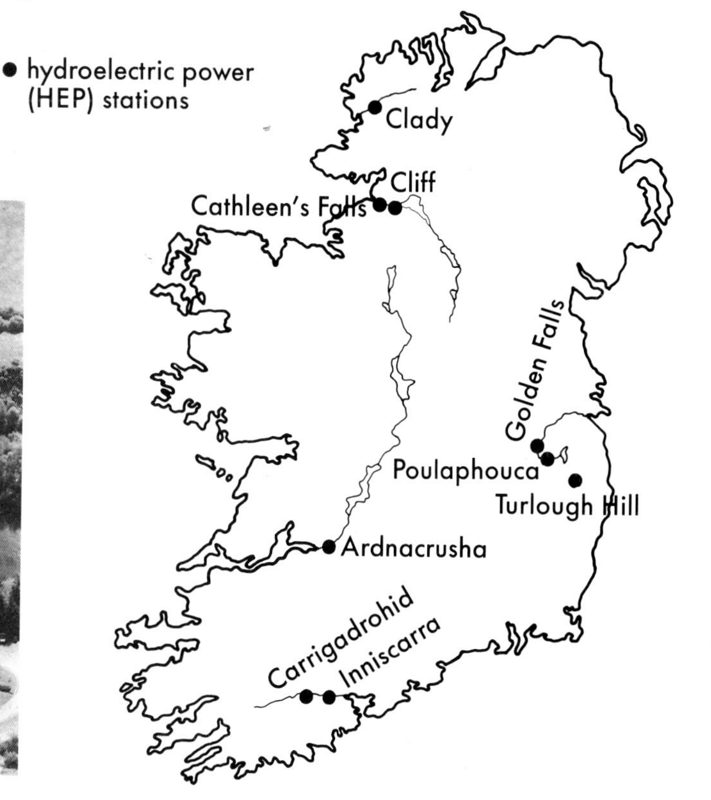

△ The HEP station at Ardnacrusha.

Benefits of hydro-electric dams	*Disadvantages of hydroelectric dams*
☐ They enable hydroelectric power to be generated.	☐ Artificial lakes flood settlements and valuable farmland.
☐ Artificial lakes can be used as reservoirs (storage areas) for urban water supplies.	☐ The survival of salmon and other river creatures may be threatened if their natural habitat is interfered with.
☐ Artificial lakes can be used for fishing and other water sports.	

Human activities sometimes result in river pollution

- ☐ ***Pesticides*** (poisonous chemicals used by farmers to kill insects) seep into the ground and eventually into the rivers.
- ☐ ***Fertilisers, animal slurry and domestic sewage*** enter rivers and cause a rapid growth of plant life. This plant life clogs up the water and uses up the oxygen which river animals need to breathe.
- ☐ ***Industrial waste*** is dumped into rivers.
- ☐ Some electric ***power stations*** pump large quantities of warm water into rivers. Warm water contains less oxygen than cool water. So a rise in river temperatures can mean suffocation for many river animals.

Some effects of river pollution

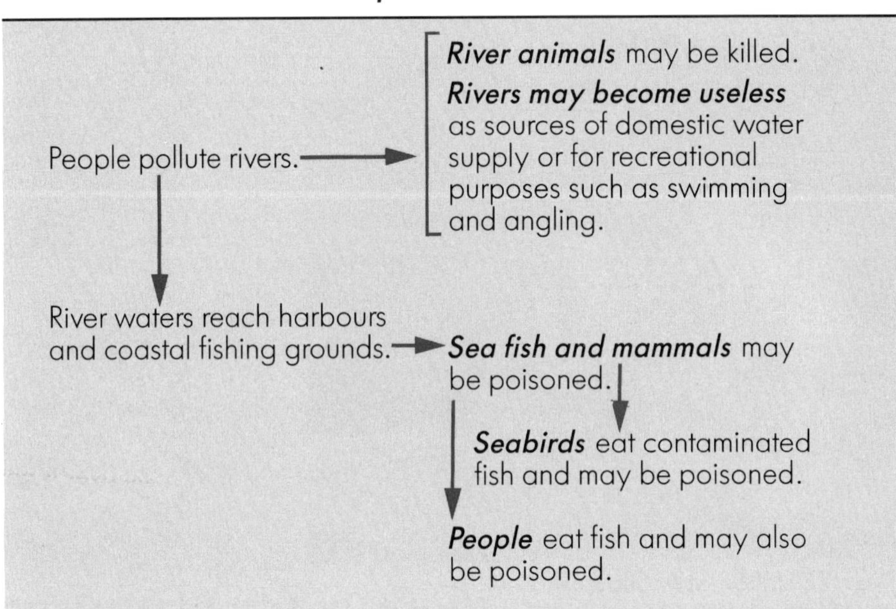

Activities

1. Study the river features in figure 7.12. Then do the following.
   (a) Identify each of the following features: source, tributary, confluence, interlocking spurs, meander, levee, ox-bow lake, delta.
   (b) Say whether the landscape between W and X has been shaped mainly by river erosion or by river deposition.
   (c) Say whether the landscape between Y and Z has been shaped mainly by river erosion or by river deposition.
   (d) In the case of each of the features lettered F, G, H and J, say whether the feature was formed mainly by: (i) river erosion; (ii) river deposition; (iii) a combination of both river erosion and river deposition.

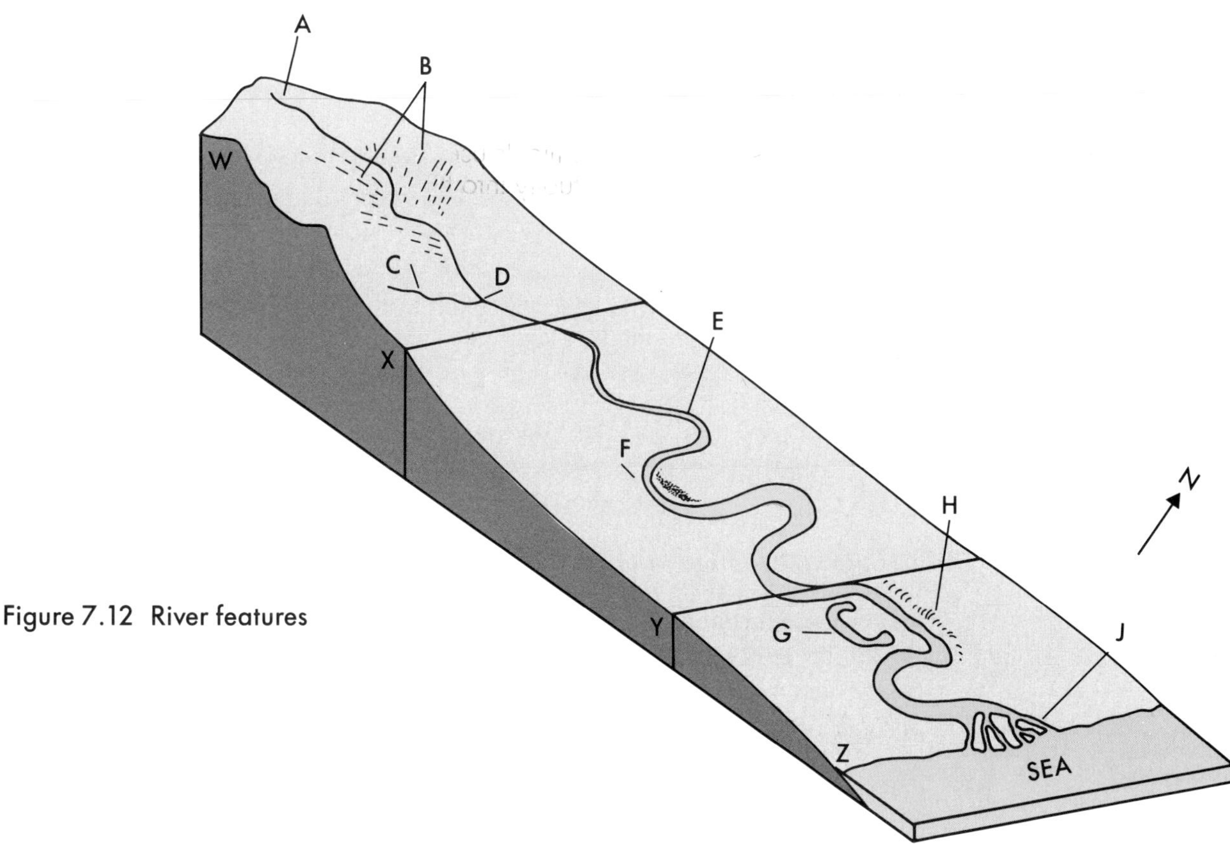

Figure 7.12 River features

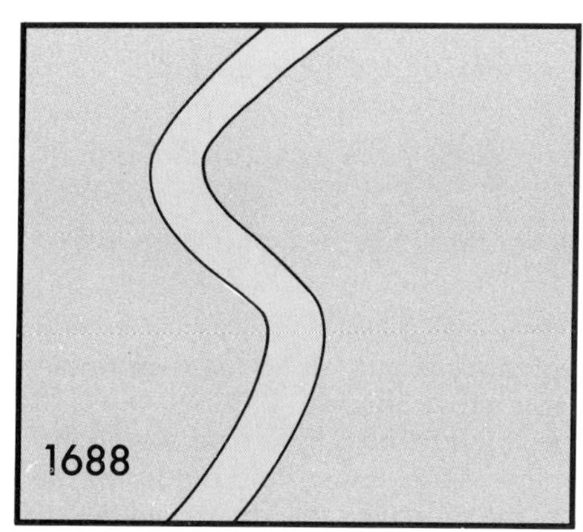

△ Figure 7.13a

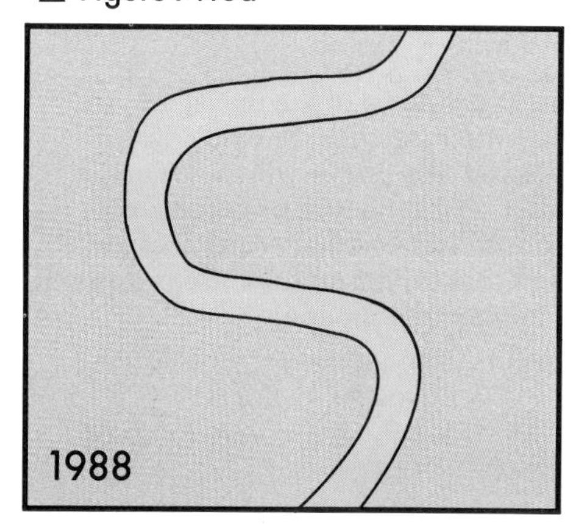

△ Figure 7.13b

(e) As a river such as the one in figure 7.12 flows from its source to its mouth, which of the following is most likely: (i) its gradient increases and its channel widens; (ii) its gradient decreases and its channel widens; (iii) its gradient decreases and its channel narrows?

(f) Observe the arrow which indicates north. Then state the general direction in which the river is flowing.

2. (a) Explain each of the following terms which relate to rivers: source, meander, flood plain.
   (b) Suggest ways in which rivers are of benefit to people.
   (c) Describe some methods by which people have tried to control or prevent the destructive forces of a river.
   (d) Describe some ways in which rivers in Ireland are being polluted. Refer to pollution by urban areas, by agriculture and by industry.

3. Locate the Emlagh River on the Tralee Bay OS map in the Map Supplement which accompanies this book.

   Make a tracing of this river and the coastal area into which it flows. On your tracing, show and label the following: the source of the Emlagh; a tributary; a confluence; the mouth of the Emlagh; the Emlagh river basin; the watershed which separates the Emlagh river basin from those of other rivers to the southeast of it.

4. Copy figures 7.13a and 7.13b. Label your diagrams so that they explain the changes which have taken place in the course of the river over the past 300 years.

   Draw a third diagram and label it to illustrate what further changes might be expected to take place in the river's course over the next 300 years.

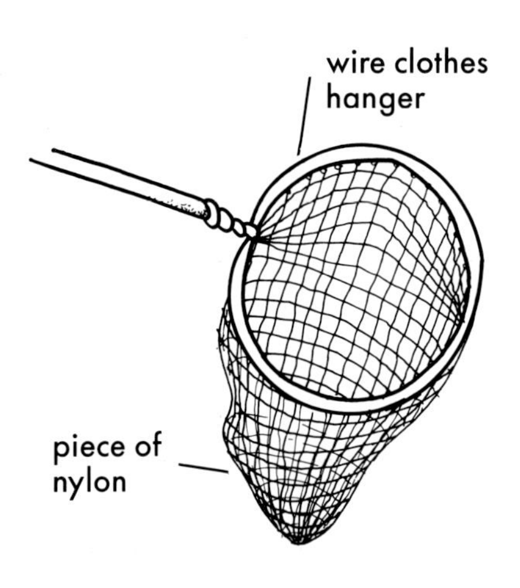

5. *Fieldwork:* To test pollution levels in a local river or stream

- ☐ Suitable for teams of two or three students
- ☐ Equipment:
  Small bottles or jars for water samples.
  A shallow dish (preferably with a white bottom) in which to examine insects.
  A small magnifying glass with which to identify insects.
  A simple net with which to collect insects. The net should be very tight-meshed. It could be made from a piece of nylon and a wire clothes hanger (see diagram).
  Some universal indicator paper and some lead acetate paper (ask your science teacher for some of each).
- ☐ *CAUTION!* Never work alone on fieldwork activities!
  Never enter deep water, no matter how tempting!

Choose your test site carefully. Do any local factories, sewage plants or other works discharge their waste into the river? If so, carry out tests both above and below these locations. By comparing the water quality of both tests, you should be able to judge the polluting effects, if any, of these works.

There are three main ways in which you can check for water pollution.

1. *Look carefully at the river.*
   (a) Is the water clear?
   (b) Does the surface contain any scum or foam—both of which may indicate pollution?
   (c) Is there a healthy weed growth present, or has the vegetation completely choked the river?
   (d) Is there any evidence of rubbish or waste in the water or along the banks?
   (e) Are there any signs of fish, birds or other animals along the river?

2. *Survey the insect life.*
   Use your net to sweep through the weeds and mud along the river banks or bed. Sift carefully through what you have collected. Pick out any insects which you find and place them, along with a little river water, in a shallow dish. Try to identify any of the insects, using field guides or the illustrations in figure 7.14. What does the insect life tell you about the river? When you have finished, return the insects to the water.

3. *Test the water*
   Fill your bottle or jar with river water. Dip a piece of the universal indicator paper into the sample. If the water is *acidic*, the indicator paper will turn red. If the water is *alkaline*, the paper will turn blue. If the water is *pure*, the paper will remain almost neutral or greenish.
   Now dip a piece of lead acetate paper into another water sample. If the paper darkens in colour, it is a sign that *hydrogen sulphide* is present in the water. This is usually a sign of industrial pollution.

Recording your findings

Make a copy of the chart shown in figure 7.15. Use it to record your findings. Discuss the results of your study with your class.

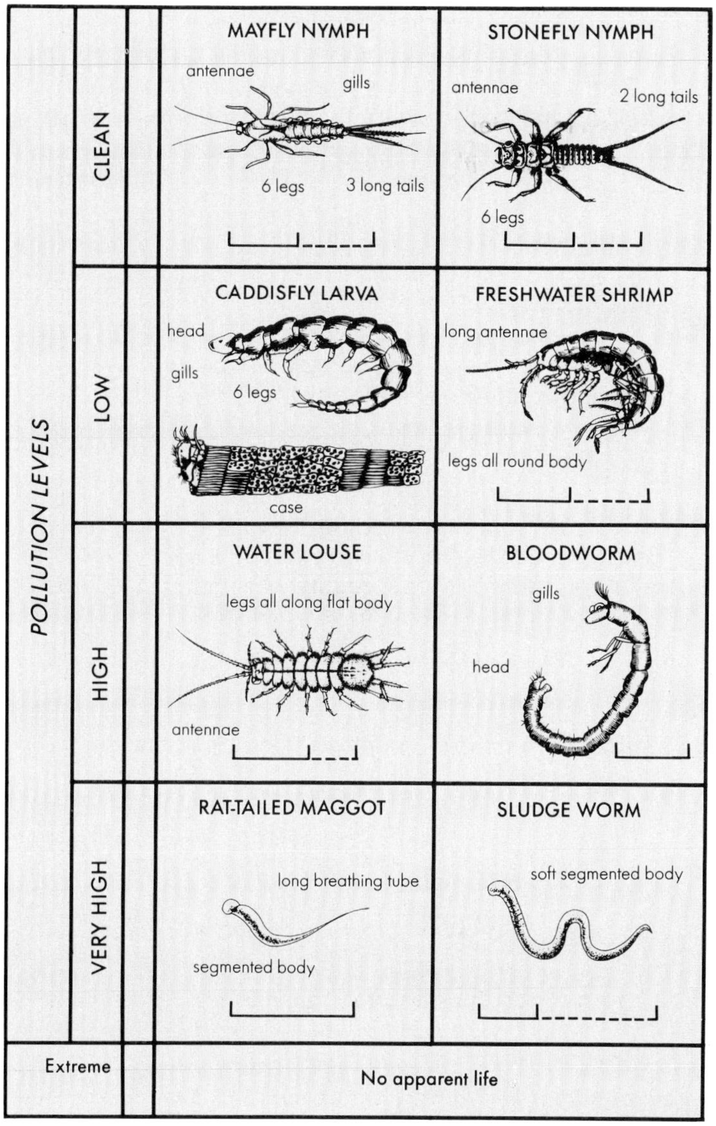

POLLUTION LEVELS		
CLEAN	MAYFLY NYMPH antennae, gills, 6 legs, 3 long tails	STONEFLY NYMPH antennae, 2 long tails, 6 legs
LOW	CADDISFLY LARVA head, gills, 6 legs, case	FRESHWATER SHRIMP long antennae, legs all round body
HIGH	WATER LOUSE legs all along flat body, antennae	BLOODWORM gills, head
VERY HIGH	RAT-TAILED MAGGOT long breathing tube, segmented body	SLUDGE WORM soft segmented body
Extreme	No apparent life	

Approximate size ranges
minimum maximum

Figure 7.14 Some insects which indicate levels of water purity or pollution

Names of fieldworkers ______________________
____
____

River: ______________________
Site of test: ______________________
Date: ______________________

Description of river
How clear is the water? ______________________
Surface scum or foam? ______________________
Weed growth? ______________________
Rubbish? ______________________
River life? ______________________

Insect life
Species found ______________________
Indication of pollution level ______________________

Water test
Acidity/alkalinity ______________________
Hydrogen sulphide ______________________

Overall conclusions
____
____

△Figure 7.15

8 THE WORK OF MOVING ICE

Just over one million years ago, the climate of countries such as Ireland began to change, gradually becoming colder and colder. Eventually, the winter snows no longer melted when summer came. Mountain snow piled up and became compressed by its own weight until it turned to ice. The ice moved slowly onto the lowlands in great rivers of ice known as ***glaciers***. Ireland became a cold wasteland, like much of the Arctic region is today.

A great Ice Age had arrived.

A glacier on the move

The picture shows a glacier in the Swiss Alps. Study the picture and identify each of the following features by their appropriate letter.

☐ ***Snowfields*** or areas of permanent snow cover. Snow accumulates in these areas and provides a constant supply of ice for the glacier.

☐ A ***glacier*** making its way slowly down the valley. (The inset shows ***crevasses***, which are long, deep, cracks in the glacier. Crevasses form when the often-brittle ice descends steeply over obstacles in its path, and so splits open.)

☐ The ***snout*** or front of the glacier, where melting is considerable and from which a river of meltwater flows.

☐ A steep-sided, flat-floored ***glaciated valley*** to the foreground of the glacier. This valley was formed at the height of the Ice Age by a glacier which was much larger than the one shown in the picture.

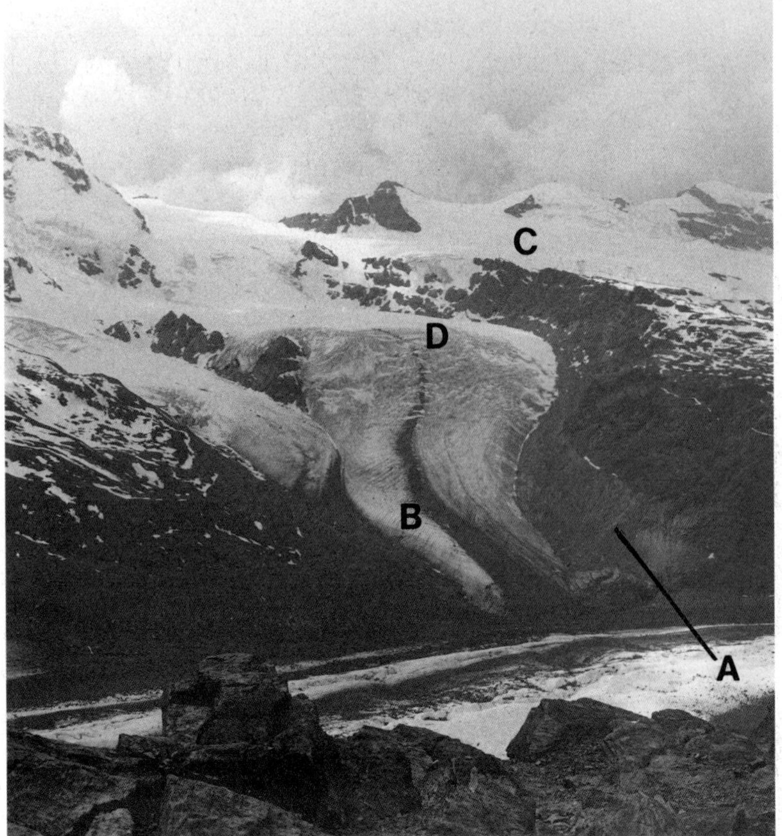

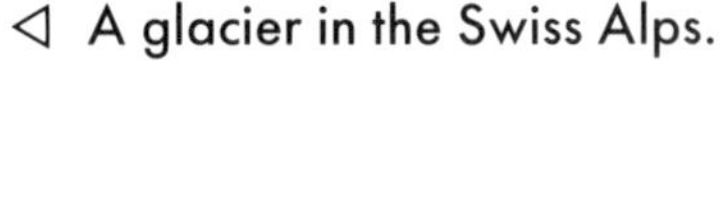

◁ A glacier in the Swiss Alps.

Crevasses in a glacier. ▽

Figure 8.1 Using a pencil, fill in the blanks with the appropriate terms: snowfield; glacier; crevasses; snout of glacier; steep side of glaciated valley; flat floor of glaciated valley; meltwater river; long, narrow lake on valley floor.

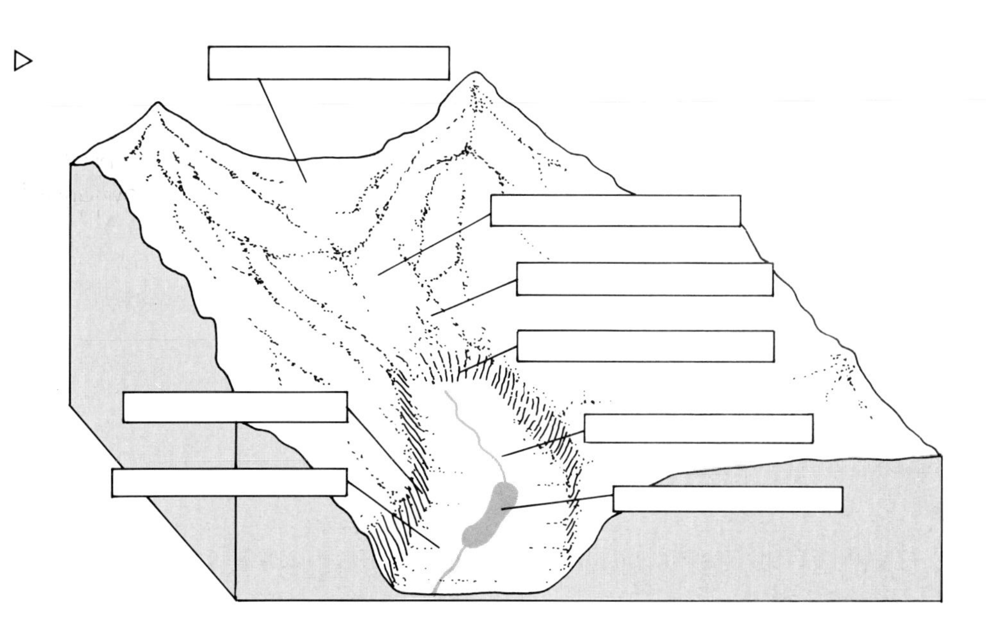

Erosion by moving ice

(mainly on highlands)

How ice erodes

1. *Plucking*

   Sometimes the base of a glacier may melt into the ground. It may then freeze again and, as the glacier moves forward, it may 'pluck' chunks of rock away with it.

2. *Abrasion*

   The plucked rocks become embedded in the base of the glacier. As the glacier moves, these rocks scour and smooth the surface over which they pass.

Features of Erosion
1. cirque (corrie)
2. arete
3. glaciated valley
4. ribbon lake
5. hanging valley

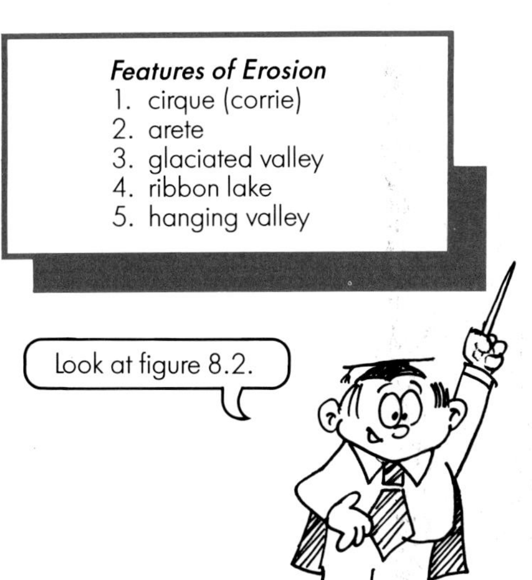

Features of glacial erosion

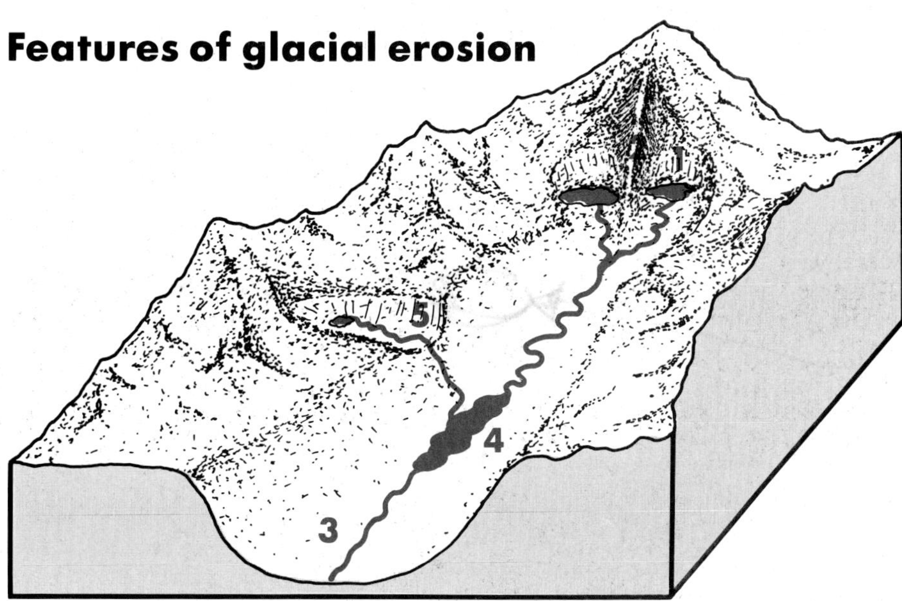

Figure 8.2 Featues of glacial erosion

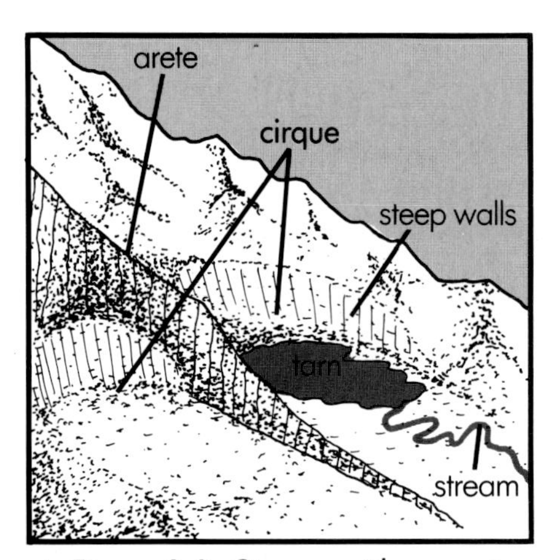

△ Figure 8.3 Cirques with an arete

Examples are at Coumshingaun in the Comeragh Mountains and the Devil's Punchbowl in Killarney, Co. Kerry.

Cirques or corries (figure 8.3)

These are large, basin-shaped hollows in mountains. They are steep-sided on three sides. Corries sometimes contain round lakes called ***tarns***.

Formation

Each cirque was once the birthplace of a glacier. Snow accumulated in these areas and became compressed into ice. The ice ***eroded*** (plucked and abraded) deep hollows as it began to move slowly downhill in the form of a glacier.

Basin-shaped hollow or corries often contain lakes called tarns. △

Examples occur at the south side of Coumshingaun and at the east side of the Devil's Punchbowl.

Aretes (figures 8.2 and 8.3)

Aretes are very narrow ridges occurring between adjacent cirques.

Formation

When ***two cirques*** developed side by side, the land between them became eroded until only a ***narrow ridge*** separated them.

An arete in the Himalayas. ▷

Glaciated valleys (figure 8.4)

Examples are Glendalough in Co. Wicklow and the Gap of Dunloe in Co. Kerry.

Glaciated valleys are deep, straight, U-shaped valleys with flat floors and steep sides.

Formation

Glaciated valleys were originally river valleys which were occupied by glaciers during the Ice Age. The powerful ***glaciers*** deepened and straightened the valleys. The glaciers ***steepened the valley sides and flattened their floors*** so that the valleys became U-shaped.

Ribbon lakes (figure 8.4)

Examples are in Glendalough in Co. Wicklow and the Gap of Dunloe in Co. Kerry.

Ribbon lakes are long, narrow lakes in glaciated valleys.

Formation

Some parts of a valley floor may have ***less resistant*** rocks than others. The softer parts of the valley may be ***eroded more deeply*** by the glaciers, leaving long hollows which later become filled with water to form lakes.

Hanging valleys (figure 8.4)

An example can be seen at Lugduff Brook which enters Glendalough by means of a hanging valley.

Hanging valleys are tributary valleys which 'hang' above the levels of the main glaciated valleys. Rivers often flow from hanging valleys into the main glaciated valleys by means of waterfalls.

Formation

Many valleys are deepened greatly by the powerful glaciers which once occupied them. But their ***tributary valleys were not deepened to the same degree***. When the glaciers melted, the floors of the tributaries remained high above the floors of the main valleys.

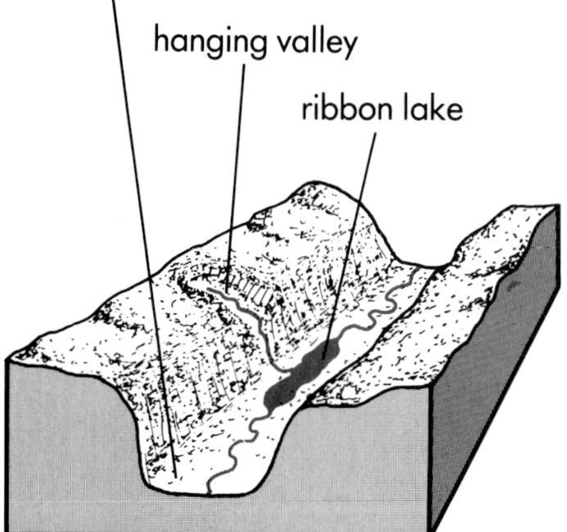

△ Figure 8.4 Glaciated valley, hanging valley and ribbon lake

Can you see a hanging valley in figure 8.4?

Glendalough, Co. Wicklow. △

- ☐ What evidence is there in the photograph that the valley shown is a glaciated valley?
- ☐ Identify a hanging valley.
- ☐ How does the photograph suggest that this glaciated area is visited by tourists?
- ☐ Suggest how Glendalough got its name (hint: Irish placenames).
- ☐ Draw a simple cross-section of the valley. Show and label: steep valley sides; a flat valley floor; a ribbon lake.

Transport by moving ice

In general, moving ice erodes highland areas. Then, as it moves ***downslope***, it transports the eroded material to lowland areas and deposits it there (figure 8.5).

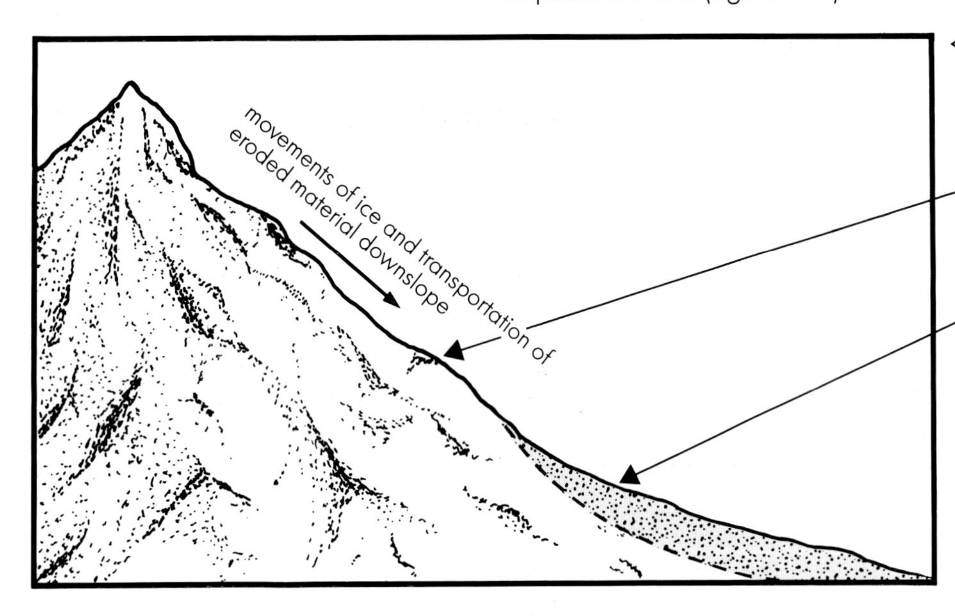

◁ **Figure 8.5**

Highland areas were eroded by the moving ice.

Eroded materials were usually deposited on lowlands.

How a glacier transports material

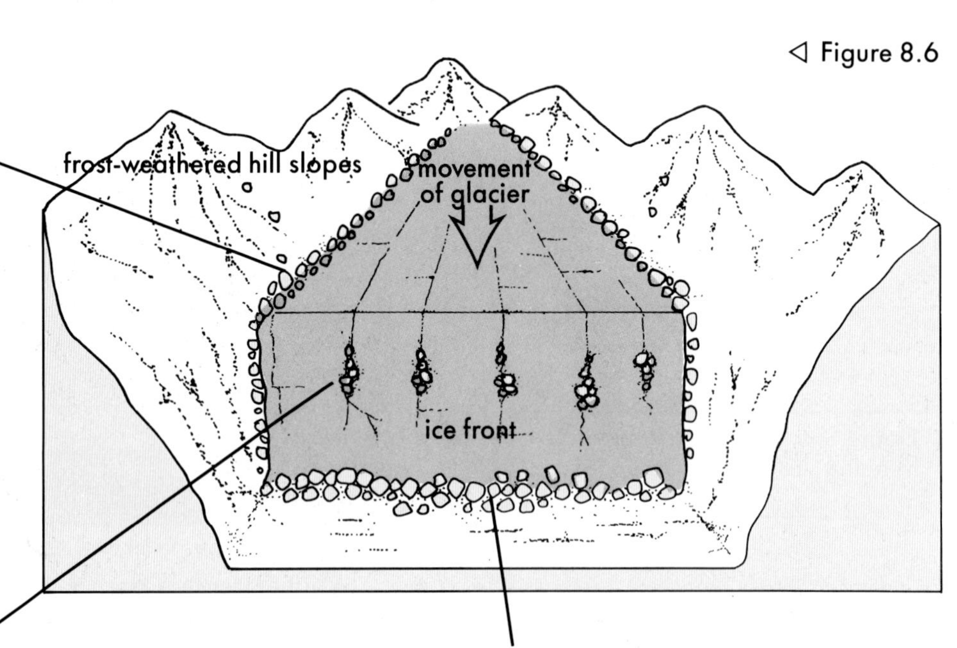

◁ **Figure 8.6**

1. Some material is carried along the surface of the glacier. This is called ***onglacial material***. Most of it consists of screes which were weathered by frost action on the hill slopes overlooking the glacier and which fell down these slopes onto the glacier's surface. (Onglacial materials often form lateral and medial moraines; see figure 8.7)
2. More material is embedded deep in the glacier. This is called ***englacial material.*** Some of it was plucked from the valley sides by the moving ice. Some of it was formerly onglacial material which fell through crevasses into the ice.
3. ***Subglacial material*** lies between the base of the glacier and the valley floor. It moves as the glacier plucks and grinds it against the rock below. Some subglacial material is pushed along before the moving ice front like litter before a giant sweeping brush.

Glacial deposition

(mainly on lowlands)

As the glaciers moved slowly from the mountains to the lowlands, they carried with them the materials which they had eroded. This material was deposited on the lowlands by the melting ice.

Features of glacial deposition

Moraines

△ Identify a lateral moraine in the photograph. What other glacial features can you see?

The word ***moraine*** is generally used to refer to any material laid down by ice. This material usually includes large boulders, small stones and fine soil particles, all mixed together.

There are special types of moraine which take the shape of long, narrow heaps of glacial deposits. These include lateral, medial and terminal moraines (figure 8.7). They are deposited on the glacial valley floors when the glaciers melt.

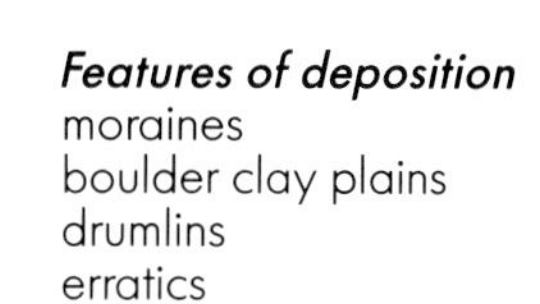

Features of deposition
moraines
boulder clay plains
drumlins
erratics

Materials fell from hill slopes down on to moving glaciers. These materials formed long, narrow ridges ***at the sides*** of the glaciers. Such ridges are called ***'side'*** or ***lateral moraines.***

Sometimes two glaciers met and moved along side by side. When this happened, two lateral moraines sometimes combined to form a single ***'middle'*** or ***medial moraine***.

Materials carried along by the glacier were often deposited at the ***farthest point reached by the ice*** – the place where the front of the moving ice finally melted. These materials formed a ridge called a ***'frontal'*** or ***terminal moraine.***

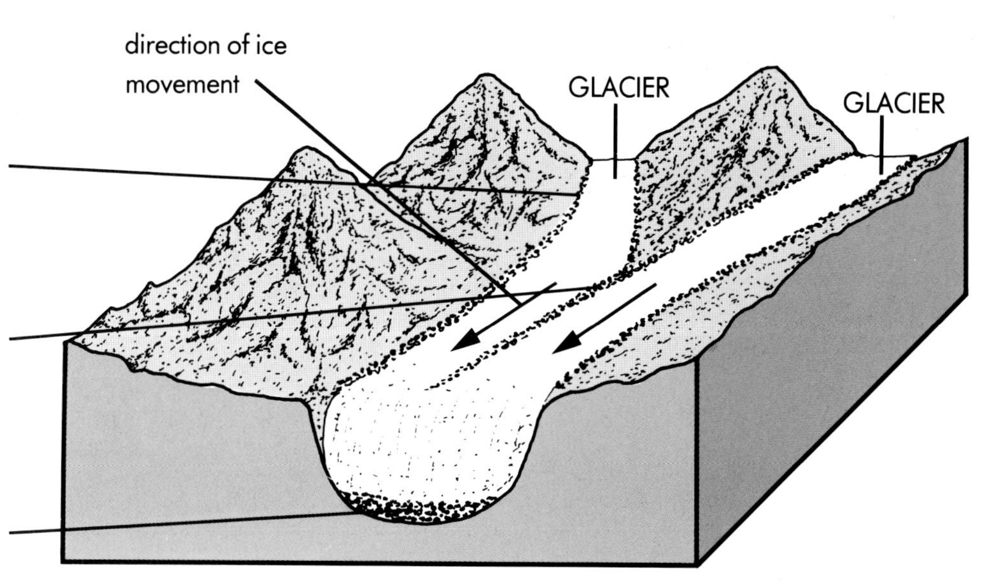

△ Figure 8.7 Lateral, medial and terminal moraines

Boulder clay plains

During the Ice Age, many lowland areas became covered by vast masses of slow-moving ice called ***ice sheets***.

The ice sheets laid out huge quantities of moraine on the lowlands. This material is called ***boulder clay*** because it contains large boulders and fine clay mixed together.

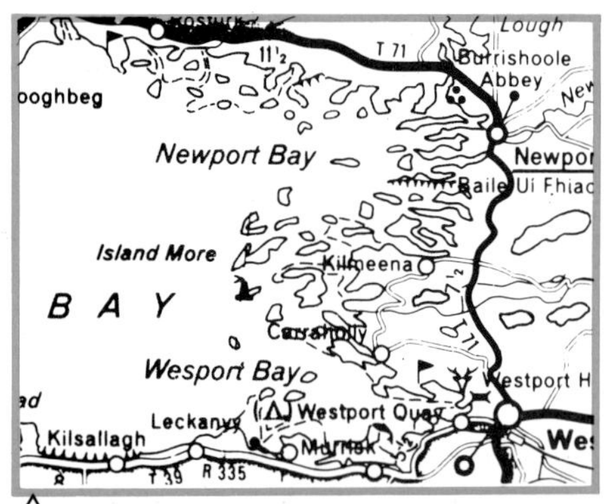

Figure 8.8 Drumlins in the Clew Bay area. Identify drumlins on land as well as partly submerged drumlins.

Drumlins

Drumlins are rounded, oval-shaped hills. They are often about 60 metres high and usually occur in large numbers known as ***swarms***.

Formation

Most drumlins were formed when the ice sheets laid down boulder clay in hill-sized heaps. Others were formed when the ice moved over existing deposits, smoothing and shaping them into oval-shaped hills.

These islands in Clew Bay are drumlins which became partly covered when sea levels rose at the end of the last Ice Age. Why do you think sea levels rose at this time?

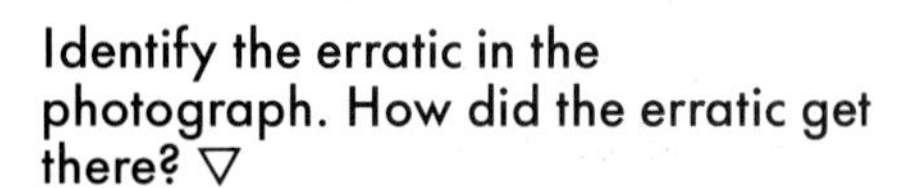

Examples: Drumlins are common in the Clew Bay area of Co. Mayo, where many have been partly covered by the sea to form small islands.

Identify the erratic in the photograph. How did the erratic get there? ▽

Erratics

The ice sheets sometimes carried boulders over great distances, dumping them in places where the rock type was quite different. Such boulders look out of place in the landscape in which they now stand, so they are called erratics.

Examples: Scottish ice sheets have dropped erratics in Co. Dublin, while granite boulders from the Mourne Mountains have been deposited in Co. Cork.

Meltwater from the ice

Towards the end of the Ice Age, vast amounts of water were released from the melting ice. This water often flowed from the ice in the form of *meltwater streams.*

These tables show the amounts of water which flow in a present-day meltwater stream in the Swiss Alps. Table A shows the varying discharge during a 24-hour period. Table B shows how the discharge varies throughout the year.

Study Table A and answer the following questions.

1. Draw a line graph to show how the discharge of the stream varies over a 24-hour period. Use an enlarged copy of the graph axes below Table A to help you.
2. Describe the pattern shown by your graph.
3. Attempt to explain the pattern shown.

Discharge of the meltwater stream over a 24-hour period on 12 May 1988

Hour of discharge	*Discharge in cubic metres per second*
0300	0.2
0600	0.3
0900	0.9
1200	2.4
1500	4.5
1800	2.6
2100	0.6
2400	0.3

Table A

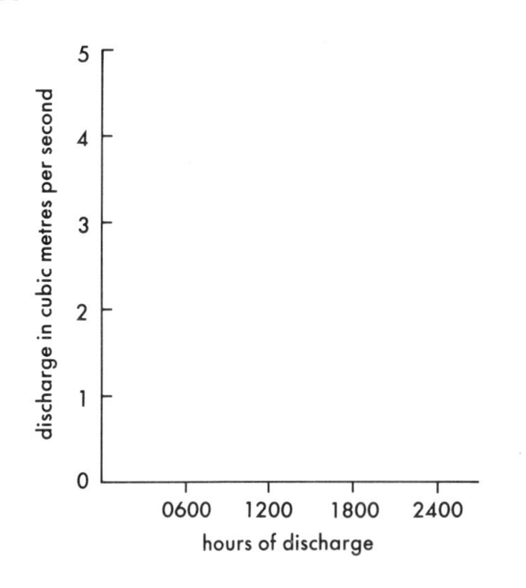

Annual discharge of a meltwater stream

Season of discharge	*Average discharge in cubic metres per second*
Winter	0.3
Spring	0.5
Summer	2.8
Autumn	2.7

Table B

Study Table B and answer the following questions.

1. Draw a series of bar graphs to illustrate the changes in the meltwater discharge.
2. Describe the pattern shown by the bar graph.
3. Attempt to explain the pattern shown.

Meltwater features
outwash plains
eskers
fiords

Meltwater features

The meltwater from the ice formed the features listed in the box.

Outwash plains (figure 8.9)

Outwash plains are areas of sand and gravel usually found near terminal moraines.

An example is the Curragh in Co. Kildare.

The Curragh in Co. Kildare. △

Figure 8.9 The formation of an outwash plain

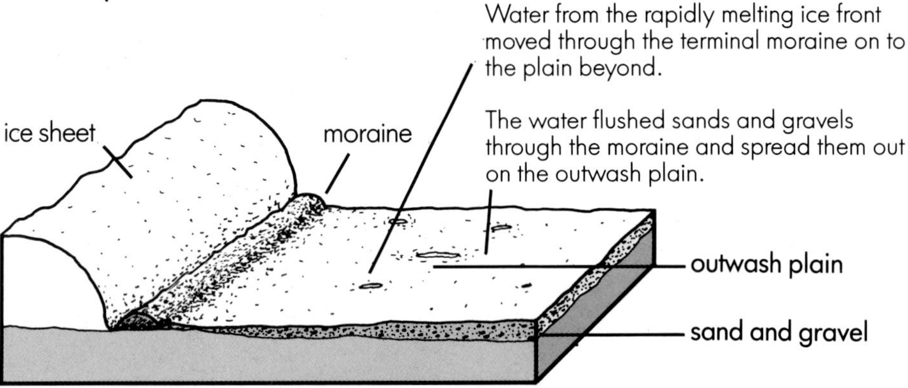

Formation

Water from the rapidly melting ice front moved through the terminal moraine on to the plain beyond.

The water flushed sands and gravels through the moraine and spread them out on the outwash plain.

Eskers (figure 8.10)

Eskers are long, narrow ridges of sand and gravel.

Examples occur between Athlone, Co. Westmeath and Athenry, Co. Galway.

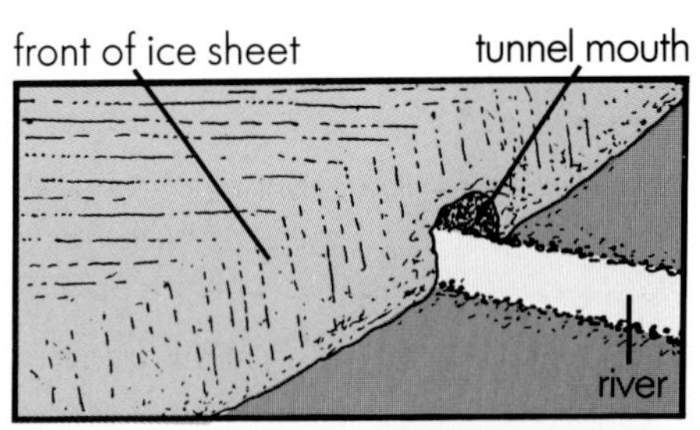

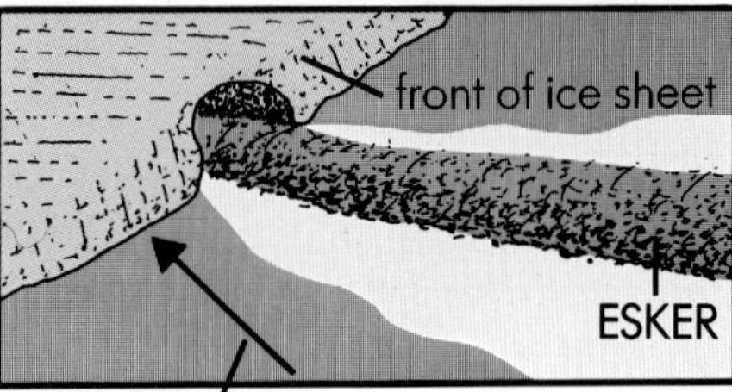

Figure 8.10 The formation of an esker

Formation

1. ***Rivers*** flowed through tunnels under the ice sheets. They flowed rapidly and carried with them large quantities of materials.
2. As each river left the ice sheet it lost its force. It then ***deposited*** its material near the mouth of its tunnel.
3. As the ice melted, the position of the ***tunnel's mouth retreated*** gradually . . .
4. . . . leaving a ***ridge*** of deposited material along the line of its retreat.

Fiords (figure 8.11)

Fiords are drowned U-shaped valleys which take the form of deep, steep-sided sea inlets.

Formation

Fiords were once glaciated valleys which were drowned by rising sea levels at the end of the Ice Age. The level of the sea rose because of the vast amounts of meltwater which entered it.

▽ Figure 8.11 A fiord

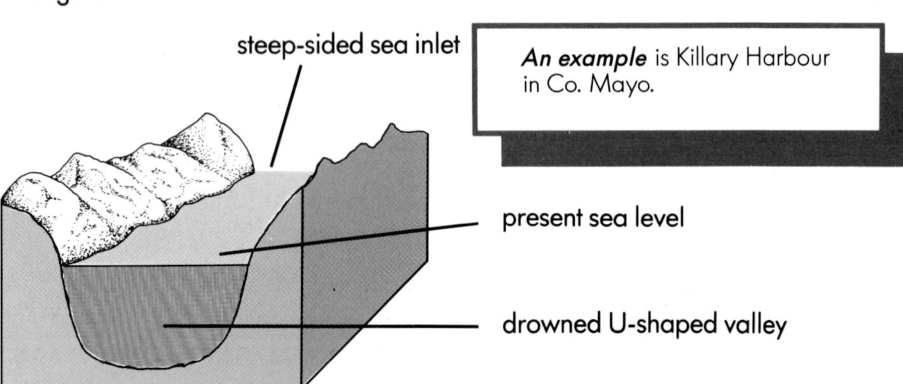

An example is Killary Harbour in Co. Mayo.

Glaciation and people

Use these newspaper cuttings to help write a brief account entitled 'The Effects of the Ice Age on the Economy of Ireland'.

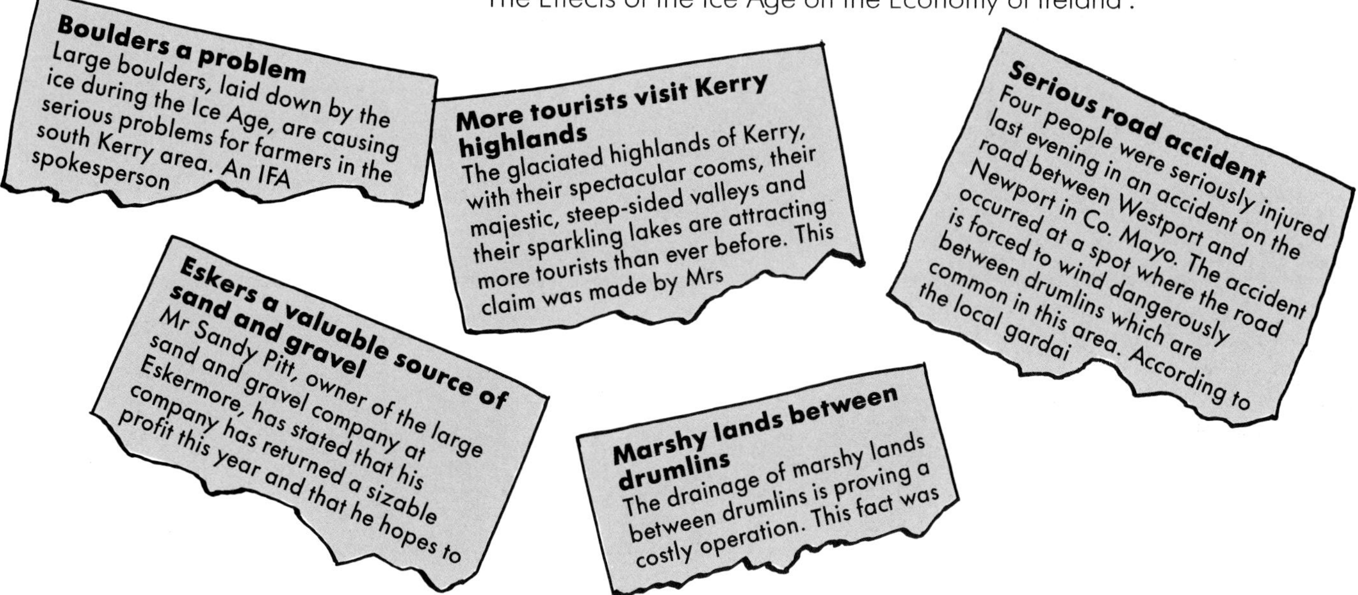

Boulders a problem
Large boulders, laid down by the ice during the Ice Age, are causing serious problems for farmers in the south Kerry area. An IFA spokesperson

More tourists visit Kerry highlands
The glaciated highlands of Kerry, with their spectacular cooms, their majestic, steep-sided valleys and their sparkling lakes are attracting more tourists than ever before. This claim was made by Mrs

Serious road accident
Four people were seriously injured last evening in an accident on the road between Westport and Newport in Co. Mayo. The accident occurred at a spot where the road is forced to wind dangerously between drumlins which are common in this area. According to the local gardai

Eskers a valuable source of sand and gravel
Mr Sandy Pitt, owner of the large sand and gravel company at Eskermore, has stated that his company has returned a sizable profit this year and that he hopes to

Marshy lands between drumlins
The drainage of marshy lands between drumlins is proving a costly operation. This fact was

Activities

1. On a page in your notebook, make a box like the one below. Use the information from this chapter to fill in the box.

Features of glacial erosion	Features of glacial deposition	Features formed by meltwater

2. Make six statements about glaciation by matching each of the 'heads' in Column A with its appropriate 'tail' in Column B.

Column A	Column B
A glacier is . . .	. . . a well known Irish glaciated valley.
A corrie is . . .	. . . a ridge of material marking the farthest advance of an ice front.
A terminal moraine is . . .	. . . a river of ice that moves slowly downslope.
A drumlin is . . .	. . . a ridge of material which was deposited by a meltwater stream.
Glendalough is . . .	. . . a large, basin-shaped hollow in a mountain.
An esker is . . .	. . . a rounded, oval-shaped hill.

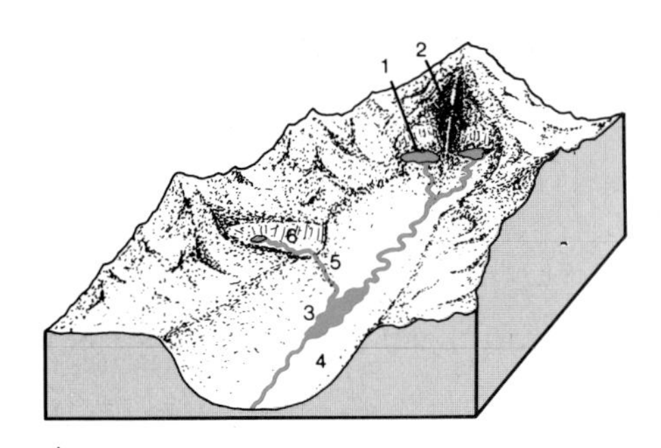

△
Figure 8.12 A glaciated upland landscape

3. (a) Examine figure 8.12 and identify the features marked 1 through 6.
   (b) Choose any three of these features and explain how each was formed.
   (c) Explain briefly each of the following terms: moraine, erratic, outwash plain.

4. *Mapwork*

   Study figure 8.13 which shows how various glacial features may be identified on Ordnance Survey maps. Then examine the Clew Bay OS map extract in the Map Supplement which accompanies this book.

   (a) Identify the type of glacial feature at each of the following locations on the Clew Bay map: L 89 69; L 90 71; L 91 65; L 91 67; L 81 68; M 03 90; M 000 648.
   (b) Identify and give the location on the Clew Bay map of *one other* example of each of the types of features referred to in (a).

Figure 8.13 Glacial features on an Ordnance Survey map ▷

Cirque: The contours in a cirque are very close (indicating very steep sides) and are almost circular in shape.

Arete: An arete may be found between two cirques which were formed close together. The arete appears as a very narrow ridge with very close contours or cliff symbols at its sides.

Hanging valley: It is a small valley which enters the main glaciated area. Note how the contours of a hanging valley tend to straighten out as they approach the glaciated valley.

Drumlins: These are clusters of small, oval-shaped hillocks. They are usually between 100 and 200 feet high.

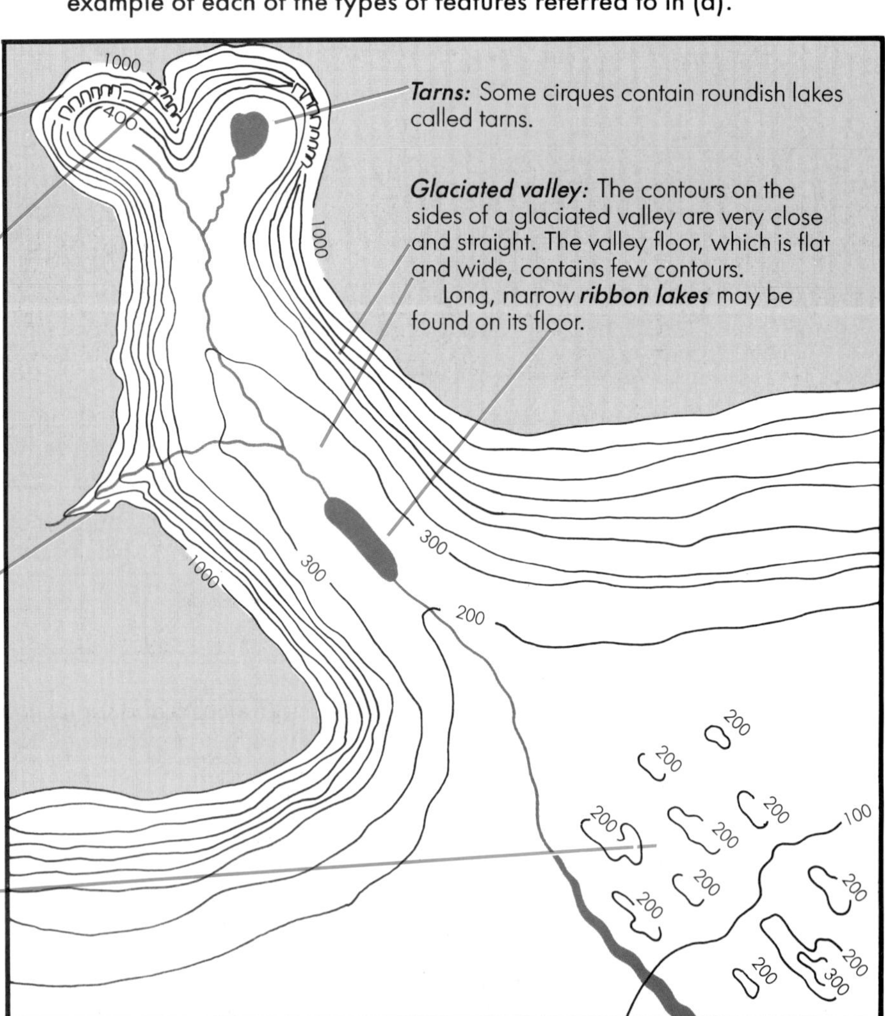

5. *Fieldwork*

   Organise a class trip to a glaciated region near your home or school. Prepare worksheets before you go and compile a report on the results of your trip. In your report, discuss the origins, characteristics and human effects of the glacial features which you have seen. Illustrate your report with diagrams, photographs and maps.

9 THE WORK OF THE SEA

The work of the sea along our coasts includes erosion, transport and deposition.

Study figure 9.1 to discover the meaning of each word: coast, shore, beach, swash, backwash.

Figure 9.1 The work of the sea ▷

Waves are formed when wind moves over the surface of the water. This movement causes the particles of water to make a circular motion which forms a wave shape. It is this ***wave shape*** (and not the actual water) which moves forward on the path of the wind.

When waves reach shallow water they ***break.*** Only then does the water rush towards the shore.

swash: the rush of water ***up the shore*** following the breaking of a wave

backwash: the return movement of water ***down the shore***

CLIFF

BEACH

COAST

HIGHEST TIDE LEVEL

LOWEST TIDE LEVEL

SHORE

BREAKING WAVE

Sea erosion

Erosion is usually carried out

- ☐ by ***strong waves***
- ☐ on ***exposed or steeply-sloping shorelines***

The strength of waves depends on two things . . . .

Is sea erosion greater on the west coast or on the east coast of Ireland? Before answering, study figure 9.2 and consider the following: (a) the direction of Ireland's most common or ***prevailing*** winds; (b) wind fetches on the east and west coasts; (c) the appearance of the west and east coasts.

Figure 9.2 ▷

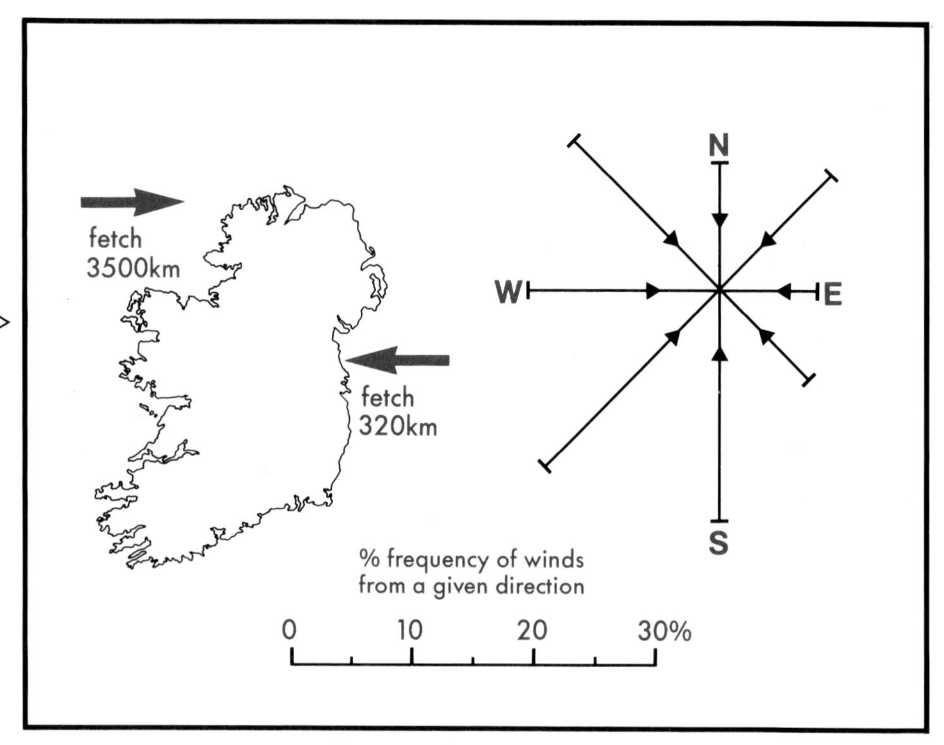

How waves erode

Water crashes against the coast, wearing and cracking the rocks.

Air becomes trapped and compressed in rock cracks as each wave crashes against the coast. As the waves move back, the compressed air expands quickly, causing tiny explosions which eventually shatter the rock.

Stones and sand are hurled against the coast by the waves. These help break down the coast and are themselves broken down.

Features of sea erosion

- bays and headlands
- sea cliffs
- sea caves
- sea arches
- sea stacks
- blow holes

Bays and headlands

A ***bay*** is a large, curved opening into the coast. A ***headland*** is a high piece of land jutting out into the sea.

Examples: Dublin Bay and Wexford harbour are examples of bays. Howth Head in Co. Dublin and Slyne Head in Co. Galway are examples of headlands.

Formation

▽Figure 9.3a Areas of hard and soft rock lie exposed to the sea

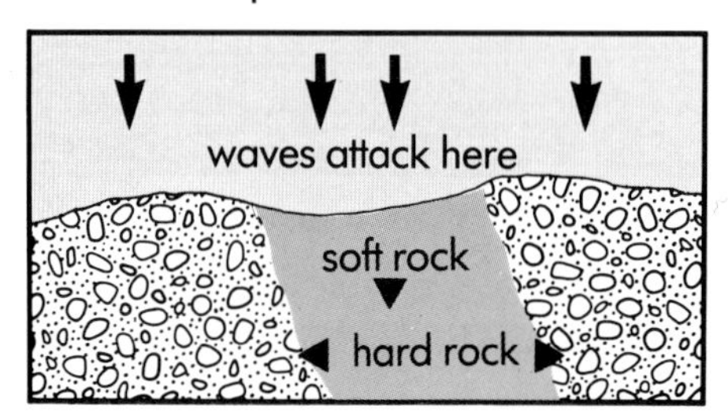

▽Figure 9.3b The soft rock is eroded back quickly to form bays. The hard rock resists erosion and juts out to form headlands.

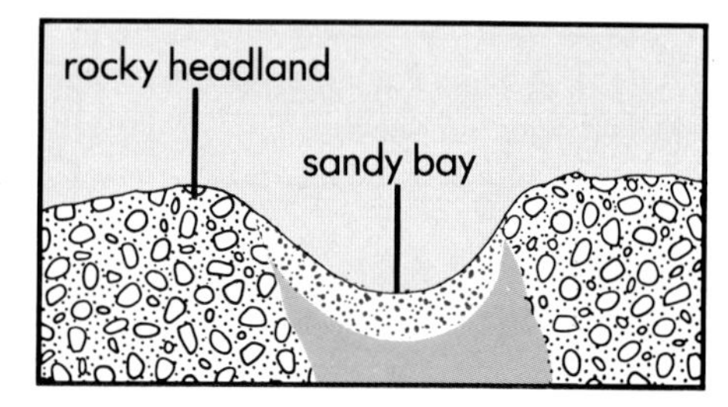

Sea cliffs

A ***sea cliff*** is a high rock face which slopes very steeply inland from the shore.

Examples can be seen at the Cliffs of Moher in Co. Clare and the cliffs of Achill Island, Co. Mayo.

Formation

Figure 9.4 The formation of sea cliffs

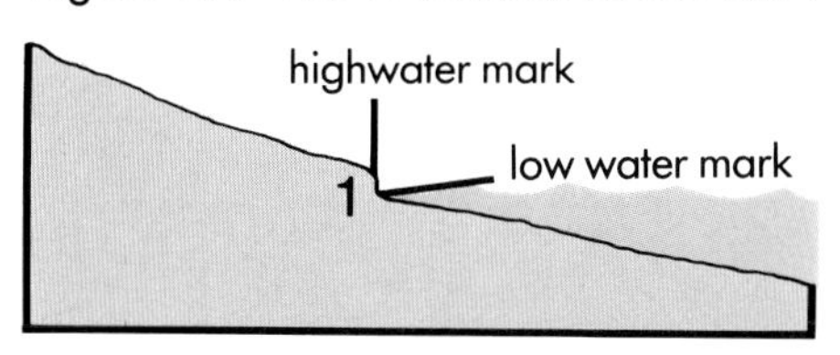

1 Waves erode a ***notch*** in the coast.

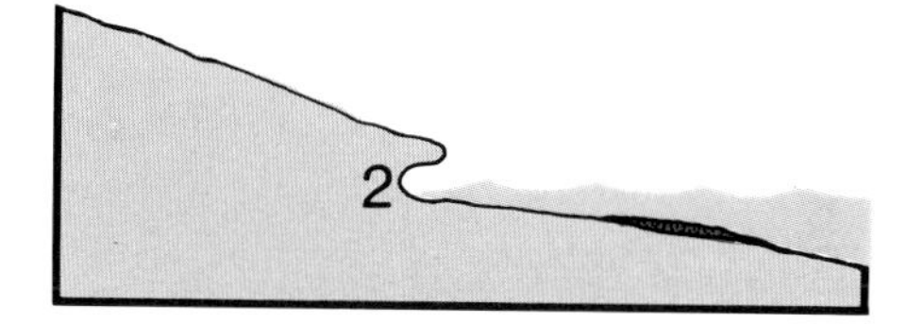

2 The notch gradually becomes deeper.

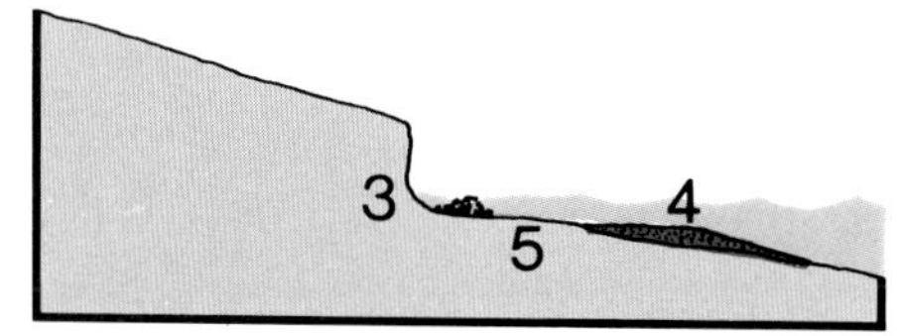

3 The area above the notch becomes so undercut that it collapses, forming a cliff.
4 Material eroded from the cliff builds up to form a wave-built terrace.
5 The former base of the cliff remains as a wave-cut platform.

△ Sea erosion on the west coast of Ireland. Identify a cliff, an undercut part of the cliff and a wave-cut platform.

Sea caves, arches and stacks

▽ Figure 9.5 The formation of sea caves, arches and stacks

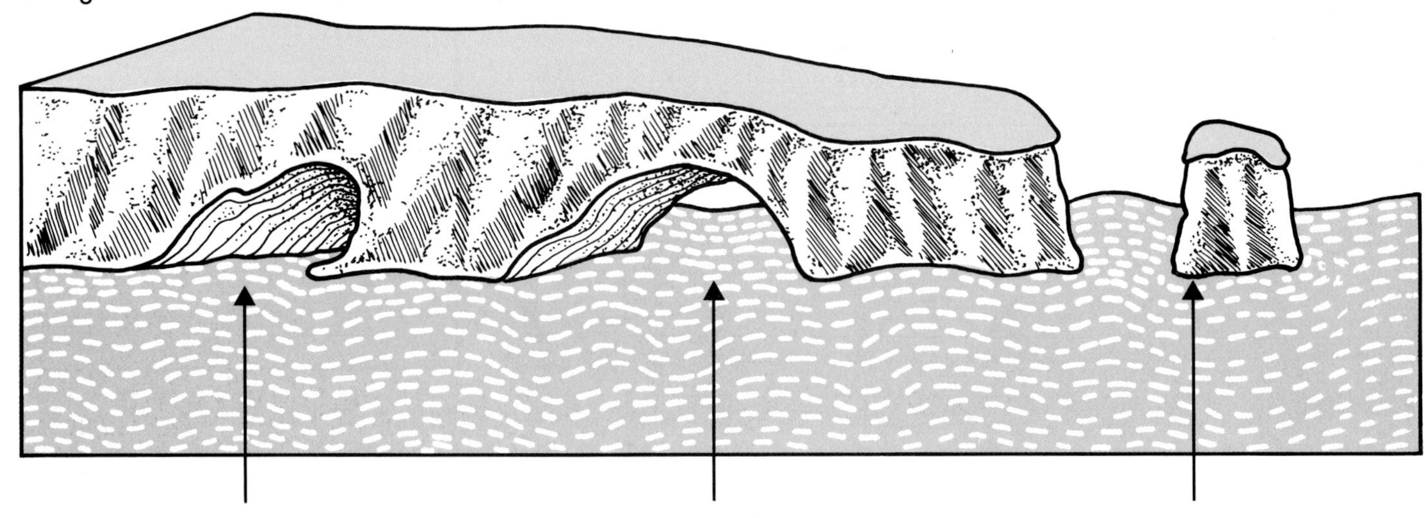

Sea cave

A ***sea cave*** is a long, cylindrical tunnel in a cliff. It usually decreases in width away from the entrance.

Formation

Waves find a weak spot at the base of a cliff and gradually erode it until a cave is formed.

> ***Examples*** may be seen at the Cliffs of Moher and along the west coast of Achill Island.

Sea arch

A ***sea arch*** is a natural archway in a rocky headland.

Formation

If a ***cave erodes right through a headland*** (or if two caves erode through from either side), an arch may form.

> ***Examples*** include the Old Head of Kinsale in Co. Cork and Portsalon, Co. Donegal.

Sea stack

A ***sea stack*** is a pillar of rock jutting out of the sea near the coast.

Formation

If an ***arch collapses***, its outer wall may stand out as a pillar of rock.

> ***Examples*** can be seen at Ballycastle, Co. Antrim and Ballybunion, Co. Kerry.

△
The Cliffs of Moher, Co. Clare. Identify a sea cave and a sea stack.

△
What features of sea erosion can you identify?

Blow holes

Examples occur at Kilkee, Co. Clare and Ballybunion, Co. Kerry.

A ***blow hole*** is a hole which joins the roof of a cave with the surface above. It is called a blow hole because sea spray may be blown up through it in stormy weather.

Formation

Figure 9.6 The formation of a blow hole ▽

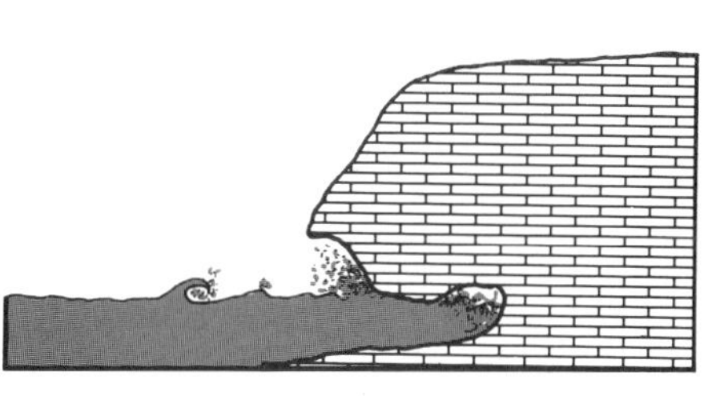

1 When powerful waves crash into a cave, ***air becomes trapped*** inside the cave.

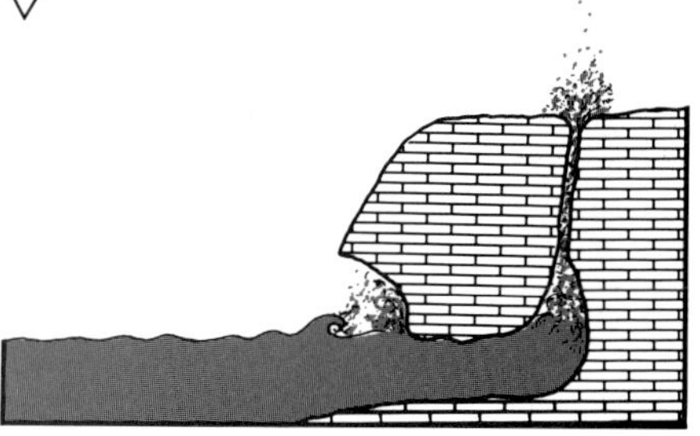

2 The ***air pressure breaks a hole*** in the roof of the cave.

Transport by the sea

The mud, sand and shingle (pebbles and stones) carried along by the sea are called its ***load***.

How waves move their loads

The load can be moved:

- ***Up*** the shore by the ***swash***, and ***back down*** the shore by the ***backwash***.
- ***Along*** the shore by the process known as ***longshore drift***. This process takes place where the waves reach the shore at an angle.

Figure 9.7 shows how this works.

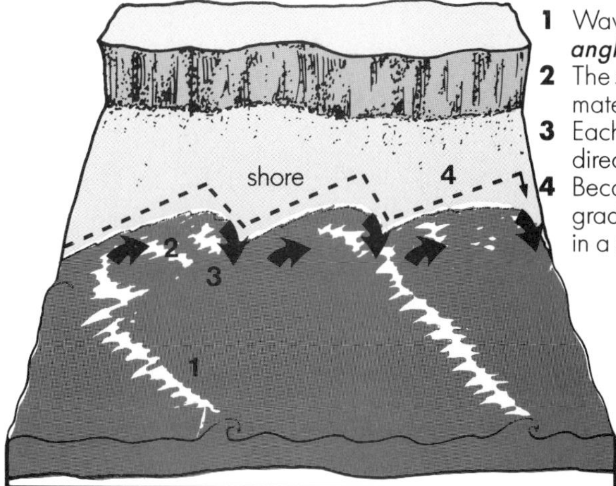

1 Waves approach the shore at an acute ***angle***.
2 The ***swash*** of each wave pushes material up and along the shore.
3 Each ***backwash*** drags the material directly down the shore.
4 Because of these things, the material is gradually ***transported along*** the shore in a zig-zag manner.

◁ Figure 9.7 How longshore drift works

Groynes are low walls built out into the sea to stop the movement of longshore drift. How do groynes prevent longshore drift? Why do port authorities or resort owners wish to prevent longshore drift? ▷

Sea deposition

How deposition takes place

The sea drops its load where the ***force of waves*** or sea currents is ***reduced***. This may occur, for example, in ***sheltered bays*** or in ***gently-sloping*** coastal areas.

Features of deposition

beaches
spits
bars
tombolos

Beaches

A ***beach*** is an area of sand or shingle which occurs between high and low tide levels.

Examples are found at Bray, Co. Wicklow and at Salthill in Co. Galway.

Formation

When ***waves break*** they ***lose their force*** and ***deposit the material*** they are carrying. The ***swash and backwash*** push and drag these materials up and down the shore. The beach material is usually sorted (figure 9.8).

△
Describe how the sea has sorted the beach material shown in the photograph.

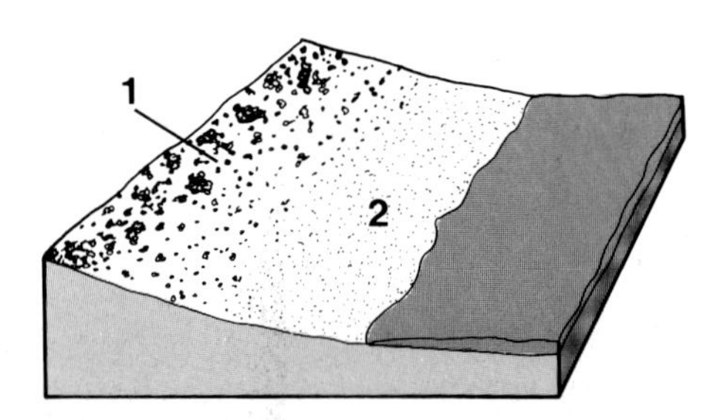

△ Figure 9.8 How beach material is sorted by the sea

1 The ***swash*** is usually powerful. It can even ***push heavy stones*** to the top of the shore.
2 The ***backwash*** is usually less powerful. It can drag ***only finer sand and gravel*** particles to the bottom of the shore.

Spits

A ***spit*** is a narrow ridge of sand or shingle. It projects into the sea but is connected to the land at one end.

Examples are at Portmarnock, Co. Dublin and Bannow Bay, Co. Wexford.

Formation

Longshore drift stops when it reaches a bay or other ***sheltered place***. The material carried by the longshore drift is ***deposited*** at these places and may build up gradually to form a ***spit*** (figure 9.9).

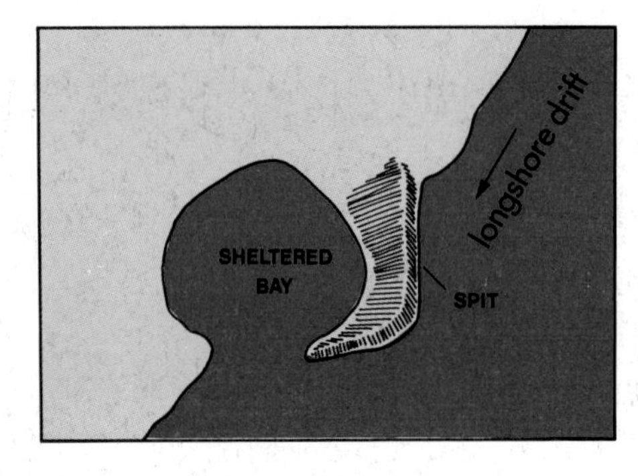

△ Figure 9.9 A spit

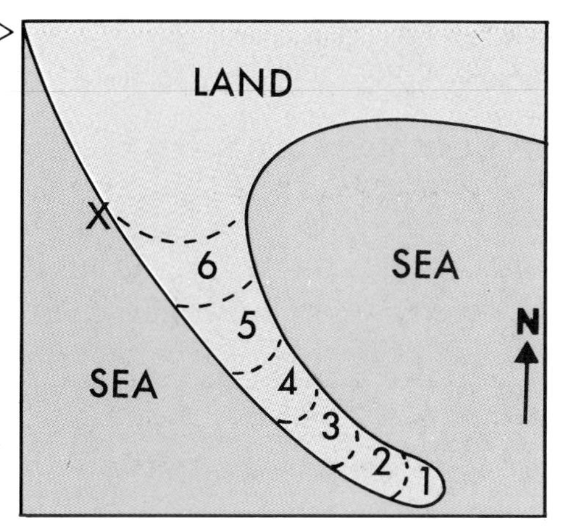

Figure 9.10 ▷

Study figure 9.10 which shows a sand spit.

1. At X, are waves most likely to come from a northwesterly, southwesterly or southerly direction?
2. Did most of the material making up the spit come from the south, the southeast, the northwest or the east?
3. Did the spit grow from 1 towards 6, from 6 towards 1, or from neither of these directions?
4. What measures could be taken to halt the growth of this spit?

Bars

A ***bar*** is a narrow ridge of sand or shingle which seals off the mouth of a bay.

Examples occur at Lady's Island Lake, Co. Wexford and at Roonagh Lough, Co. Mayo.

Formation

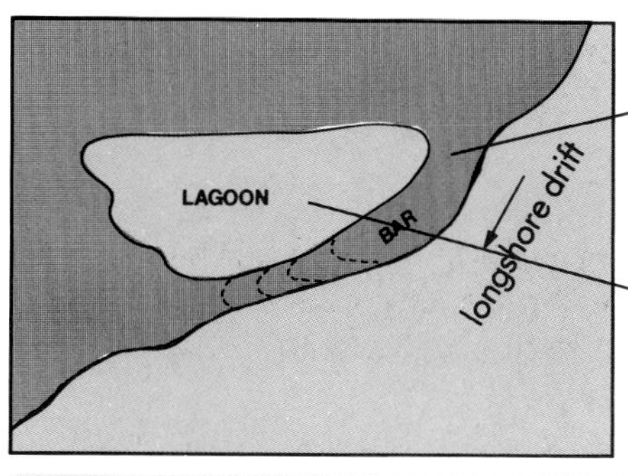

A ***spit*** may grow in length until it completely ***seals off a bay***. The former spit is then referred to as a ***bar.***

A lagoon is a small lake ***formed behind a bar.*** It was originally part of the bay which was sealed off by the bar.

◁ Figure 9.11 A bar

Tombolos

A ***tombolo*** is a narrow ridge of sand or shingle which joins an offshore island to the mainland.

Examples occur at Sutton, Co. Dublin and at Castlegregory, Co. Kerry.

Formation

A ***spit*** may grow in length until its seaward end reaches an offshore island. The former spit is then referred to as a ***tombolo***.

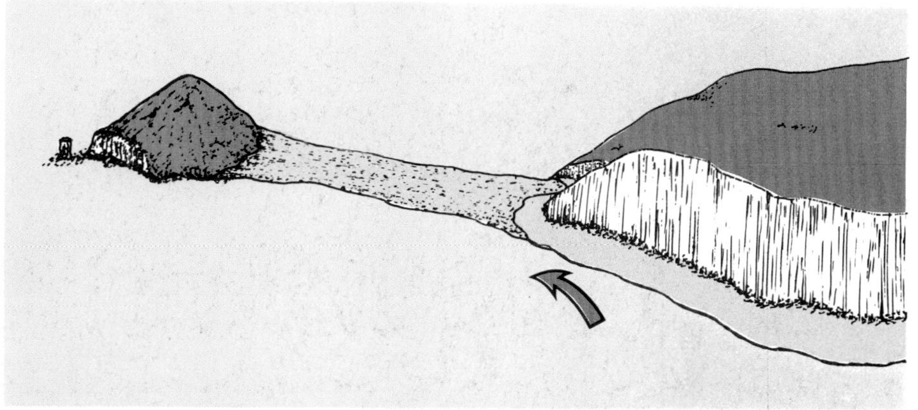

△ Figure 9.12 A tombolo

The tombolo at Brow Head, Co. Kerry. ▷

The Destructive Power of the Sea

A Case Study of Ballycotton, Co. Cork

For centuries, the sea has provided livelihoods for the fishermen in the East Cork village of Ballycotton. But the sea has sometimes been an enemy as well. Its powers of erosion and deposition have inflicted great damage on Ballycotton and the surrounding area (figure 9.13 and Box 1).

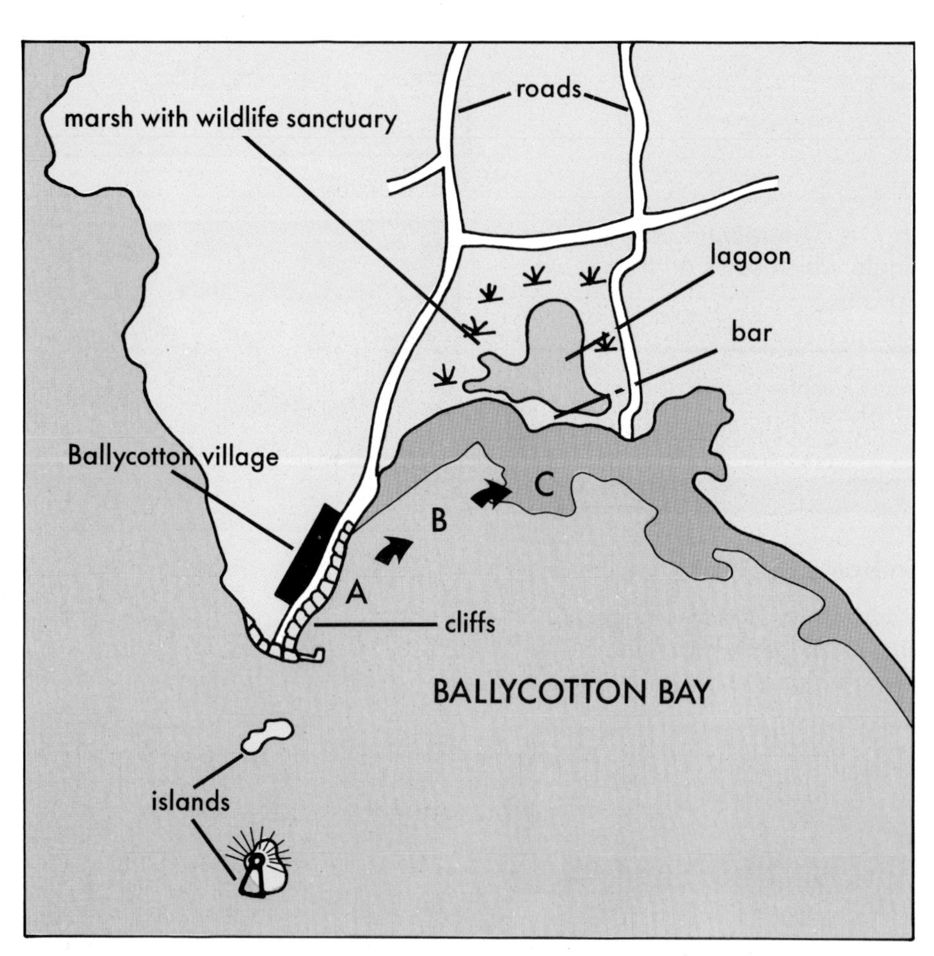

Figure 9.13 The Ballycotton Bay area ▷

▽ Box 1 Marine processes and problems at Ballycotton

Processes		Problems
A. ***Sea erosion*** occurs here on soft boulder clay cliffs.	▶	Valuable farmland, several houses and some roadways have been eroded away.
B. ***Longshore drift*** is removing the eroded material towards the head of Ballycotton Bay.		
C. ***Deposition*** has resulted in a bar blocking the head of the bay.	▶	The bar has blocked drainage to the bay. A lagoon and marshes have formed and have flooded farmlands and roads. The marsh, however, is an important wildlife sanctuary.

The people of Ballycotton have repeatedly requested Cork County Council to carry out coastal defence work in the area. Box 2 contains six possible options for dealing with Ballycotton's problems. (Letters A, B and C refer to figure 9.13.)

Consider each option carefully. Decide on the option which you favour most and the one you favour least. Give reasons for your choices. Remember that the cost of any option would have to be met by taxpayers throughout Co. Cork, not just by the people of Ballycotton.

Box 2 ▽ Coastal defence options for Ballycotton

Possible option	*Likely cost*	*Positive comment*	*Negative comment*
1. Build ***timber groynes*** at A and B.	£30,000	Will prevent longshore drift, and thus ***reduce erosion*** at A and deposition at C	Strong waves would destroy groynes within a few years.
2. Build ***reinforced concrete groynes*** at A and B	£150,000	Should last for several decades.	
3. Build long, reinforced ***sea wall***.	£850,000	Will prevent erosion at A.	Beach at A and B would be removed by longshore drift.
4. Build ***concrete groynes plus sea wall***.	£1,000,000	Will prevent erosion and preserve beach.	
5. Dump huge ***boulders*** on beach at A and B.	£30,000	Will protect cliff from wave attack.	Unsightly boulders would cover much of the attractive beach.
6. Let ***nature*** take its course. ***Compensate owners*** of property threatened by erosion.	£50,000 (over the next 20 years)	This is the only option which would preserve the wildlife sanctuary.	The problems outlined in Box 1 would continue.

Activities

1. (a) **Match each of the *coastal features* shown in figure 9.14a with the appropriate word selected from the box.**

beach	headland	tombolo	arch
cliff	longshore drift	lagoon	cave
trapped air	spit	bay	stack
waves	bar	blowhole	

▽ Figure 9.14a

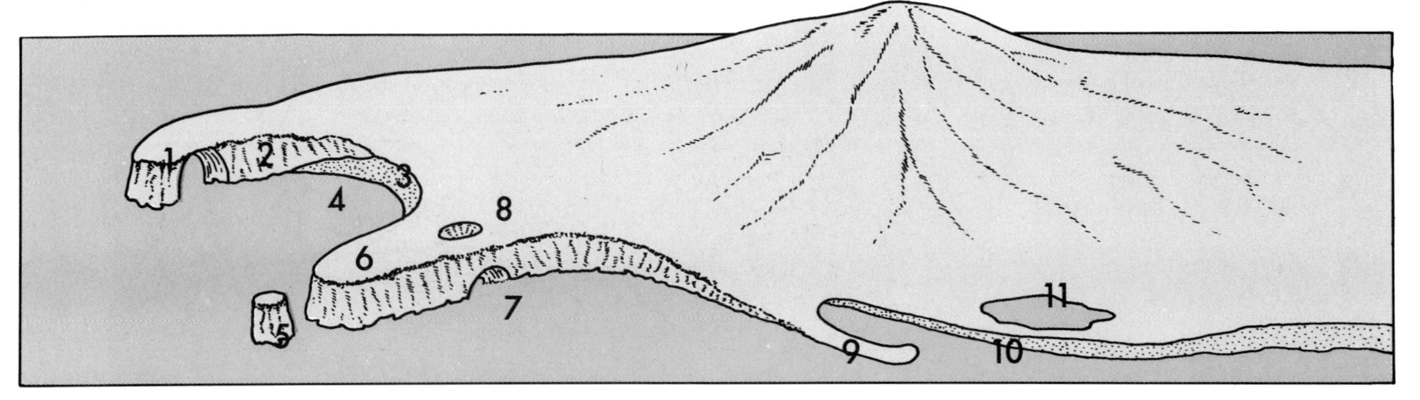

(b) Copy the flow diagram in figure 9.14b which shows the *coastal system*. Fill in the blanks in your diagram by selecting appropriate terms from the box.

▽ Figure 9.14b

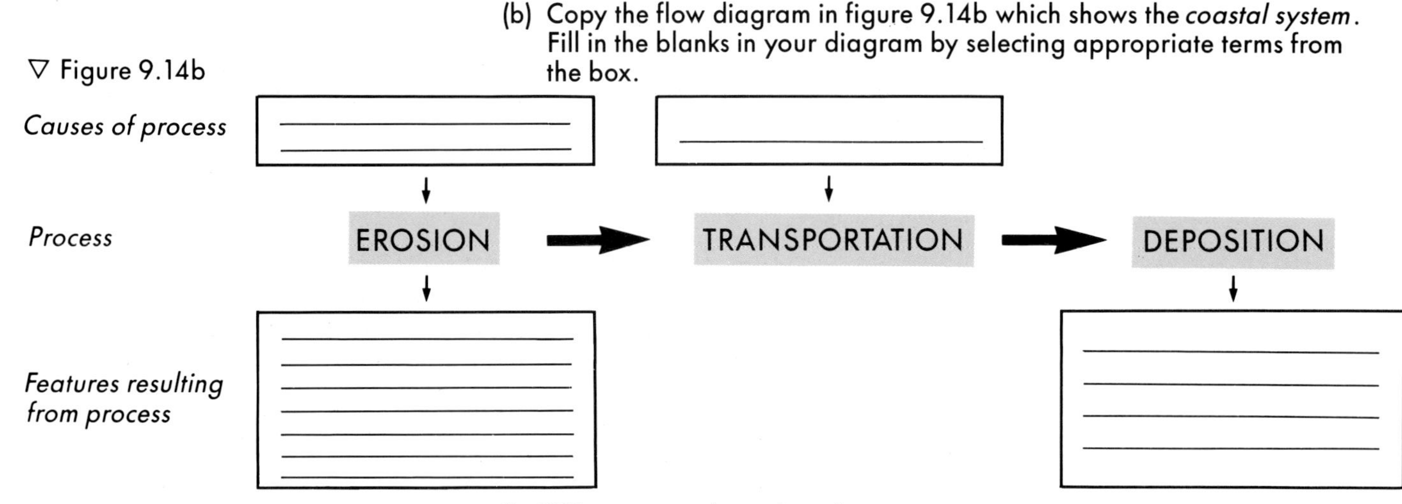

2. *Cliff, cave, sand spit, beach*

All of these features are found along the Irish coastline.

(a) Identify which of these features are formed by erosion, and which are formed by deposition.

(b) Choose *two* of the features mentioned above – *one* formed by erosion and *one* formed by deposition. For *each* one you choose, explain how it was formed, using diagrams.

(c) Human use of the coastline can also cause erosion – in sand dune areas, for example. Explain how this happens and describe the efforts that can be made to improve the situation.

3. *Mapwork*

Figure 9.15 shows a fragment of the Tralee Bay map extract which is in the Map Supplement which accompanies this book. The numbers 1 through 11 on figure 9.15 show the Ordnance Survey symbols for features formed by sea erosion and deposition. Study these features on the Tralee Bay map extract. Then identify similar features found elsewhere on the Tralee Bay or Clew Bay map extracts found in the Map Supplement.

Figure 9.15 Features of sea erosion and deposition on an Ordnance Survey map

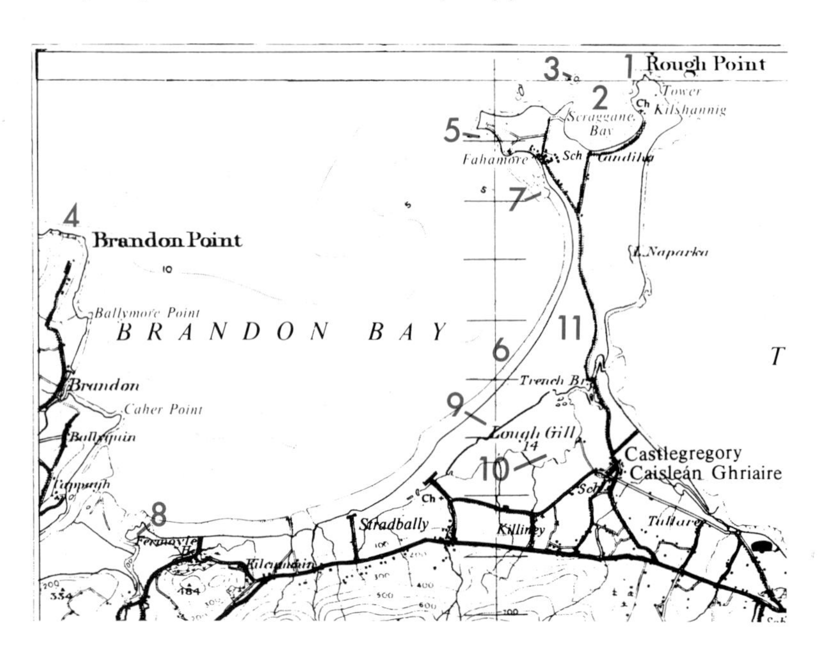

4. *Project idea*

Using newspaper cuttings, old calendar pictures, postcards etc., mount a class wall display of coastal features. Write a suitable caption for each picture displayed.

5. *Fieldwork*

At low tide—*and in the company of an adult*—visit part of the coastline and carry out the following field activities.

- ☐ *Sketch map* – Draw a large sketch map and/or picture sketch of the area visited. Show and name some of the main coastal features.
- ☐ *Nature of the coastline* – Is the coastline steep or gently sloping? Is it exposed or sheltered? From your observations, would you conclude that it is more likely to undergo erosion or deposition?
- ☐ *Erosion* – List the features of coastal erosion which you can see. Draw sketches and write descriptions of them. Describe how each feature is formed.
- ☐ *Movement of load*
  - (a) Do waves approach the shore from a direct or oblique angle? Can you draw any conclusions from this about the probable movement of the sea's load?
  - (b) Place an orange on the beach at the water's edge. Mark its position with a large stone. Then observe the movements of the orange over a period of 15 minutes. Do your observations confirm the conclusions you reached in (a)?

    (*Note:* This experiment is best carried out when the tide is coming in.)
- ☐ *Deposition*
  - (a) Examine the pebbles found on the beach. Are they smooth or jagged? Why?
  - (b) Contrast the size of beach material at the top of the beach with that at the bottom of the beach. Account for any contrast you have observed.
  - (c) List the main features of sea deposition which you find. Draw a sketch and write a description of each feature. Describe how each feature was formed.
- ☐ *People and the sea*
  - (a) Describe any damage which sea erosion or deposition have done to the locality. What could local people do to prevent this damage? Has anything been done already?
  - (b) Describe any benefits which the work of the sea may have provided for the local people. What could the people do to use these benefits to the best advantage?

10 ORDNANCE SURVEY MAPS: AN INTRODUCTION

A map is a scaled-down plan of all or part of the earth's surface. All maps are drawn to a particular ***scale***.

Figure 10.1a

☐ Small scale maps show large areas with little detail.

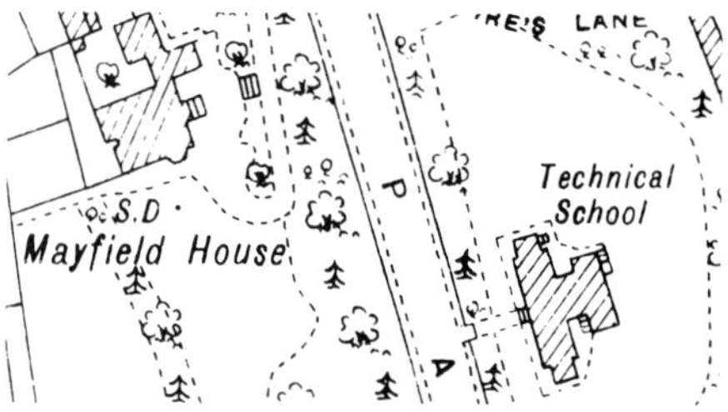

Figure 10.1b

☐ Large scale maps show small areas in great detail.

Classwork

Study the Waterford city street map and the plan of Waterford in the Map Supplement which accompanies this book. Note that the area covered by the Waterford plan is shown on the C7 and D7 grid of the street map of Waterford.

☐ Which of these two maps — the street map or the plan — has the larger scale?

☐ Which of the maps shows the greater detail? Name any five points of detail shown on one map but not on the other.

☐ Which map shows the greater area?

☐ Which of the two maps would be more useful: (a) for people wishing to find their way around the city as a whole; (b) for people wishing to study house sizes on Broad Street?

Do Activity 1, page 74.

Scale

Scale is the relationship between a distance on a map and its corresponding distance on the ground.

For example, if the scale of a map is one centimetre to one kilometre (1 cm:1 km), a length of six centimetres on the map would represent six kilometres on the ground.

How scale is shown

In the Ordnance Survey (OS) maps which we will study, scale is shown in three different ways.

Linear scale is a divided line which shows map distances in kilometres and miles.

Statement of scale, such as ½ inch to one mile.

Representative fraction (RF) such as 1:126,720. This presents distance on the map as a ratio or fraction of the corresponding distance on the ground. For example, the RF 1:126,720 tells us that any unit of measurement on the map corresponds to 126,720 similar units on the ground.

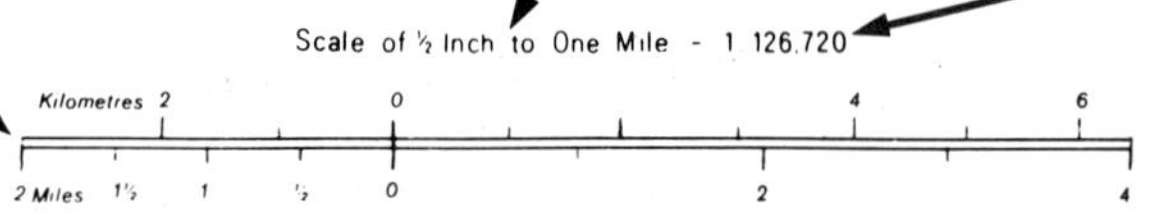

Scale of ½ Inch to One Mile - 1 126,720

Kilometres 2 0 4 6

2 Miles 1½ 1 ½ 0 2 4

Figure 10.2 Three ways of showing scale.

Examine each of the following Ordnance Survey map extracts in the Map Supplement which accompanies this book.

- ☐ Tralee Bay
- ☐ Dublin District
- ☐ Waterford (street map)
- ☐ Waterford (plan)

Write down the statement of scale and the representative fraction for each of these map extracts.

Measuring distances on maps

The ***straight line*** or ***shortest distance*** between points is sometimes called the distance 'as the crow flies'. This distance is measured as follows.

Do Activities 2 and 3, page 74.

1. Place a straight ***edge of paper*** between two points on the map. Mark precisely where each point touches the edge of the paper (figure 10.3a).
2. Now place the paper's edge on the map's ***linear scale*** and measure carefully the distance between the two marks (figure 10.3b).

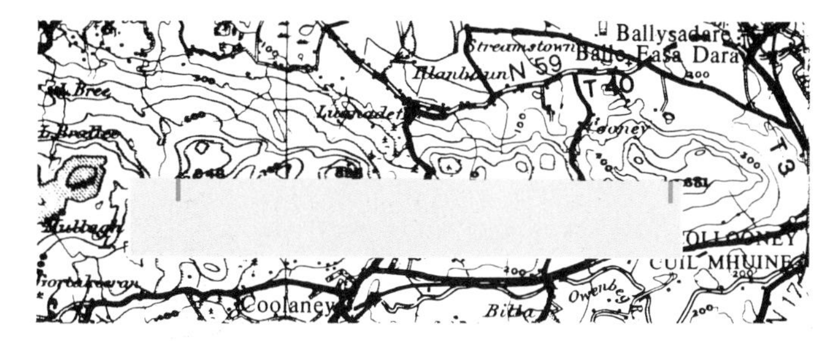

Figure 10.3a

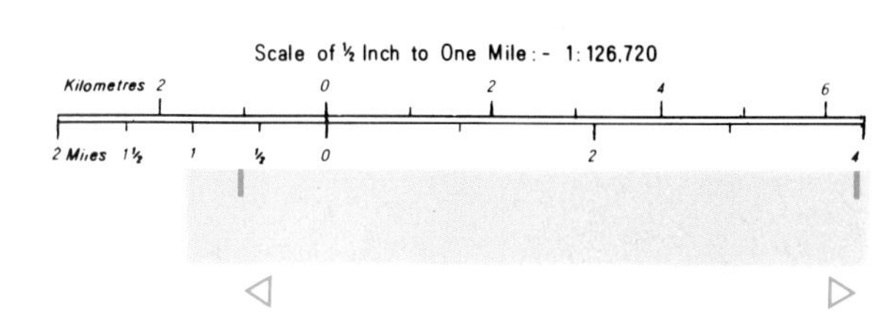

Figure 10.3b

Curved line distances, such as those along roads or railways, may be measured as follows.

1. Lay the edge of a ***strip of paper*** along the centre of the road, marking the starting point on the paper's edge.
2. Press a pencil point on the paper at each place where the road curves. Use this pencil point to ***pivot*** the edge of the paper along each succeeding section of road, until the finishing point on the road is reached. Mark this point on the paper's edge.
3. The length of the curved road is now represented on the straight edge of paper. Use the map's ***linear scale*** to measure this length.

Do Activity 4, page 74.

Calculating map areas

For higher level only!

Regular areas

To calculate a regular area represented by all or part of a map:

1. Measure the ***length*** of the area (a straight line measurement).
2. Measure the ***breadth*** of the area (a straight line measurement).
3. ***Multiply*** the length by the breadth to calculate the area.

Note! When you multiply miles by miles, your answer will be in ***square miles***.

Example
To calculate the area (in square miles) of the map shown in figure 10.4:

length	=	6.5 miles
breadth	=	4.9 miles
area (6.5 x 4.9)	=	31.8 square miles

Do Activity 5, page 74.

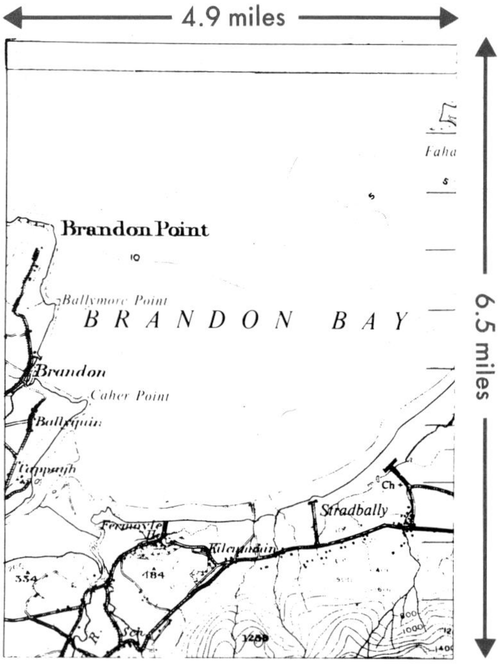

4.9 miles

6.5 miles

Brandon Point

BRANDON BAY

Brandon

Stradbally

2 Miles 1½ 1 ½ 0 2

Figure 10.4

Irregular areas

An ***irregular area*** (such as that of a wood, a lake or an island) can be calculated roughly from a map by using the method outlined in the box.

First, ensure that the irregular area is covered by squares of equal size (see figure 10.5). Then carry out the following steps.

1. ***Mark off and count*** the number of squares which are ***completely covered*** by the irregular area.
2. ***Mark off and count*** the number of squares ***partly covered*** by the irregular area. ***Divide*** this number by 2.
3. ***Add*** the answers obtained in steps 1 and 2. This gives the approximate ***total number*** of squares covered by the irregular area.
4. ***Calculate the area of one square.*** (Each square made by joining Eastings and Northings is $1km^2$ in area.)
5. ***Multiply*** the total number of squares (step 3) by the area of one square (step 4) to calculate the ***irregular area***.

Example
To calculate the area of Lough Splash (figure 10.5):

1. Number of completely covered squares = 10
2. Number of partly covered squares ÷ 2 = $^{19}/_2$ = 9.5
3. Total number of squares (10 + 9.5) = 19.5
4. Area of each square = $1km^2$
5. Total area ($19.5 \times 1km^2$) = $19.5 km^2$

X = completely covered square
O = partly covered square

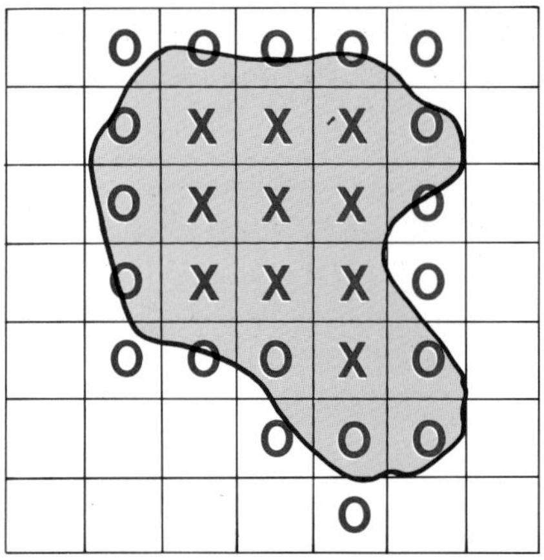

Do Activity 6, page 74.

Figure 10.5

Directions

Directions are usually given in the form of compass points. Figure 10.6 shows the sixteen main points of the compass. Learn these points and study their positions.

Figure 10.7 shows that Position 2 lies to the northeast of Position 1.

In which direction would one travel in going: from 1 to 3; from 1 to 4; from 1 to 5; from 5 to 4; from 4 to 1?

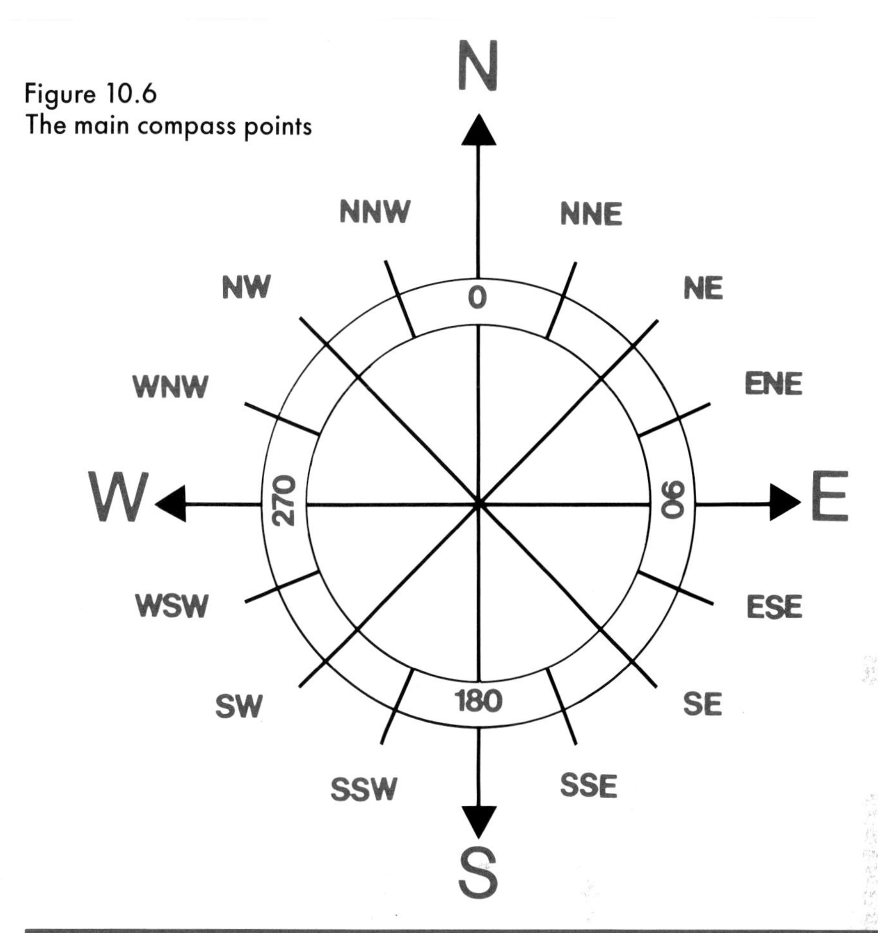

Figure 10.6
The main compass points

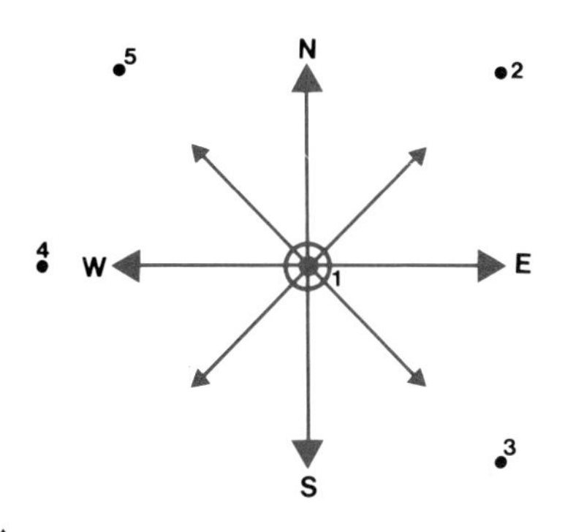

△
Figure 10.7

Do Activities 7 and 8, page 74.

Two North Poles

Did you know that there are two North Poles? There are, as explained below.

True north points to the North Star. This direction is shown by a gothic cross on the margin of each OS map.

Magnetic north (the direction towards which compass needles point) is west of true north.

The *angle of difference* between true north and magnetic north is called the *magnetic variation*.

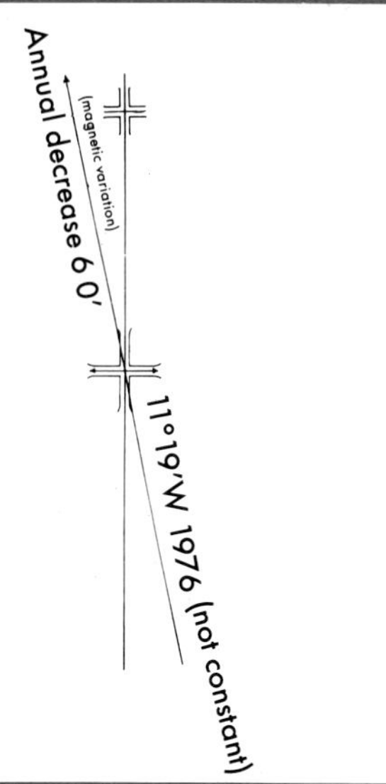

Figure 10.8 True north and magnetic ▷ north

Activities

1. Study the Dublin District map and the Central Dublin street map in the Map Supplement which accompanies this book. Which of these maps is drawn to the larger scale? Which shows the larger area? Which map shows more detail?

2. Measure in centimetres the length and width of your school desk or table. Then draw a plan of your desk or table to a scale of 1mm : 1cm.

3. (a) Measure in metres the shortest distance between the tourist office and the clock tower which are shown near Meagher's Quay in the Waterford street map in your Map Supplement.
   (b) Measure in miles and tenths of a mile the shortest distance between the peak of Kilmashogue Mountain and the peak of Glendoo Mountain. Refer to the Dublin District OS map in your Map Supplement.
   (c) Measure in miles and tenths of a mile the shortest distance between the youth hostel (red triangle) near Corraun Hill and the hostel near the mouth of Killary Harbour (Clew Bay map).

4. (a) Using the Waterford street map, measure in kilometres and metres the length of that part of the N25 road which is north of the River Suir. This road includes Sally Park, Dock Road, Fountain Street and Ross Road.
   (b) Using the Tralee Bay map, measure in kilometres the length of the L103 roadway between Castlemaine and its junction with the T68 roadway near Anascaul.
   (c) Using the Clew Bay map, measure in miles the length of the T71 roadway between Westport and Newport.

5. (a) Calculate in square miles the area represented by the Dublin District map.
   (b) Calculate in square kilometres the area represented by the Tralee Bay map.
   (c) Calculate in square metres the area represented by the Waterford plan.

6. Calculate the approximate area of that part of the Dublin District map which lies to the northeast of the N11 (or T7) roadway. This roadway is shown on the map by a green line.

7. Use the Tralee Bay map to do the following.
   (a) In which direction does one travel: (i) from Tralee to Milltown; (ii) from Milltown to Castlegregory; (iii) from Castlegregory to Lispole?
   (b) In which general direction does the coast run between Tralee and Castlegregory? In which general direction does the Owencashla River run as it flows into Tralee Bay?

8. Refer to the Waterford plan. What is the general direction from Meagher's Quay to Michael Street? What is the general direction from the tower to the warehouse on Jenkin's Lane?

11 SMALLER-SCALE ORDNANCE SURVEY MAPS

Three smaller-scale OS maps can be found in the Map Supplement which accompanies this book. These are maps of:

- ☐ Clew Bay (scale = ½ inch to 1 mile)
- ☐ Tralee Bay (scale = ½ inch to 1 mile)
- ☐ Dublin District (scale = 1 inch to 1 mile)

Locating places — the National Grid

The National Grid is used to locate places on OS maps.

On the margin of each of your one-inch and half-inch OS maps, you will find a tiny map of Ireland which is divided into lettered squares called ***sub-zones*** (figure 11.1a).

Each OS map has printed on it, in red, the letter(s) of the sub-zone(s) from which that map extract was taken (figure 11.1b).

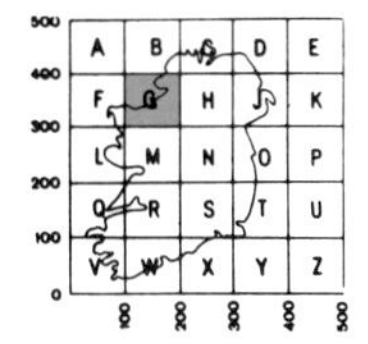

△ **Figure 11.1a Sub-zones**

▽ **Figure 11.1b**

Each sub-zone is divided into smaller squares by a grid of lines called ***co-ordinates***.

Some of these lines are ***vertical*** and are called ***Eastings.***

The other lines are ***horizontal*** and are called ***Northings***.

Both Eastings and Northings are numbered from 00 to 99. Eastings increase in value from left to right. Northings increase in value from base to top (figure 11.2).

All Eastings and Northings are drawn in on one-inch maps, such as the map of Dublin in your Map Supplement.

Figure 11.2 Eastings and Northings ▽

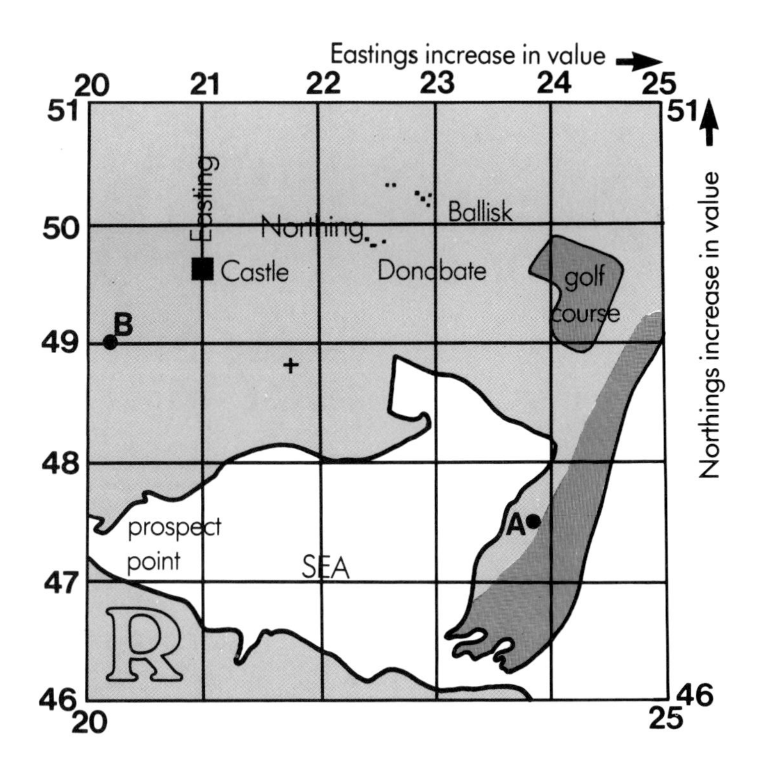

On half-inch maps, only one Easting and Northing in ten is shown. The positions of other lines are indicated by black markings along the margins of the map. These can be drawn in by carefully joining the appropriate markings (see the Clew Bay and Tralee Bay maps).

Grid references

The locations of places on OS maps are expressed in the form of ***grid references***.

Four-figure grid references

Four-figure grid references are used to locate an entire square, giving the ***general location*** of objects or features within that square.

How to get a four-figure grid reference

Step 1: Give the sub-zone letter.

Step 2: Give the two-digit ***Easting*** number which forms the ***left side*** of the square.

Step 3: Give the two digit ***Northing*** number which forms the ***base*** of the square.

Remember!
Give ***all*** grid references in this order.
1. Letter
2. Easting
3. Northing

Example
The four-figure grid reference for the church (marked with a cross) on figure 11.2 is R 21 48.
What is the grid reference for the golf course in figure 11.2?
Which square is indicated by the reference R 23 50?

Do Activity 1, page 87.

Six-figure grid references

Six-figure grid references are needed to give the exact location of small features.

To get a six-figure reference, we must imagine that the distance between each pair of Eastings and each pair of Northings is divided into tenths. So the church in figure 11.2 is on Easting 21.7 and on Northing 48.8. The decimal points are not shown when writing the grid reference, so the full reference for the church is R 217 488.

What are the grid references for points A and B on figure 11.2?

Locating a feature from a six-figure grid reference – such as W 226 529 (figure 11.3).

What features are located at R 202 474 and at R 210 496 in figure 11.2?

Figure 11.3

Do Activity 2, page 87.

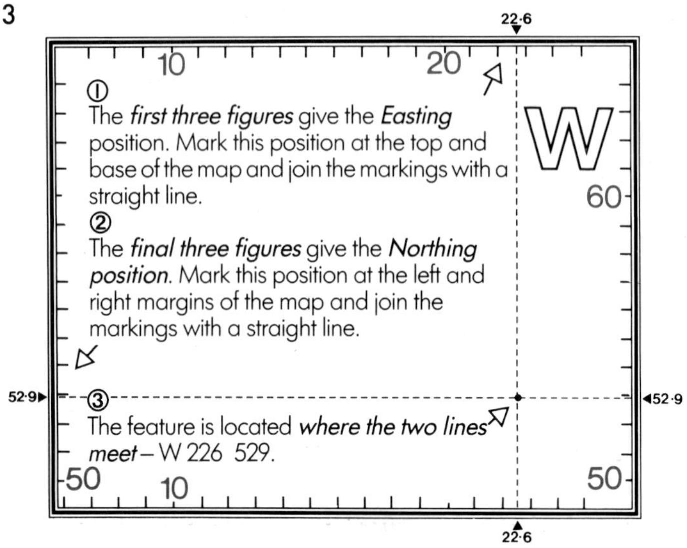

Symbols

Many features are shown on OS maps by means of ***symbols***.

Some symbols are given in the ***key*** or ***legend*** found at the base of each map. A selection of symbols found on one-inch and half-inch maps is given in figure 11.4. You will learn other symbols later in this chapter.

Figure 11.4 Some map symbols

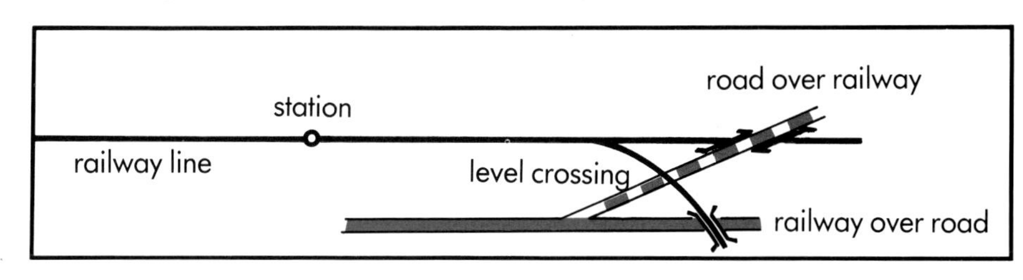

General
building •
large old house ■Ho.
school •Sch.
church +
youth hostel ▲
caravan park
ESB power stations: steam ⦿ hydro ○
county boundary – – –

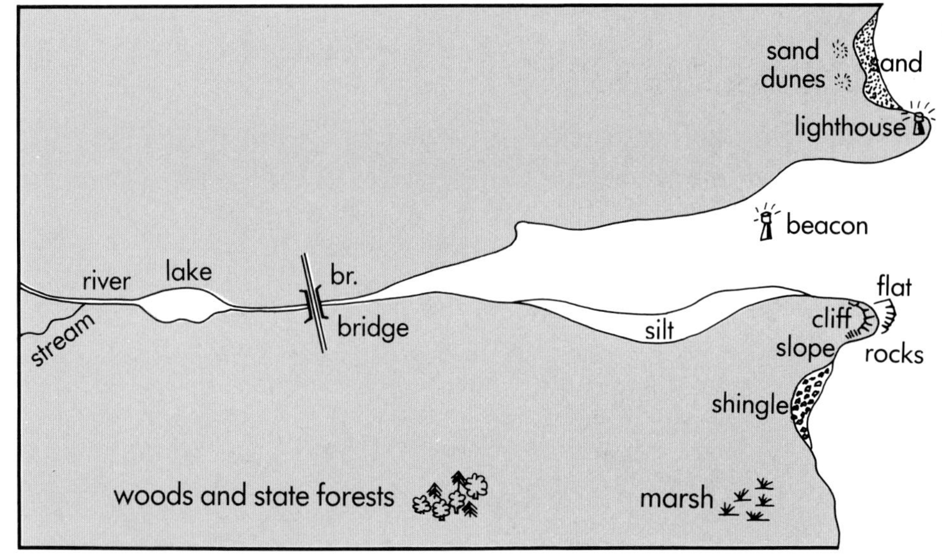

Do Activity 3, page 87.

Showing height

Do Activity 4, page 87.

Height above sea level (sometimes called height above Ordnance Datum or height above OD) is expressed in feet on most Irish OS maps.

Symbol	*Method*	*Explanation*
◬ 2273	triangulation station	It shows exact height, usually of a hill or mountain peak.
• 251	spot height	It shows exact height.
∼ 100	contour	A contour is a line which joins places of equal height. Contours occur usually (though not always) at intervals of 100 feet. (Submarine contours give the depth of seabeds in ***fathoms***. One fathom equals six feet.)
	colour key	A full colour key appears on the margin of each OS map.

Study figure 11.5.

- ☐ Match the locations A–F with the appropriate height selected from those listed in the box.
- ☐ Estimate the height of the places marked 1–6.

Do Activity 5, page 87.

Figure 11.5 ▷

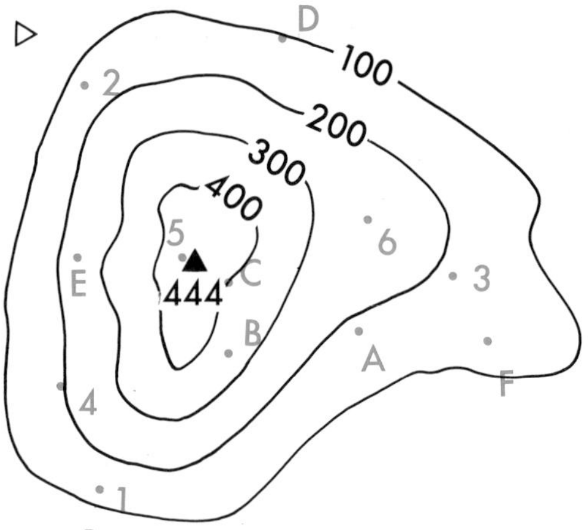

200′ OD
410′ OD
150′ OD
2000′ OD
110′ OD
350′ OD
440′ OD
91′ OD
228′ OD

Showing slope

Steepness of slope is shown by the closeness of contours (figure 11.6).

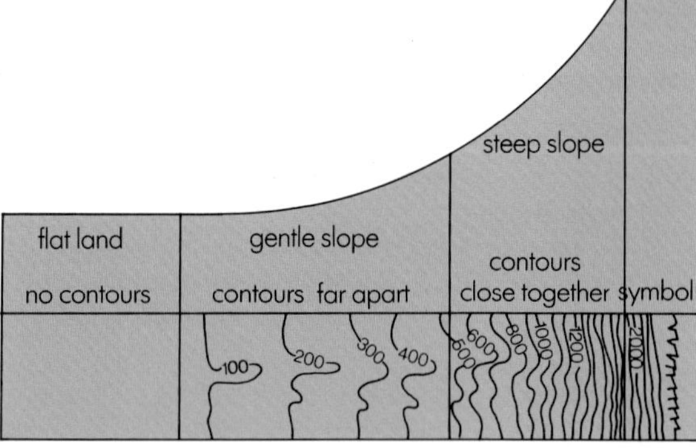

Figure 11.6 Contours show steepness of slope

Do Activity 6, page 87.

Some types of slope

Figure 11.7a shows a ***concave slope***. It is gentle near the base and steep near the top.

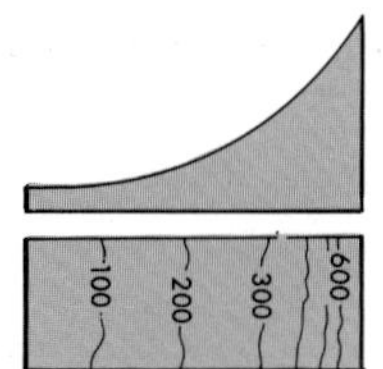

△ **Figure 11.7a A concave slope**

Contours are widely spaced near the base and close together near the top.

Figure 11.7b shows a ***convex slope***. It is steep near the base and gentle near the top.

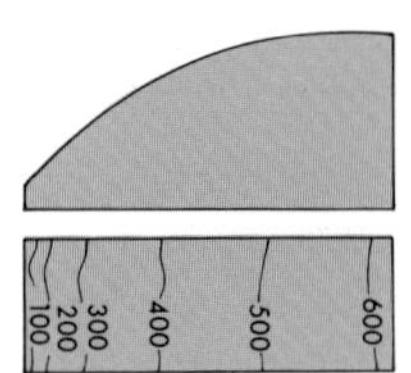

△ **Figure 11.7b A convex slope**

Contours are close together near the base and widely spaced near the top.

Figure 11.7c shows an ***even slope***.

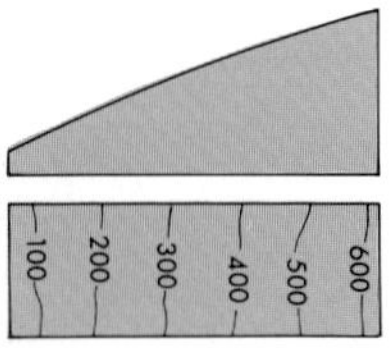

△ **Figure 11.7c An even slope**

Contours are evenly spaced throughout the slope.

Do Activity 7, page 88.

Gradient

Gradient is the slope expressed by means of a ratio. A gradient of 1:10 means that the slope rises or falls one unit of measurement for every ten similar units of measurement travelled horizontally (figure 11.8).

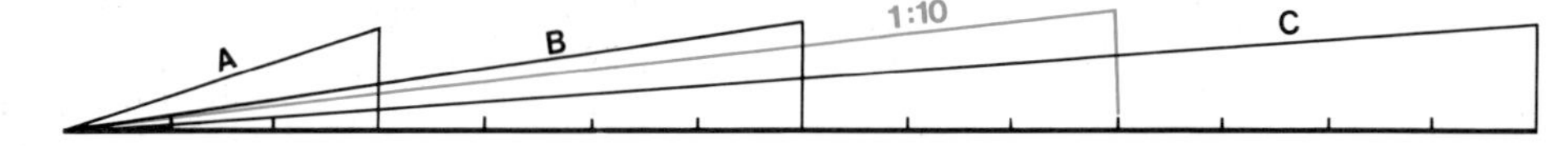

Figure 11.8 What are the gradients of the slopes marked A, B and C? Which gradient is the steepest? Which is the most gentle gradient?

How to draw a cross section

Example: Between A and B, figure 11.9a.

1. Lay a straight ***edge of paper*** along the line of the section on the map.
   Mark on the paper's edge the beginning and end of the section, indicating the height of each.
   Mark also the points where contours meet the paper's edge, indicating the height of each (figure 11.9b).
2. On ***graph paper***, draw a horizontal axis to represent sea level. Draw a vertical axis to a scale of 1/10 inch (one tiny square) to 100 feet (figure 11.9c).
3. Lay the edge of paper along the horizontal axis so that the left end of the section touches the vertical axis.
   Insert points on the graph paper at appropriate heights and directly above the markings on the paper's edge (figure 11.9c).
4. Join the points with a ***smooth line*** (figure 11.9d).
5. Add to the section the names of ***important features*** as well as the horizontal scale (the scale of the map) and the vertical scale (1/10 inch to 1 mile).

Do Activity 9, page 88.

Cross sections

A ***cross section*** gives us a side view of the relief or shape of a landscape.

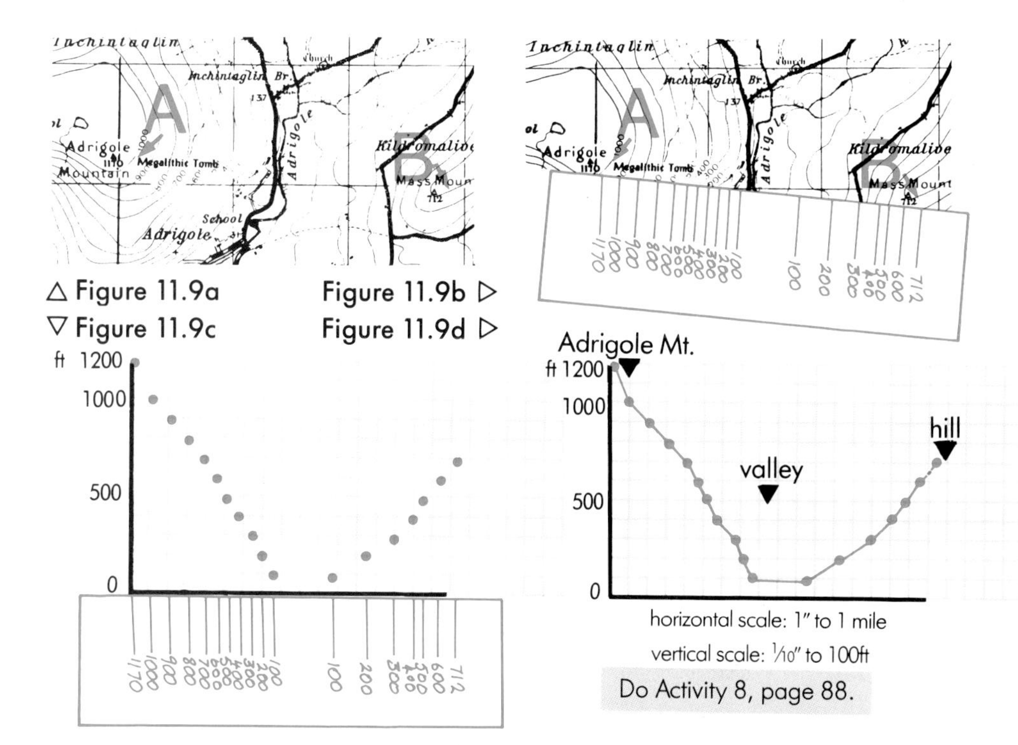

△ Figure 11.9a Figure 11.9b ▷
▽ Figure 11.9c Figure 11.9d ▷

Do Activity 8, page 88.

Sketch maps

Drawing a sketch of an OS map

1. Draw a rectangular ***frame*** of the sketch. It must be the same shape (though not necessarily the same size) as the frame of the map.
2. Using a pencil, lightly divide both the OS map and the sketch into ***segments*** (figures 11.10a and 11.10b). Use these segments to guide you when positioning features on the sketch.
3. Insert the coastline (if any) on the sketch. Insert and name other required ***features,*** omitting any unnecessary details (figure 11.10b).

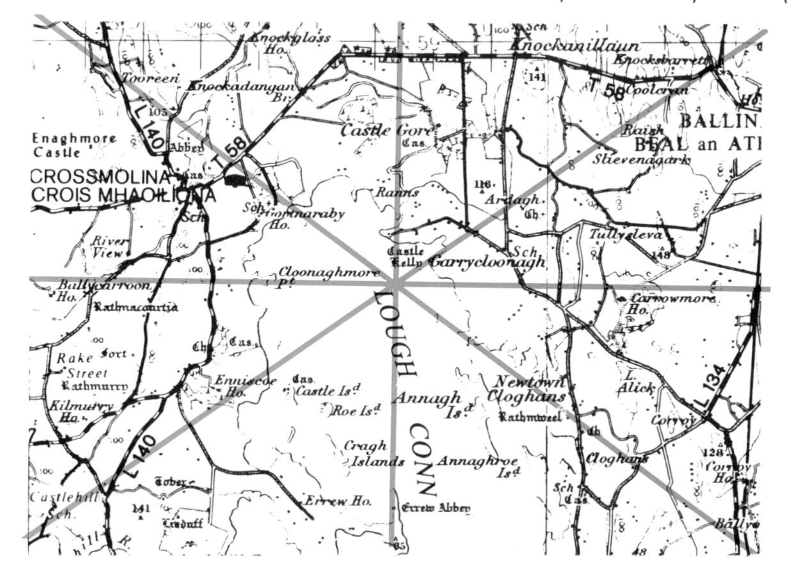

◁ **Figure 11.10a OS map of the Crossmolina area**

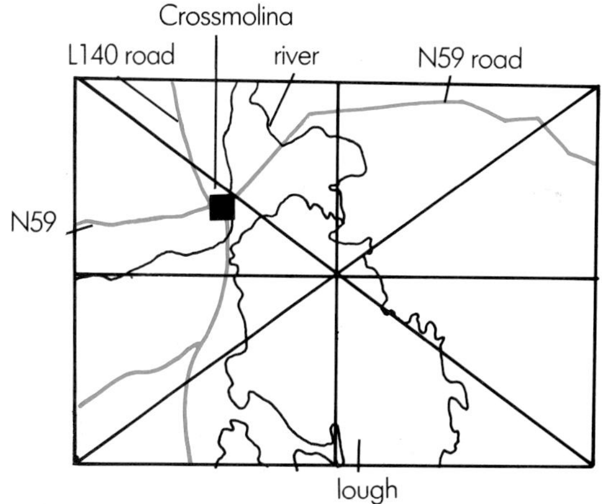

Figure 11.10b Sketch map showing △ location of Crossmolina

Relief

Relief refers to the shape of the surface of the land. Study the definitions of relief features in the box. In pencil, write the letter for each relief feature in the appropriate blank in figures 11.11a and 11.11b.

Relief Features

A	mountain	a steep-sided landform, usually over 1000 feet in height
B	hill	a steep-sided landform, usually less than 1000 feet in height
C	ridge	a long, narrow area of high land
D	col	a dip between two summits
E	spur	a protruding tongue of high ground
F	valley	a narrow depression, usually occupied by a river
G	plain	a large area of fairly flat lowland
H	drumlins	groups of small hillocks (about 100 feet high) which were laid down by ice
J	estuary	the part of a river mouth which is tidal
K	bay	a large coastal inlet
L	headland	a cliff-like area jutting out into the sea

Do Activity 10, page 88.

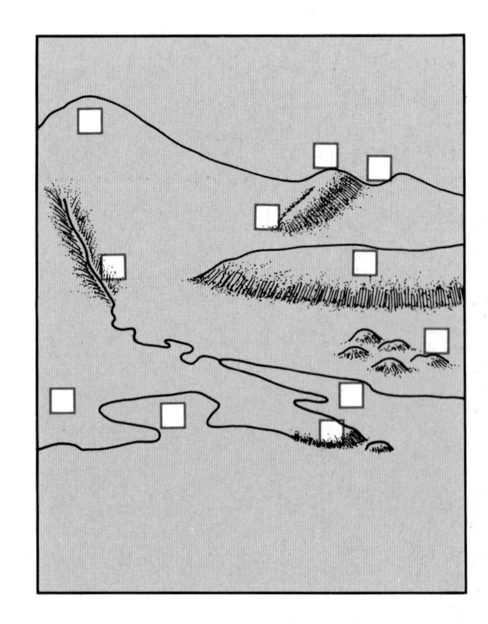

Figure 11.11a

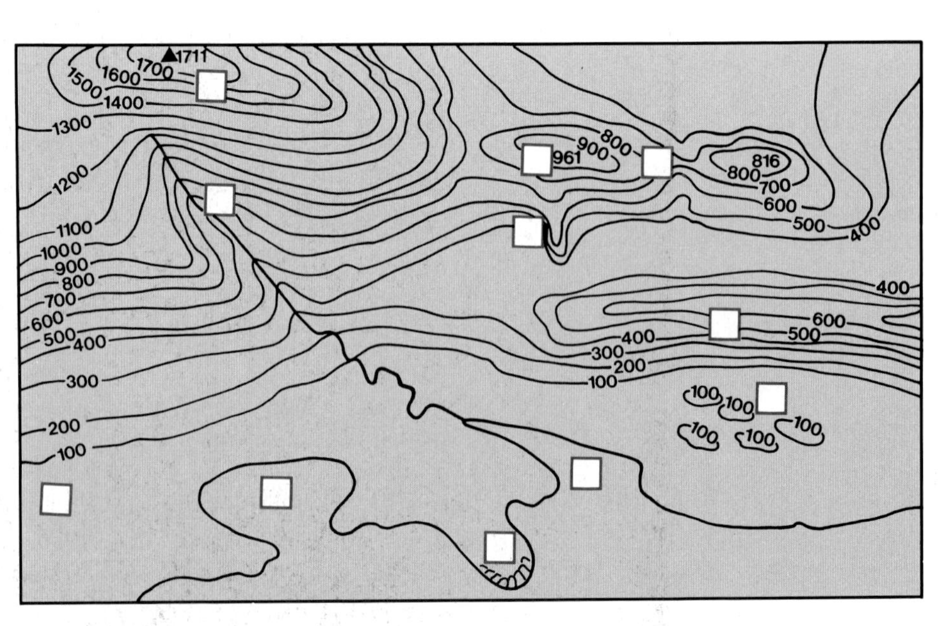

Figure 11.11b

A Relief Feature

- ☐ name and position
- ☐ height and steepness of sides
- ☐ size (approximate length and width)
- ☐ shape
- ☐ how formed
- ☐ other information

Describing individual relief features

In describing any relief features, try to refer to as many of the points in the box as possible.

Example: A description of Mweelrea (Clew Bay map L 79 66)

Mweelrea is a mountain situated north of Killary Harbour. Its peak is 2688ft OD and it has very steep sides. Two small, circular lakes occur on its eastern side. These lakes were probably formed by glacial (ice) erosion.

Do Activity 11, page 88.

Drainage

Drainage refers to the way and extent to which water flows off the land surface. It refers especially to ***rivers, lakes and marshes.***

Study the definitions of common drainage features given in the box. Write the letter for each drainage feature in the appropriate blank in figures 11.12a and 11.12b.

Do Activity 12, page 88.

Drainage Features	
A well-drained area	an area containing many relatively straight rivers
B poorly-drained area	an area with too much surface water, containing marshes or many lakes
C dendritic drainage pattern	where a river and its tributaries form a pattern resembling a tree in winter
D radial drainage pattern	where rivers flow out in several directions from a hill or mountain
E tributary	a small river which joins a larger one
F meander	large curves in the course of an 'old' river

Figure 11.12a ▽

Figure 11.12b ▷

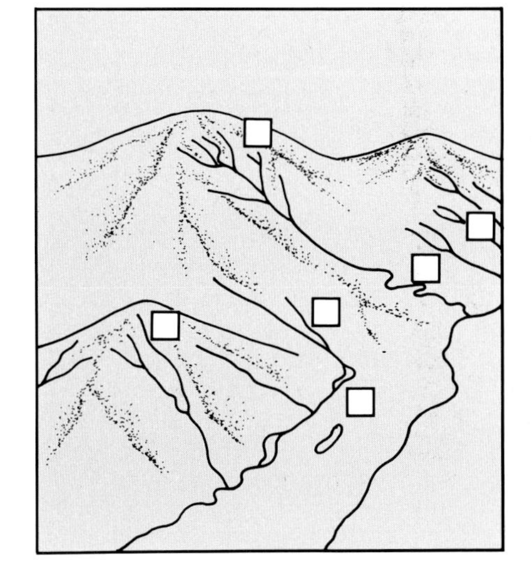

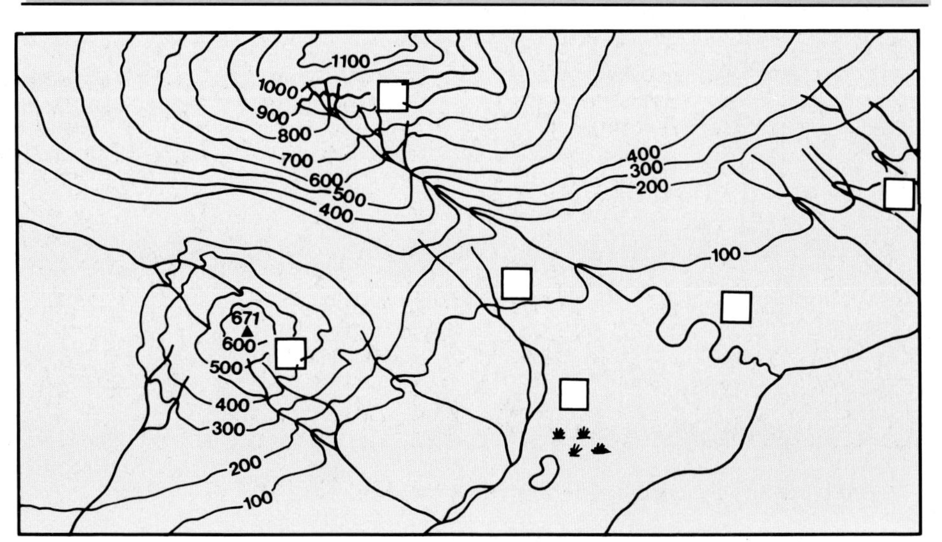

Do Activity 13, page 88.

To describe the general drainage of an area

- ☐ Do many or few rivers drain the area?
- ☐ Is the area well or poorly drained? (If poorly drained, refer to lakes and marshes.)
- ☐ Do rivers form dendritic and/or radial patterns?
- ☐ What are the main rivers?

To describe one particular river

- ☐ What is its name?
- ☐ Where does it rise?
- ☐ In which direction does it flow?
- ☐ Where does it reach the sea?
- ☐ Do many tributaries join it?
- ☐ Do these tributaries form a dendritic pattern?
- ☐ Does the river flow quickly (down a steep slope) or does it meander slowly?

Example: A description of the Bunowen River (Clew Bay map)
The Bunowen River rises in the Sheeffry Hills. It flows in a northwesterly direction. It reaches the sea at Cloghmoyle Harbour. Many tributaries join it and they form a dendritic pattern. The river flows quickly down the steep slopes of the Sheeffry Hills. Farther north, it meanders slowly on flatter land.

Communications

The most important type of communications are roads. Other types of communication include railways, air routes and seaways.

The main classes of roads and their functions (what they do) are outlined in figure 11.13.

Figure 11.13 The main classes of roads and their functions

Map symbols	Classes	Functions
N1	National Primary	Connect main towns
N81	National Secondary	
T1	Trunk	
L87	Link	Connect major roadways and/or serve smaller towns or villages
	Third class	Join rural areas to larger roads
	Other (by-roads)	

Do Activity 14, page 88.

How to describe the roadways in an area

Number	Are there many (as in heavily populated areas) or few (as in lightly populated areas)?
Classes	Of what classes are they and what are their functions?
Straightness	Are the roads relatively straight or winding?
Gradients	Do they avoid steep slopes by: (a) following valleys or cols; (b) avoiding hills, drumlins etc.; (c) using 'corkscrew' bends?
Flooding	Do they stay a little back from meandering rivers to avoid flooding?

Do Activity 15, page 88.

Settlement

△ **The Poulnabrone Dolmen in the Burren, Co. Clare.**

> *Location*
> Ancient settlements were located in places which provided for the basic needs of the inhabitants — water, fuel, building materials, farmland.

Do Activity 16, page 88.

We will study settlement under these headings.

- ☐ former settlement
- ☐ rural settlement
- ☐ urban settlement
- ☐ settlement patterns
- ☐ placenames

Former settlement

Evidence of former settlement is shown on OS maps by the presence of ***antiquities***. These are indicated in gothic lettering. The most common antiquities are listed in the box.

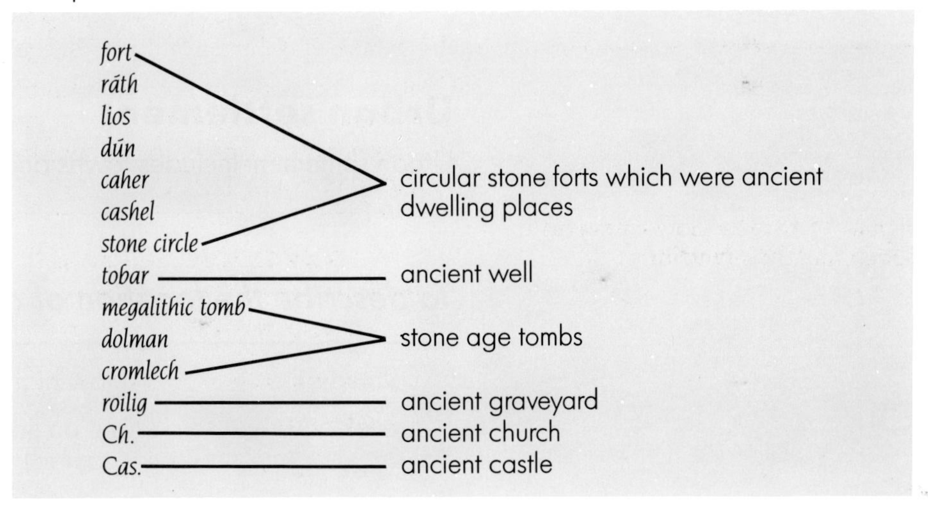

Rural settlement

Rural settlement includes ***isolated houses and small villages***.

The ***density*** of settlement in any area refers to the number of houses per square kilometre (or square mile).

Some factors affecting density of settlement are outlined in the box.

Factors affecting density

Dense settlement in	*Factor*	*Sparse settlement in*
(sheltered) lowlands	◀ altitude ▶	(cold, exposed) highlands
gently-sloping or flat areas	◀ slope ▶	steeply-sloping areas
(warmer) south-facing slopes	◀ aspect ▶	(colder) north-facing areas
well-drained areas	◀ drainage ▶	poorly-drained areas
areas with many small farms	◀ farm size ▶	areas with large farms

Rural settlement may occur in these ***patterns*** or ***forms*** (figures 11.14a, 11.14b and 11.14c).

Dispersed: houses are isolated from each other.

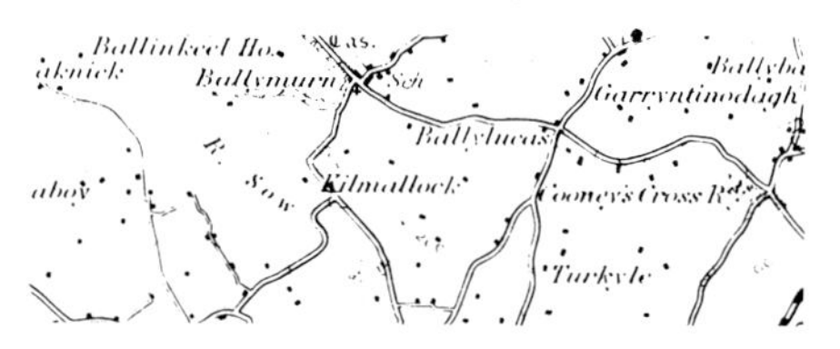

△ **Figure 11.14a**

Nucleated: houses cluster together, as at the meeting point of roads.

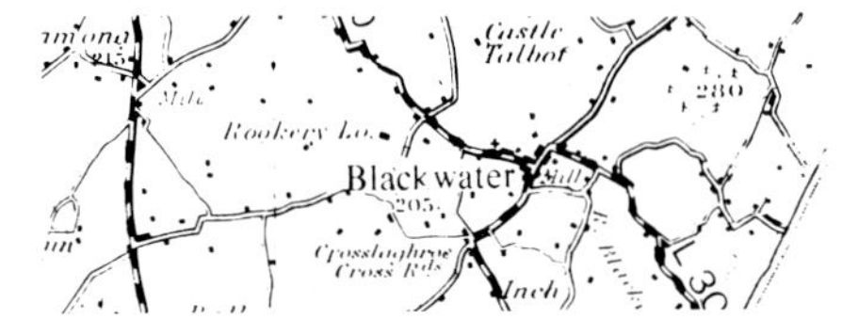

△ **Figure 11.14b**

Ribboned: village houses form a long, narrow line, usually along a road.

△ **Figure 11.14c**

Do Activity 17, page 88.

Urban settlement

Urban settlement includes towns and cities.

Use some of these points.

To describe the location of a town

altitude	How high is it above sea level?
relief	Is it on flat or gently-sloping lowland or in a (named) valley?
sea	How far is it from the sea? Is it in a (named) sheltered estuary or harbour?
rivers	Is it built: (a) near a river which provides water for domestic purposes or industry; (b) near a bridging point of a river; (c) a little way back from a meandering river which might be liable to flooding?
roads	Is it built at the meeting point of (named) roads? Do the roads meet to cross a bridge where (named) valleys meet?
shelter	Is it sheltered by (named) mountains? From which direction is it sheltered?

Do Activity 18, page 88.

Functions

The functions of a town refer to the ***services and benefits*** which the town provides for people. Most towns perform several functions, although these are not easily identified on one-inch and half-inch maps.

Here are some past and present functions of towns.

Functions of Towns

Function	*Explanation*	*Clues on 1″ or ½″ maps*
defence (former function)	Some towns originated near castles or towers which were easily defended.	Castles or towers in or near the town
ecclesiastical (former function)	Some towns originated near old abbeys or monasteries which once provided alms, education and health services.	Monasteries or abbeys in or near the town
market	Many towns once provided markets for the produce of the surrounding countryside. Some still have co-operative marts.	Well-drained lowlands nearby
port	Commercial (cargo) ports tend to be large. Fishing ports tend to be smaller.	Deep, sheltered bays or estuaries; piers, docks, lighthouses or beacons
tourist resort	Some towns have developed largely because of the tourist opportunities which they provide.	Scenic: mountains, lakes, woodlands Coastal: beaches, dunes, cliffs Cultural: antiquities Facilities: caravan parks, youth hostels, golf courses
dormitory	These towns develop as residential centres for people who work in nearby cities.	A city nearby.

Do Activity 19, page 88.

Patterns

The shapes, forms or patterns of towns are often related to the ***relief*** of the area.

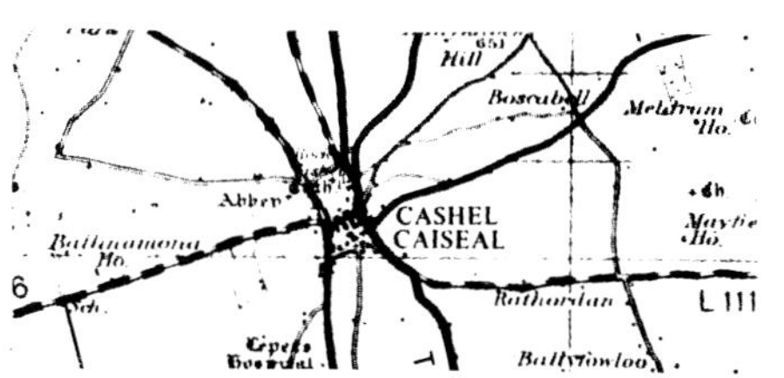

△ Figure 11.15a

Nucleated towns are common in flat or gently-sloping sites where there is no restriction on building.

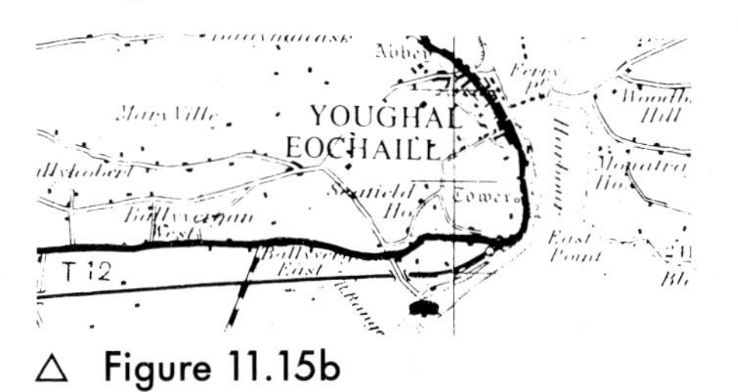

△ Figure 11.15b

Ribboned or linear (long and narrow) towns often develop along coastlines or at the foot of steep slopes where only narrow strips of land are available for easy building.

Placenames

The origins of some places are easily discovered from their names. The name Newbridge, for example, would suggest that the settlement was established near what was then a new bridging point on a river.

To understand the origins of many placenames, we must first know the meanings of the Irish words which make up the names. Here is a list of words commonly used in Irish placenames.

Words used in Irish placenames

Gaelic words

word	meaning
sliabh (slieve/sleve)	mountain
cnock (knock)	hill
beann (ben/bin)	head
drom (drum)	back of hill
abhainn (owen)	river
cúm (coom)	mountain hollow (cirque)
inis (inish/inch)	island
inch	flood plain of river
carraig (carrick)	rock
baile (bally)	small settlement
mainistir (monaster)	monastery
teampall (temple), **cill (kil/kill)**	church
coill (kil/kill), **ros**	wood
ráth, **líos (lis)**, **dún (doon)**, **caher**, **cashel**	ancient stone fort
cúl (col/cole)	behind
...ard	high
...mór (more)	large
...beag (beg)	small
...ín (een)	little
bun	beneath
...bán (bawn/bane)	white ***or*** permanent pasture
...dubh (duv)	black
...glas (glass)	green
...gorm	blue
...buí (boy)	yellow
...sean (shan/shane)	old

English words

word	meaning
borough	fortified place
...ton	town
...ford	crossing place of a river (or the Viking word for inlet)

The placename Lismore is made from ***lios*** (an ancient stone fort) and ***mór*** (large). The settlement probably grew near such a stone fort.

Suggest origins for each of these placenames: Carrick-on-Shannon, Knockglass, Foxford, Inishbeg, Coombawn.

Do Activity 20, page 88.

Tourism and maps

Many tourist attractions are shown on OS maps. Figure 11.16 shows several tourist possibilities for the Rosses Point/Glencar area of Co. Sligo.

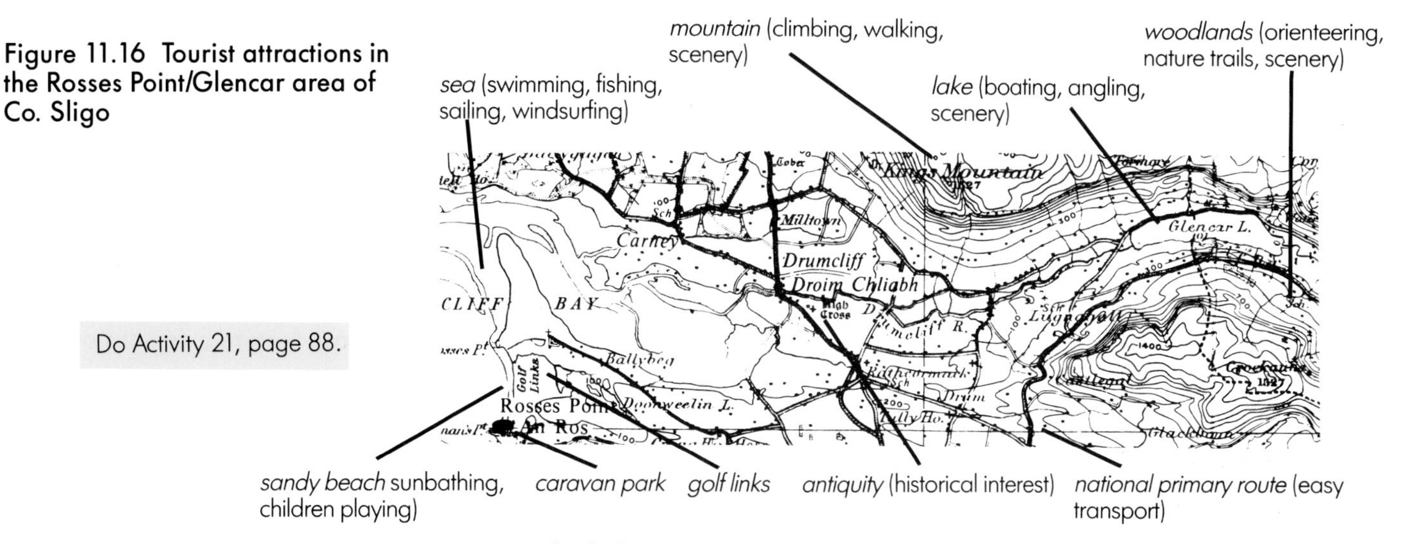

Figure 11.16 Tourist attractions in the Rosses Point/Glencar area of Co. Sligo

Do Activity 21, page 88.

Activities

1. (a) Look at the Dublin District map. Locate Rathgar and Sandyford using four-figure grid references. Identify the settlements at O 13 30 and O 19 23.
   (b) Use the Clew Bay map to locate Croagh Patrick and Corraun Hill by means of four-figure grid references. Identify the main features at L 90 71
   and at L 99 84.

2. (a) Use the Dublin District map to identify the feature at each of these grid references: (i) O 200 258; (ii) 0 134 142; (iii) O 176 278.
   (b) Use the Tralee Bay map to identify the features at: (i) Q 646 122; (ii) Q 715 149; (iii) V 527 999.
   (c) Give six-figure grid references for each of these places on the Dublin District map: (i) the peak of Glendoo Mountain; (ii) the peak of Kilmashogue Mountain.

3. (a) On the Dublin District map, identify the symbols at each of the following places: (i) O 210 288; (ii) O 197 278; (iii) O 150 161; (iv) O 155 190.
   (b) On the Tralee Bay map, identify the symbols at: (i) Q 640 124; (ii) Q 555 125; (iii) Q 590 070.

4. On the Clew Bay map, identify examples of four different methods of showing height. Which height is shown by each example you identify? Name and give the height of the highest mountain shown on the Clew Bay map.

5. Calculate the height above sea level of each of the following places on the Dublin District map: (a) O 210 210; (b) O 178 180; (c) O 210 195; (d) O 140 190; (e) O 150 220.

6. Rank according to steepness of slope (steepest slope first, gentlest slope last) each of the following locations on the Clew Bay map: (a) L 79 70; (b) L 80 80; (c) L 80 70; (d) L 90 70.

7. What type of slope exists at each of the following places on the Clew Bay map: (a) the southern side of the hill of 1259′ (L 82 63); (b) the northern side of Croagh Patrick; (c) the northern side of the hill of 462′ (M 02 94); (d) M 03 65.

8. Using the Dublin District map, draw a cross section between the peak of 1760ft (O 147 145) and the peak of 1825ft (Prince William's Seat, O 178 182). Show and name the Glencree River and a third class road on your cross section.

9. Make a sketch map of the Tralee Bay map. On your sketch, show and name: Tralee Bay; Rough Point; the River Maine; Tralee town; the T68 roadway; land over 200 feet OD.

10. On the Clew Bay OS map, find one example of each of the relief features listed on page 80. Identify the position of each feature by means of a four-figure grid reference.

11. Describe each of these features which can be found on the Clew Bay map: Croagh Patrick, the valley of the River Erriff, Old Head, Clew Bay.

12. On the Clew Bay map find one example of each of the drainage features listed on page 81. Identify the position of each by means of a four-figure grid reference.

13. Using the questions on page 82 as guides, describe: the general drainage of the Sheeffry Hills (Clew Bay map); the River Carrowhisky (Clew Bay map).

14. Give examples of four different classes of roads on the Tralee Bay map. Name the class and give the function of each roadway you choose.

15. Describe the roadways in the area shown on the Clew Bay map south of Northing 80.

16. Identify the antiquities which appear on the Clew Bay map at: L 762 738; L 795 797; L 836 808. What does the presence of these antiquities tell you about the people who once lived in the area?

17. Describe the density and pattern of rural settlement at each of the following places on the Tralee Bay map: Q 80 10; Q 63 09; Q 59 01. Explain the factors which influence the settlement density of each place.

18. Use map evidence to describe and account for the locations of Castlemaine (Tralee Bay map) and Westport (Clew Bay map).

19. Using only evidence from the Tralee Bay map, describe the functions of Tralee and Fenit. List other everyday functions which you think Tralee performs but which are not apparent from the map.

20. Suggest possible origins for the following placenames on the Dublin District map: (a) Glendoo — O 14 20; (b) Ballyedmonduff — O 18 22; (c) Galloping Green — O 20 27; (d) Ballyross — O 17 15; (e) Dundrum — O 16 27.

21. Using map evidence only, describe the attractions of Westport (Clew Bay map) and Castlegregory (Tralee Bay map) as family holiday centres.

12 A CLOSER LOOK AT CITY STREET MAPS

The Map Supplement which accompanies this book includes two city street maps. One is of ***Central Dublin***. The other is of ***Waterford city***.

- ☐ **What is the scale of each of these maps (as shown by the representative fraction)?**
- ☐ **Which map has the larger scale?**
- ☐ **Which map shows the larger area?**

Central Dublin street map

Locating places

The ***letter and number*** grid reference system is used to locate places on the Central Dublin street map (1:7500).

The map has been divided into squares. The squares are lettered A–E across the top and base of the map. The squares are numbered 1–5 along the sides of the map.

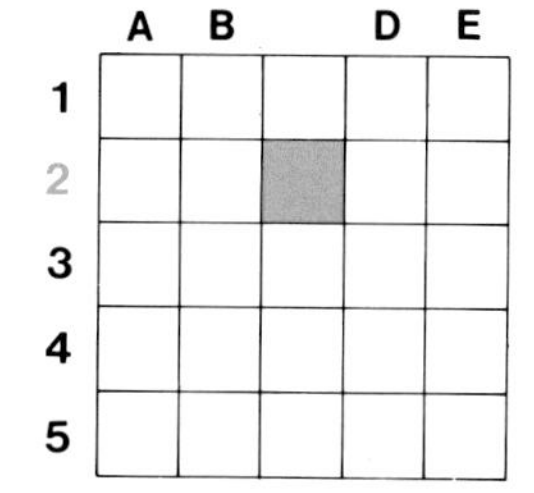

These squares can be used to locate places on the map. The grid reference for the GPO, for example, is C2. Here is how we locate the GPO.

- ☐ Refer to the letter C at the top or base of the map.
- ☐ Refer to the number 2 at the east or west side of the map.
- ☐ Identify the grid in which C and 2 meet. The GPO will be found within this square.

Activities

1. **Name the squares in which each of the following is located: St. Stephen's Green; St. Patrick's Cathedral; O'Connell Street Upper.**
2. **Name the theatres found in B4 and in C5.**
3. **Which parks are found in squares A5 and E5?**

Symbols

Study the symbols contained in the legend accompanying the Central Dublin street map (1:7500). Name the symbols used for each of the following:

- ☐ a built-up area
- ☐ a railway station and a railway line
- ☐ a garda station
- ☐ a one-way street
- ☐ a cinema or theatre

City journeys – Some activities relating to Central Dublin Street Map (1:7500)

1 Car Journey

Imagine that you wish to take the most convenient car route from the Garden of Remembrance (square C1) to the dental hospital at Lincoln Place (located near College Park at the southeast portion of the map).

1. **List the streets along your route.**
2. **State the general direction in which you would travel along each street listed.**
3. **Name one theatre and two public buildings which you would pass on your journey.**

2 Taxi journey

Imagine that you are a taxi driver. You have been asked to take the shortest permissible route from Pearse Station (at E4) to the Carlton Cinema (at C2).

1. **Name the streets along which you would travel.**
2. **Calculate the length of your route in kilometres and metres.**

3 Tourist trail

Dublin holds many attractions for the tourist.

1. **Make a tracing of the frame of that section of the map showing places south of the River Liffey. On your tracing, show and name five places which would be of interest to tourists.**
2. **Beginning and ending at Pearse Station (E4), plan the shortest walking route which would allow tourists to pass by the five places of interest which you selected in 1 above. Show this walking route on your tracing. Name the streets on your route. Use arrows to show tourists the directions they should take.**

Central Dublin — Street map and photograph

Figure 12.1 Central Dublin: a vertical ▷ aerial photograph

Figure 12.1 shows an aerial photograph of part of Central Dublin. It is a ***vertical aerial photograph***. This means that the photograph was taken when the camera was pointing directly down over the area shown.

Trinity College, the River Liffey and St Stephen's Green are shown and named on the photograph. These places are also shown and named on the map.

1. *Relate the map to the photograph.* Then use the information given on the map to identify the buildings named A, B, C and D on the photograph. Name the streets indicated by letters E, F, G, H, J, and K on the photograph.
2. Suggest reasons for the names given to streets J and K.
3. In which direction would you travel from building C to Trinity College? (Consult your map before answering.)

Comparing maps and photographs

Maps and photographs each have certain advantages and disadvantages. Compare the street map of Waterford with the aerial photograph.

1. **Name some kinds of information which the street map gives, but which the photograph does not.**
2. **Name some kinds of information which the photograph gives, but which the map does not.**
3. **Which would be more useful for doing the following, the map or the photograph?**
   - ☐ planning a bus route through the city
   - ☐ showing the location of a building
   - ☐ helping tourists to find their way around Waterford
   - ☐ giving tourists a first impression of what Waterford looks like

Waterford City Street Map

Locating places on the Waterford Street Map

A ***letter and number grid reference system*** is used to help locate places on the Waterford street map (1:9000). The grid reference for Manor Street, for example is D7.

☐ **Find each of the following places on the Waterford street map.**

Place	*Grid reference*
Gaelic Park	E7
Courthouse	D8
Penrose Lane	C6
Coal Quay	C7

☐ **Give the location reference for: (a) the bridge over the River Suir; (b) Bishopsgrove, on the north side of the river.**

Symbols

Study the symbols in the legend which accompanies the Waterford street map.

1. **Name and quote a grid reference for:**
   - ☐ a national primary route shown on the map
   - ☐ two regional roads
   - ☐ two smaller thoroughfares (streets)
   - ☐ two public buildings (excluding schools)
2. **Name two antiquities to be found at C8. What does the presence of these antiquities suggest about the history of Waterford city?**

Waterford street map and photograph

Figure 12.2 Waterford city: an oblique aerial photograph

Figure 12.2 shows an aerial photograph of part of Waterford city. It is an ***oblique aerial photograph***. This means that it was taken when the camera was pointing at an acute angle to the ground when the photograph was taken. The photograph was taken over the River Suir at the place marked X on the map. The photograph is facing towards Broad Street.

Relate the photograph to the map. Then do the following.

1. Use the information on the map to name the following places shown on the photograph.
   - ☐ the street marked A
   - ☐ the tower marked B
   - ☐ the buildings marked C, D and E
2. Suggest how the street marked A got its name.
3. Make a tracing of the photograph. Show and name the following on your tracing.
   - ☐ the River Suir
   - ☐ Meagher's Quay and Coal Quay
   - ☐ Barronstrand Street, Broad Street and Michael Street
   - ☐ Patrick Street and the cinema on Patrick Street

Follow the trail

The passage below describes a journey through Waterford. The journey begins at Cork Road near the Holy Ghost hospital (E6) and ends where Barronstrand Street meets Meagher's Quay (C7).

1. Rewrite the passage, using the information on the *map* to fill in the blanks.
2. Calculate the length of the journey in kilometres and metres.

Travelling in a northeasterly direction along Cork Road, we pass the grounds of the Holy Ghost Hospital and Home on our left. At the end of these grounds, we pass ______________ Street on the right and ______________ Street on the opposite side.

As we go along Manor Street, we pass ______________ Park, which is directly opposite ______________ Square. We continue past the old city walls and cross the junction with the regional road named ______________ Street.

We then travel along Parnell Street in a ______________ direction. To our right are three public buildings; a ______________, a ______________ and a technical institute. Out of our range of view and to the southeast of these public buildings, the River ______________ meanders in a generally ______________ direction until it enters the River Suir at grid ______________.

Taking the next left at C______________ Street, we travel only ______________ metres before meeting a T-junction. There we take a ______________ direction along Lady Lane. On our right are two public buildings, a ______________ and a ______________.

Upon entering Michael Street, we turn in a ______________ direction towards the River Suir. We pass two streets on our right. They are ______________ Street and ______________ Street. Then as we enter Meagher's Quay, the ______________ Tower stands directly in front of us. Behind it, the River Suir flows in a ______________ direction towards the Mooring Posts on our right.

13 A CLOSER LOOK AT TOWN PLANS

Large-scale Ordnance Survey maps called ***plans*** are generally used to study towns.

Plans of parts of Roscrea and Waterford are included in your Map Supplement book.

Scale

Most modern town plans are made to a scale of 1 : 1000 (1mm to 1m). The linear scales of these plans show measurements in both metres and feet.

Do Activity 1, page 101.

8-figure grid references can also be used to locate places. These consist of 4-figure Eastings, followed by 4-figure Northings.

Study the Eastings and Northings on the plan of ***Roscrea***. ***Rosemount House*** is located at Easting 1384 (138.4) and at Northing 8931 (893.1). Its grid reference is therefore 1384, 8931.

Locating places

The location of prominent places on a town plan can be described by referring to the approximate area of the plan in which these places are found.

Look at the plan of Roscrea in your Map Supplement. You will note the following.

- ☐ A tennis ground is shown on the northeast corner of the plan.
- ☐ A factory is shown on the south side of Gaol Road.
- ☐ A park is located at the meeting place of Glebe View and Carroll's Row, on the south.

Do Activity 2, page 101.

Showing height

Height is shown by those methods described in the box.

Method	*Explanation*	*Example*
minor altitude	Gives the height above sea level of a particular point.	+22 (22 metres* above OD)
bench mark	Gives the height of a spot marked on a wall or building. The apex of the arrow points to the spot.	BM 56.4 (56.4 metres* above OD)

*Town plans of 1:1000 give height in metres.

Symbols

Most features on town plans are shown by symbols or initials. The most commonly used ones are shown in figure 13.1.

Figure 13.1 Symbols used on town ▽ plans

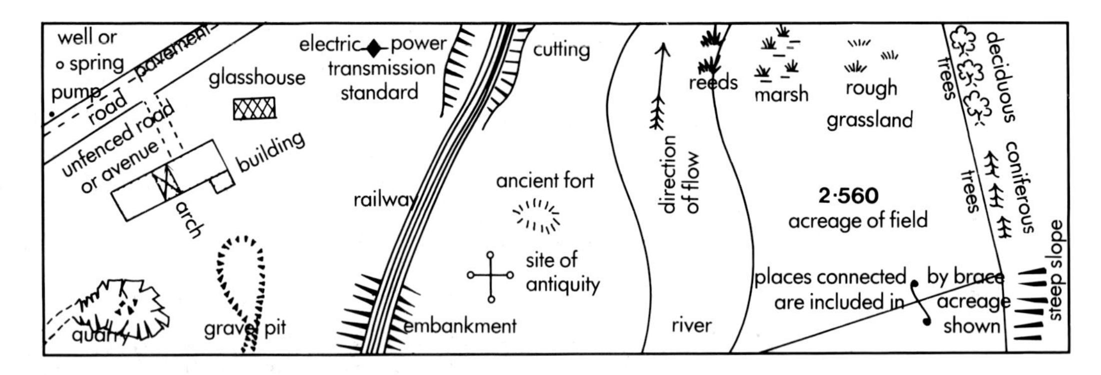

Boundaries

county	— — —
barony	— — — —
parish	- - - - - -
townland	●●●●●●●●●●●
change of boundary	●●●/- - -

Positions of boundaries

CR	centre of road or river
CS	centre of stream
SR	side of road/river
CW	centre of wall
CF	centre of fence
FW	face of wall
FF	face of fence
6′FW	6ft from face of wall
und.	undefined

Other initials indicating features

MP	mile post (on a roadside) or mooring point (on a dockside)
SP	signal post
SB	signal box
GP	guide or sign post
PO	post office
LB	letter box
LS	lamp standard
TB	telephone box
TK	telephone kiosk
GS	Garda station
WB	weigh bridge
WM	weigh machine
MH	manhole
LC	level crossing
FB	footbridge
FP	footpath
Chy	chimney
WT	watertap
P	pump
Fn	fountain
FS	flag staff
T	transformer
ES	electricity station
H	hydrant
HWM	high water mark
LWM	low water mark

Location of towns

Town plans can provide much information about the ***site*** (precise location) of a town.

Example: Describing the site of Roscrea (see Roscrea plan)

Roscrea is sited on lowlying land which is just under 100 metres OD.

The land slopes from the east (102.23m near the east end of Gaol Road) to the west (89.7m at the Mall).

Do Activity 3, page 101.

To describe a town site from a town plan

Altitude	Do bench marks and minor altitudes indicate the town's height above sea level?
Relief	Do heights show the town to be on flat or gently-sloping ground? In which direction is it sloping?
Sea	Is the town on the coast? What kind of coast (rocky, sandy, indented)?
River	Is the town situated on a river? On which side? Do buildings avoid marshes or reedbeds, which indicate the possibility of flooding? Is the town on a bridging point where roads meet?

An aerial view of Roscrea. Is the town regular or irregular in layout? Identify: a castle; a school; a large 'green' area; a terrace of houses; the main meeting place of roads; the place where the residential area merges into the countryside. ▷

Do Activity 4, page 101.

Shape — form, layout and pattern

We have already seen that towns may be ***nucleated*** or ***ribboned*** in shape.

Towns may also be ***regular*** or ***irregular*** in layout.

A ***regular layout*** is more usual in planned urban areas, as shown in figure 13.2a.

An ***irregular layout*** is a sign of unplanned — and usually old — urban growth (figure 13.2b).

Figure 13.2a A planned urban area ▽

Adjacent buildings and gardens tend to be ***uniform*** in size and plan (shape).

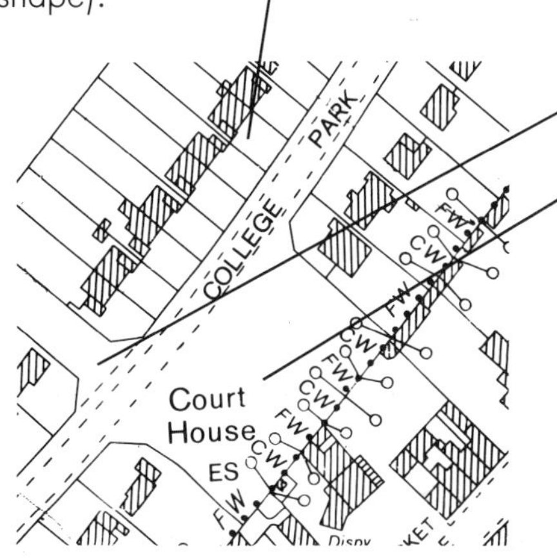

Roads are straight and usually join at ***right angles***.

Squares or ornamental parks may occur.

▨ building

Figure 13.2b An unplanned urban area ▽

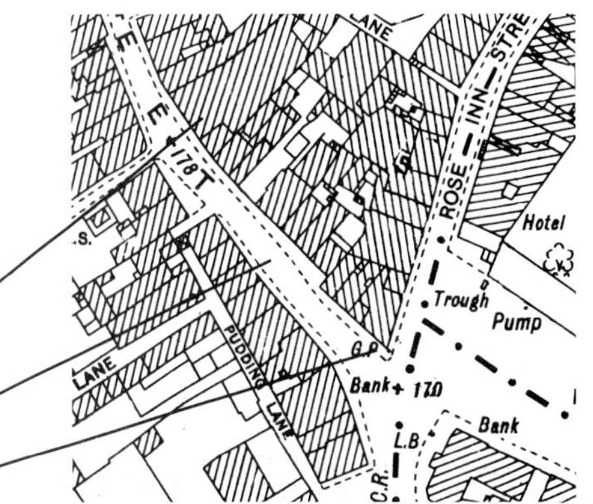

Adjacent buildings ***vary*** in size and plan.

Individual roads may ***wind*** and vary in width.

Road may meet at ***acute angles***.

Functions and services

Clues about present and former functions and services of towns can be found in a number of places.

- ☐ ***Street names*** — Market Lane, for example, is a name which suggests that the place once served as a market for the produce of the surrounding area.
- ☐ ***Named buildings*** and other structures such as those listed in the box. When describing the locations of such buildings, refer to the streets on which they are found.

Functions	*Some buildings and structures which give clues to functions*
defence	castle, tower, town wall, town gate (all refer to former function)
religious	monastery, abbey, priory, presbytery ('Presby'.), church, chapel, meeting hall, convent (those in gothic print refer to former function)
manufacturing	factory, plant, works, tanyard, creamery (mill or mill race suggest a former function)
market	market square, market house, fair green (all former functions), co-operative mart, creamery (present functions)
port	quay, dock, wharf, mooring post (MP), mooring ring, crane, warehouse, boat slip, lifeboat station, lighthouse, beacon (The presence of sand or mud might suggest that the port is not now in use.)
other commercial	bank, post office, coalyard, hotel
financial	bank, post office, credit union
administrative	corporation, county council or urban council buildings, town hall, municipal building
legal	courthouse, garda station (GS), bridewell (former police station)
medical	hospital, infirmary, dispensary (dispy)
educational	school, college, academy, library
holiday resort	beach, pier, hotel, caravan park, youth hostel, golf course, antiquity
other recreational	cinema, ballroom, hall, park, ball alley, sports ground, library

Do Activity 5, page 101.

Street names

The origins of some street names may be obvious from the information given in town plans. For example, Castle Street (on the Roscrea plan) clearly got its name from the fact that Roscrea Castle stands on its southern side.

Good suggestions can also be made about the origins of many other street names. Gaol Road in Roscrea is probably so called because a jail (gaol) was once located there, even though this jail no longer exists on the Roscrea plan.

Some useful street name words to know

Shambles (as in Shambles Lane)	a place where things were once sold
...... ***gate*** (as in Water Gate)	usually refers to a former gate in an ancient town wall
Irish...... (as in Irishtown)	an area in which Irish people rather than English settlers lived
Pound (as in Pound Lane)	a place where stray animals were kept
Tanyard (as in Tanyard Lane)	a place where cattle hides were made into leather
Bridewell (as in Bridewell Street)	a former police station
High ...... (as in High Street)	may refer to a street of high altitude or a main shopping street

Do Activity 6, page 101.

Describing buildings

When describing housing, you should ask yourself the following questions.

- ☐ ***buildings*** — Are they large or small? Are they uniform or varying in size and shape?
- ☐ ***front and back gardens*** — Are they large, small, or non-existent? Are they uniform or varying in size and shape?
- ☐ ***density*** — How many buildings are found in a given area? Are these buildings terraced (high density), semi-detached (medium density) or detached (low density)?

△
These photographs illustrate the types of housing shown in figures 13.3a and 13.3b. What extra information do the photographs give? Is there any information which the plans give which the photographs do not?

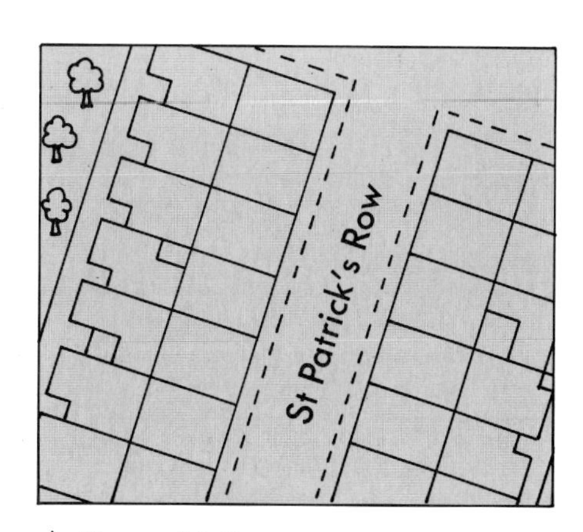

△ **Figure 13.3a**

Contrast the housing in figure 13.3a with that in figure 13.3b.

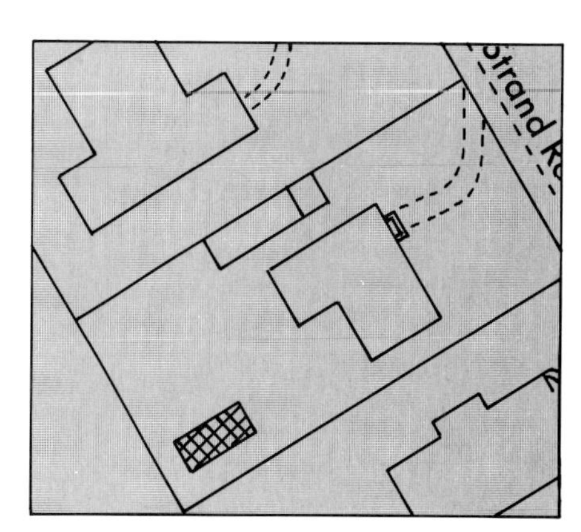

△ **Figure 13.3b**

Do Activity 7, page 101.

Planning the location of large buildings

Locating large, modern buildings calls for careful planning. Study the locational factors in the chart (page 100) and then carry out Activity 8 on page 101.

Locational Factors

Location	Factory	Hospital	School	Shopping area
large, well-drained site	...for buildings and car parks	...for buildings and car parks	...for buildings and car parks	...for buildings and car parks
near residential areas	...which provide workers	...which provide patients	...which provide students	...which provide workers and customers
near wide, uncongested roads	...to facilitate transport	....to facilitate transport	...to facilitate transport and reduce danger of accidents	...to facilitate transport
near railways, ports or airports	...to facilitate transport	____	____	____
away from large factories	____	...to avoid noise and pollution	...to avoid noise and pollution	____

Do Activity 8, page 101.

Environmental problems and solutions

Traffic congestion may occur	*Solutions*
☐ in commercially busy streets	☐ traffic lights at busy street junctions
☐ where streets narrow suddenly	☐ parking restrictions, together with the provision of convenient car parks
	☐ a one-way street system
☐ where streets converge, such as near a bridge	☐ a by-pass or ring road to reduce traffic passing through a town

Water pollution may occur	*Solution*
☐ where houses or factories discharge untreated effluent into a river	☐ the purifying of effluent in a treatment plant outside the town

Do Activity 9, page 101.

Air pollution may be caused by	*Solutions*
☐ carbon monoxide from vehicle exhausts	☐ a by-pass or ring road to reduce traffic
☐ smoke from factories or domestic fires	☐ use smokeless fuels

For a comparison of the usefulness of maps of different scales . . .

. . do Activity 10, page 102.

Comparing maps of different scales

Your Map Supplement contains two maps of Waterford city. One is the Waterford street map, drawn to a scale of 1:9000. The other is the plan of Waterford, drawn to a scale of 1:1000.

Examine these maps together. Then do the following.

1. On the street map, find the area represented by the plan. In which grids on the street map is this area shown?
2. On the street map, carefully frame the area represented by the plan. (Do this lightly with pencil.) Then make a tracing of the area you have framed. On your tracing, show and name the two cinemas to the west of Broad Street. Show and name the cathedral.

Activities

1. (a) Measure the area of the tennis ground shown on the northeast of the *Roscrea plan*. Give your answer in square feet.
   (b) Measure the width of the Waterford plan. Give your answer in metres.
2. Describe the general location and give the grid reference of each of the following on the *Roscrea plan*: Damer House, the youth centre, the playground, the hotel.
3. Use the information given on the Waterford plan to write three sentences about the site of Waterford city.
4. Do you consider the area shown by the *Waterford plan* to be planned or unplanned? Explain your answer fully.
5. (a) Name and locate the services provided in the area shown by the *Waterford* plan under each of these headings: religious, financial, recreational.
   (b) Name and give the locations of services provided in the area shown by the *Roscrea plan* under each of the following headings: recreational, religious, commercial, manufacturing.
6. Using information from the Waterford plan, suggest explanations for each of the following placenames: Coal Quay, Broad Street, Blackfriars Lane, Lady Lane.
7. Contrast the type of houses at St. Cronan's Terrace in the *Roscrea plan* with Rosemount House.
8. (a) Would you say that the factory and school shown on the *Roscrea plan* are suitably located? Explain your answer fully, referring to actual places on the plan.
   (b) Do you consider your school to be suitably located? Explain your answer fully.
9. Study the plan of *Roscrea* in your Map Supplement.
   (a) Name a place which you think suffers from traffic congestion.
   (b) Say why you think this is so.
   (c) Carefully outline ways of easing traffic congestion at the place you have selected.

10. In the last four chapters, you have examined maps of several different scales. Copy figure 13.4 into your geography notebook. In your copy, tick the appropriate boxes to indicate the map scale which is best suited to each of the purposes listed A – E. Make a cross in the appropriate box to indicate which map scale is least suited to each of the purposes listed.

▽ Figure 13.4

Purposes	*1 : 126 720 (eg. Clew Bay map)*	*1 : 9000 (eg. Waterford city street map)*	*1 : 1000 (eg. plan of Waterford)*
A To make a detailed survey of a single street	☐	☐	☐
B To help a foreign motorist drive through the city	☐	☐	☐
C To study the location of a town in relation to a large area around it	☐	☐	☐
D To study the ground areas and shapes of public buildings	☐	☐	☐
E To draw up a tourist trail for the city	☐	☐	☐

11. **Field work — A town trail**

Plan a trail (preferably circular) of part of your local town area.

Suggested duration: two or three class periods.

Suggested equipment for each student: copy of plan of the area, pens, paper, clipboard, compass (optional).

Some suggested topics: distance, directions, use of symbols, gradient, contrast between two houses or housing estates, the location of an important building, traffic congestion.

More detailed study: Distinguish between the commercial, residential, industrial, educational and other functions of buildings in a selected street or area.

While on the trail, mark lightly on the plan the principal functions of each building.

Later, shade in each building using a different colour for each function named. Calculate the percentage of buildings given to each function. Illustrate these percentages by means of a pie chart.

14 EXAMINING PHOTOGRAPHS

Photographs are an excellent means of illustrating geographical features. This chapter will explain some of the skills needed to understand and interpret geographical photographs.

Referring to locations in a photograph

When referring to the locations of objects in a photograph, it is useful to divide the photograph into nine areas, as shown in figure 14.1.

Figure 14.1
▽

left background	centre background	right background
left centre	centre or middle distance	right centre
left foreground	centre foreground	right foreground

The ***background*** is the zone farthest from the camera when the photograph was taken.

The ***centre*** zone is between the background and the foreground.

The ***foreground*** is the zone nearest the camera.

Photograph A: Zermatt, a Swiss Alpine village. ▷

Look at photograph A.

Some houses are in the foreground.

The village church is in the left centre.

A high pyramid-shaped mountain (The Matterhorn) is in the centre background.

Distance and size in photographs

The sizes of various objects in a photograph often help us to distinguish between different features. It is important to remember the following two points when looking at any photograph.

- ☐ ***Foreground objects*** usually appear larger than objects of similar size in the background.
- ☐ The ***area*** covered by the photograph increases with the distance from the camera position.

Look at the photograph of Glendalough in Co. Wicklow (photograph B). (Page 104).

- ☐ Notice that the tree branches shown in close up at the top of the photo in the foreground appear to be larger than *entire trees* in the middle distance.
- ☐ Notice that, in the immediate foreground, the photograph covers an area of only a few metres across (ie approximately the size of a tree branch). Towards the background, the photograph covers an area over a kilometre wide.

Estimating size in a photograph

To estimate the size of objects in a photograph, it is useful to compare that object's size with that of a ***nearby*** object, the size of which is familiar or which can be easily estimated.

The heights of the fir trees in the middle distance of photograph B (Glendalough), for example, may be estimated by comparing them with heights of the two-storey buildings nearby.

Do Activity 2, page 109.

Measuring distance accurately

To measure distance accurately using a photograph, one must identify the corresponding distance on a map and then measure this distance from the map.

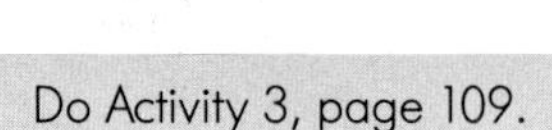

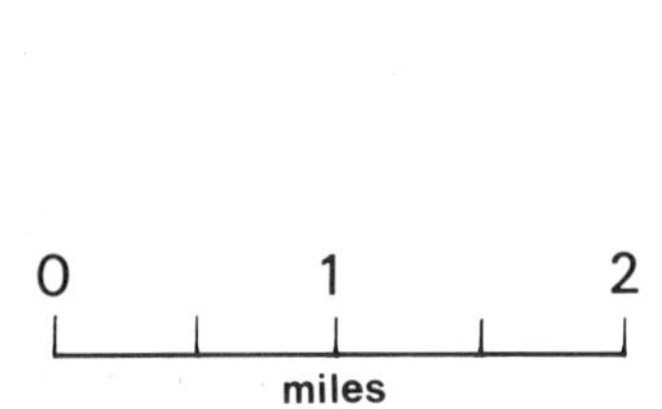

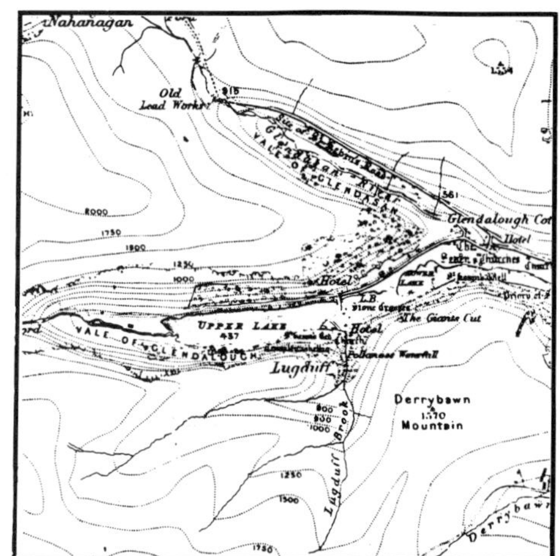

Figure 14.2 Map of Glendalough, Co. Wicklow ▷

Photograph B: Glendalough. ▷

Photograph C: The Cliffs of Moher. ▷

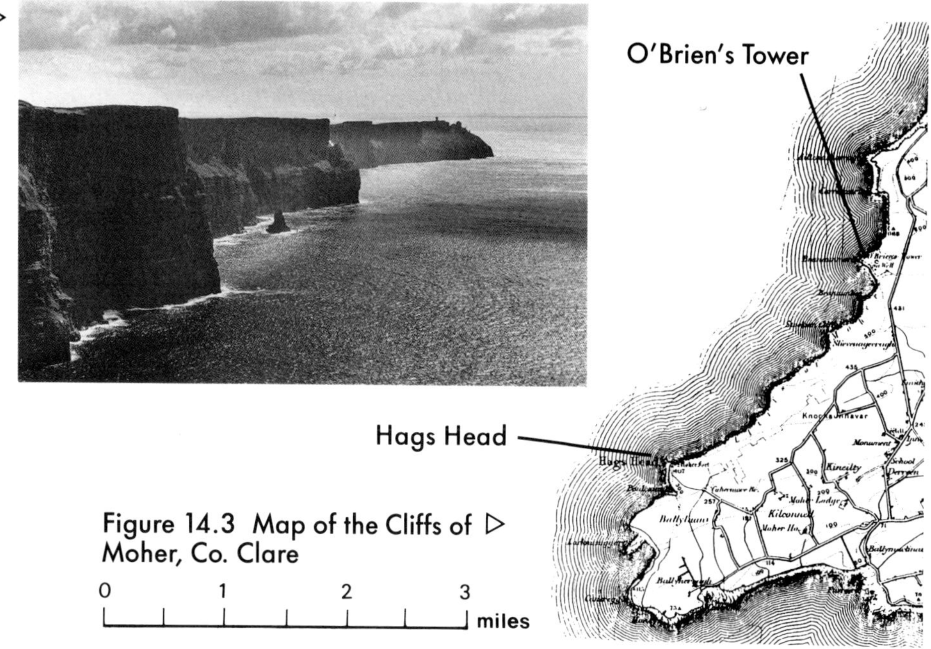

Figure 14.3 Map of the Cliffs of Moher, Co. Clare ▷

Study the photograph of the Cliffs of Moher and the 1″ OS map which accompanies it. Calculate the distance between the foreground and the tower on the cliff in the right background. The photo was taken near O'Brien's Tower and the camera was pointing towards Hags Head.

△
Figure 14.4a The 'photograph'

Photographs and direction

To judge directions in any photograph, one must first know the direction in which the camera was pointing when the photograph was taken. To discover this, we must refer to OS maps.

To find the direction in which a photograph was taken

(See figures 14.4a and 14.4b)

1. *Draw a line through the centre of the photograph from the foreground to the background.* This represents the direction in which the camera was pointing when the photograph was taken (figure 14.4a).
2. Identify two or three *prominent landmarks* through or near which the line you have drawn in 1 passes.
3. Identify these landmarks *on the OS map* of the area. Then draw a line on the map which corresponds to the line drawn on the photograph. Mark an arrow on the line so that it points *away from* the camera position on the photograph (figure 14.4b).
4. Identify the direction in which the line on the map is pointing. This is the direction in which the camera was pointing when the photograph was taken (figure 14.4b).

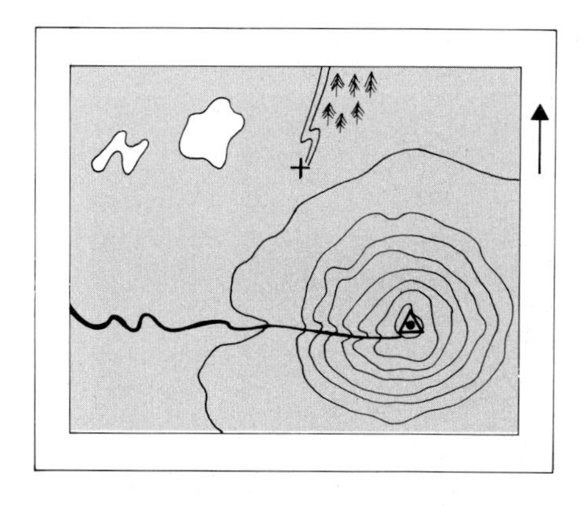

△
Figure 14.4b The map

Do Activity 4, page 109.

Recognising features in photographs

Geographical photographs often show many features. It takes a great deal of practice to recognise these features.

The shapes, sizes and shading of objects can be used to help identify objects.

Study photograph D, which shows an area near the coast.

Try to identify the following features:

- ☐ a larger beach
- ☐ a small beach, off which boats are moored
- ☐ a main road
- ☐ two fields with haystacks
- ☐ hedgerows
- ☐ small woodlands
- ☐ a row of houses
- ☐ a larger building
- ☐ farm buildings surrounded by trees

Making tracings from photographs

It is sometimes useful to make a tracing of a photograph in order to highlight the landscape shown.

Steps to follow when making a tracing

1. Place tracing paper on the photograph. On the tracing paper, mark carefully the four corners of the photograph. Join these marks to make a ***frame*** which will give you the ***boundaries*** of the photograph.
2. Make sure your tracing boundaries lie ***exactly*** on the boundaries of the photograph. Then trace lightly the ***main lines*** which appear in the photograph. These usually include the skyline or the ***horizon***, the coast and one or two other prominent features.
3. Trace the ***landscape features*** which you wish to show. Omit all ***unnecessary details***.
4. Identify the features on your tracing by means of ***arrowed labels***.
5. Give your tracing a ***title***. This will help to explain the landscape features you have shown. (Your title need not be the same as that given on the photograph.)

Do Activity 5, page 109.

How to draw a sketch of a photograph

1. Draw a ***rectangular frame*** of the sketch. It must be the same shape (though not necessarily the same size) as the frame of the photograph.
2. Using a light pencil, divide the photoaraph and the sketch into ***segments*** (figures 14.5a and 14.5b). Use these segments to guide you when you are positioning features on the sketch.
3. On your sketch, show and name the principal features of the photograph. Omit unnecessary details (figure 14.5b).
4. Give your sketch a ***title*** which helps to explain the features you have shown in your sketch.

Figure 14.5a The Lauterbrunnen valley in the Swiss Alps.
▽

Do Activity 6, page 109.

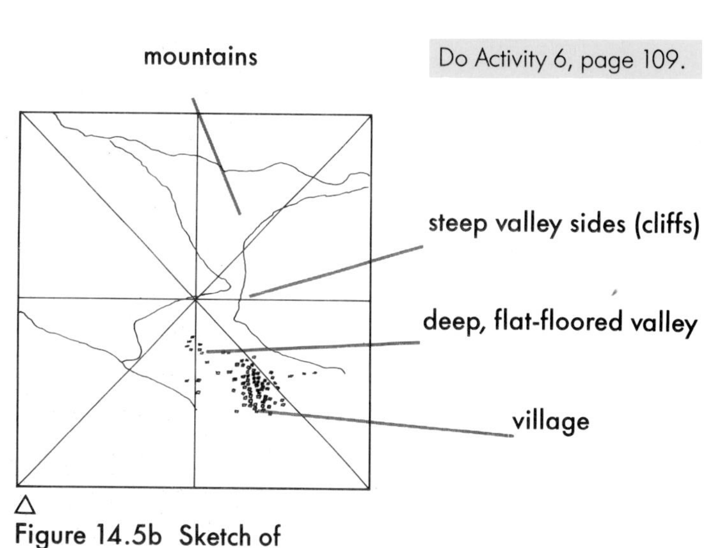

△
Figure 14.5b Sketch of Lauterbrunnen, a glaciated valley.

Photographs and the time of year

It is often possible to guess the time of year when a photograph was taken. These guidelines will help you do this.

Deciduous trees and bushes	These will be covered in foliage from around May to October. They will be bare from November to April.
Flowers in bloom	Most flowers will be in bloom from spring to early autumn.
Crops	Crops such as hay and cereal crops appear ripe during July and August.
Bare ploughed fields	Fields are usually bare between late winter and early spring.
People's clothing	Clothing such as heavy overcoats would suggest wintertime. Light clothing would suggest summer.
Crowded beaches	These are more likely during July and August, though this depends on day-to-day weather conditions.

Do Activity 7, page 109.

Maps versus photographs

Discuss the advantages of maps versus photographs using the information in the box.

Maps and photographs are two useful methods of representing landscape features on flat surfaces.

Here are some of the advantages and disadvantages of each.

	Maps	*Photographs*
Scale	Each map is drawn to a single scale and so can be used to calculate distances.	Most photographs do not show a single scale.
Realism	Maps show landscape features by means of symbols only.	Photographs can give a more realistic-looking view of landscape features.
Clarity	Symbols on all parts of a map can be shown with equal clarity.	Background features often appear vague in photographs.
Details	Maps can show some details which photographs cannot.	Photographs can show some details which maps cannot.

Activities

1. Examine the photograph of Glendalough on page 104. Give the locations of each of the following: the hotel; the round tower; the nearest lake; a clump of furze (gorse).

2. Examine the photograph of the Cliffs of Moher on page 105. Estimate the approximate height of the tower (Moher fort) on the cliff (Hag's Head) in the centre background of the photograph. Use the height of the cliff, as shown on the map, to help you.

3. Examine photograph B (Glendalough) and figure 14.2 on page 104.
   (a) Calculate the distance in miles between the two lakes in the photograph.
   (b) Calculate the length of the lake which can be seen towards the background of the photograph.

4. (a) In which direction was the camera pointing when photograph C (Cliffs of Moher) was taken? Use photograph C and figure 14.3 to obtain your answer. (Assume that the photograph was taken very near O'Brien's Tower.)
   (b) Which direction is it from the lake in the left centre of the photograph of Glendalough (B) to the hotel in the foreground?

5. Make a tracing of the Cliffs of Moher photograph (page 105). On the tracing, show and label the following: cliffs; a headland; a sea stack.

6. (a) In your opinion, at what time of year was the photograph of Glendalough (B) taken? Account for your answer.
   (b) In your opinion, at what time of year was photograph D (a coastal area) taken? Account for your answer.

15 WORLD LOCATIONAL GEOGRAPHY

To learn the locations of the important places listed in this chapter you will need:

- ☐ the ***blank map*** of the world in the Map Supplement which accompanies this book
- ☐ an ***atlas map*** of the world.

To learn the location of each place

1. Find the place on your ***atlas map.***
2. Locate its position on the ***blank map.***
3. Try to ***memorise*** its position.

Note: Do not write on your blank map!

To test your knowledge

1. Try to locate each place on the blank map.
2. Check your accuracy by consulting your atlas.

Note: You may work with a friend on this.

Some Lines of Latitude

0° — Equator
23½°N — Tropic of Cancer
23½°S — Tropic of Capricorn
66½°N — Arctic Circle

Some Lines of Longitude

0° — Prime or First or Greenwich Meridian
180°E/180°W — International Date Line

Some Sea Areas

☐ ***Arctic Ocean***

White Sea

☐ ***Atlantic Ocean***

Norwegian Sea
North Sea
Baltic Sea
North Channel
Irish Sea
St George's Channel
English Channel
Bay of Biscay
Straits of Gibraltar
Mediterranean Sea
Tyrrhenian Sea
Ionian Sea
Gulf of Taranto
Adriatic Sea
Aegean Sea
Black Sea
Gulf of Guinea
Caribbean Sea
Gulf of Mexico
Gulf of St Lawrence
Hudson Bay

☐ ***Indian Ocean***

Arabian Sea
Gulf of Aden
Red Sea
Iranian (Persian) Gulf
Bay of Bengal

☐ ***Southern Ocean***

Great Australian Bight

☐ ***Pacific Ocean***

Bering Sea

Some Islands and Headlands
(H = headland)

☐ ***Arctic Ocean area***

North Cape (H)
Newfoundland

☐ ***Atlantic/ Mediterranean areas***

Greenland
Iceland
Hebrides
Isle of Man
Isles of Scilly
Channel Islands
Balearic Islands
Corsica
Sardinia
Sicily
Malta
Crete
Cyprus
Canary Islands
Cape of Good Hope (H)
Tierra del Fuego
Cape Horn (H)
Cuba
Dominican Republic/Haiti

☐ ***Indian Ocean/Pacific areas***

Malagasay
Sri Lanka
Sumatra
Java
Borneo
New Guinea
Philippines
Taiwan
Islands of Japan
New Zealand
Tasmania
Vancouver

Some Mountain Ranges

☐ ***Europe***

Pennines
Grampians
Sierra Nevada
Cantabrians
Pyrenees
Central Massif/Cevennes
Brittany Highlands
Vosges
Jura
Alps
Black Forest
Appennines
Pindus
Dinaric Alps
Carpathian
Caucasus
Urals
Scandinavian mountains

☐ ***Asia***

Zagros
Deccan plateau
Himalayas
Plateau of Tibet

☐ ***Africa***

Atlas
Ethiopian Highlands
Drakensberg Range

☐ ***Australia***

Great Divide

☐ ***The Americas***

Rockies
Appalachians
Mexico Plateau
Andes
Brazilian Highlands

Some Rivers

☐ ***Europe***

Shannon
Severn
Thames
Elbe
Rhine
Meuse
Moselle
Main
Schelde
Seine
Loire
Douro
Tagus
Guadiana
Guadalquivir
Ebro
Saone
Rhone
Tiber
Po
Danube
Dnieper
Volga

☐ ***Asia***

Indus
Ganges
Mekong
Si-Kiang
Yangtse-Kiang
Hwang-Ho
Amur

☐ ***Australia***

Murray
Darling

☐ ***Africa***

Niger
Zaire (Congo)
Limpopo
Zambezi
Nile

☐ ***North America***

Makenzie
Yukon
Columbia
Colorado
Rio Grande
Mississippi
Missouri
St Lawrence

☐ ***South America***

Orinoco
Amazon
Parana
Paraguay

Some Lakes

☐ ***Asia***

Caspian Sea
Aral Sea

☐ ***Africa***

Lake Victoria
Lake Tanganyika
Lake Malawi

☐ ***North America***

Great Bear
Great Slave
Lake Winnipeg
Lake Superior
Lake Michigan
Lake Huron
Lake Erie
Lake Ontario

Some Canals

Suez
Panama

Some Countries and Cities

(+ denotes capital city)

☐ ***Europe***

country	*city*
Ireland	Dublin+
	Cork
	Limerick
United Kingdom	London+
	Cardiff
	Birmingham
	Liverpool
	Manchester
	Edinburgh
	Glasgow
	Belfast
Denmark	Copenhagen+
Federal Republic of Germany (West Germany)	Bonn+
	Dortmund
	Essen
	Cologne
	Frankfurt
	Munich
	Hamburg
France	Paris+
	Le Havre
	Nantes
	Bordeaux
	Toulouse
	Marseilles
	Lyons
	Strasbourg
The Netherlands (Holland)	The Hague+
	Amsterdam
	Rotterdam

Country	*City*
Belgium	Brussels+
	Antwerp
Luxembourg	
Spain	Madrid+
	Barcelona
	Seville
Portugal	Lisbon+
	Oporto
Italy	Rome+
	Milan
	Turin
	Genoa
	Venice
	Naples
Greece	Athens+
USSR	Moscow+
	Leningrad
Rumania	Bucharest+
Yugoslavia	Belgrade+
Austria	Vienna+
Hungary	Budapest+
Czechoslovakia	Prague+
Poland	Warsaw+
German Democratic Republic (East Germany)	Berlin+
Switzerland	Berne+
Norway	Oslo+
Sweden	Stockholm+
Finland	Helsinki+

☐ *North America*

country	*city*
Canada	Ottawa+
	Montreal
United States	Washington DC+
	New York
	Chicago
	Los Angeles

☐ *Central and South America*

country	*city*
Mexico	Mexico City+
Colombia	Bogota+
Peru	Lima+
Chile	Santiago+
Argentina	Buenos Aires+
Brazil	Brasilia+
	Rio de Janeiro
Venezuela	Caracas

☐ *Oceania*

country	*city*
Australia	Canberra+
	Sydney
	Melbourne
New Zealand	
Indonesia	

☐ *Middle East and Asia*

country	*city*
Turkey	Ankara+
	Istanbul
Saudi Arabia	
Iran	Tehran+
Pakistan	
India	Delhi+
Thailand	
Vietnam	Ho Chi Minh City (Saigon)+
China	Bejing (Peking)+
	Shanghai
	Hong Kong (British territory until 1997)
Korea	
Japan	Tokyo+
Philippines	

☐ *Africa*

country	*city*
Egypt	Cairo+
Libya	
Algeria	
Nigeria	Lagos+
Angola	
South Africa	Pretoria+
Mozambique	
Tanzania	
Kenya	
Ethiopia	
Sudan	
Zimbabwe	

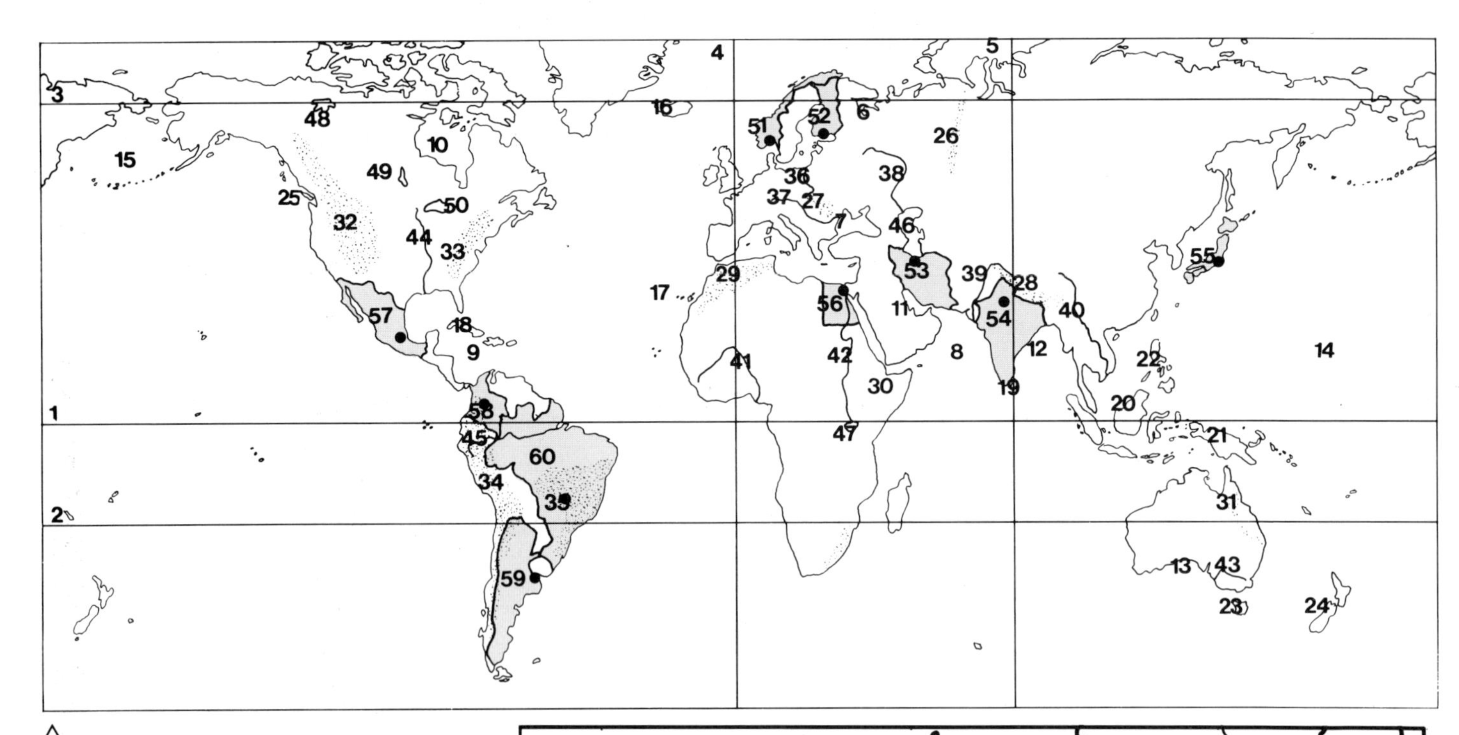

△
Figure 15.1

Figure 15.2 ▷

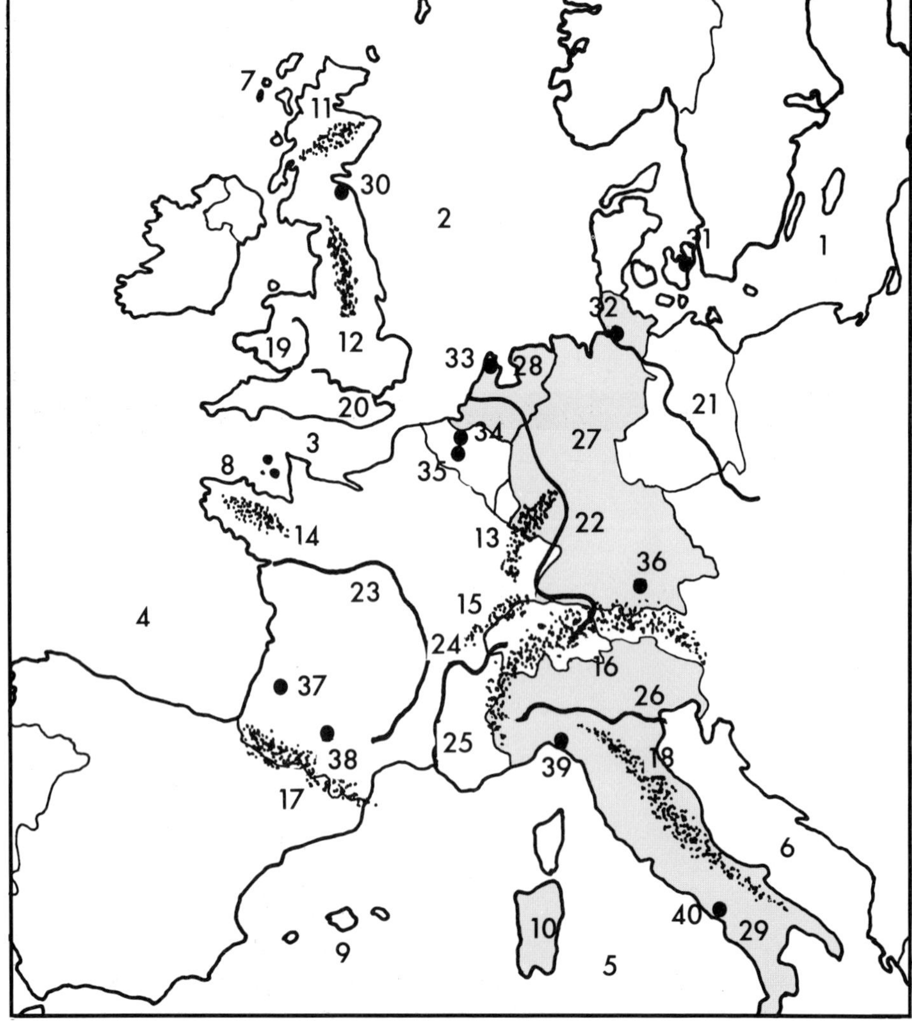

Activities

1. Study figure 15.1. Then name the following.
   (a) the lines of latitude and longitude 1–5
   (b) sea areas 6–15
   (c) islands 16–25
   (d) mountain ranges 26–35
   (e) rivers 36–45
   (f) lakes 46–50
   (g) countries 51–60, and the cities shown with a dot in each country

2. The map in figure 15.2 shows the location of 40 features. Write the names of the features in your geography notebook.
   (a) sea areas 1–6
   (b) islands 7–10
   (c) mountain ranges 11–18
   (d) rivers 19–26
   (e) countries 27–29
   (f) cities 30–40

16 IRELAND: LOCATIONAL GEOGRAPHY

To learn the locations of the places listed in this chapter, you will need:

- ☐ the ***blank map*** of Ireland included in the Map Supplement which accompanies this book
- ☐ an ***atlas map*** of Ireland

To learn these locations, and to test your knowledge of them, follow the directions on page 110.

Some headlands, peninsulas and inlets

Dublin Bay
Howth Head
Dundalk Bay
Carlingford Lough
Dundrum Bay
Strangford Lough
Belfast Lough
Fair Head
Larne Lough
Giant's Causeway
Lough Foyle
Malin Head
Inishowen Peninsula
Lough Swilly
Bloody Foreland

Gweebarra Bay
Rossan Point
Donegal Bay
Sligo Bay
Killala Bay
Erris Head
Blacksod Bay
Clew Bay
Killary Harbour
Slyne Head
Galway Bay
Loop Head
Shannon Estuary
Tralee Bay
Dingle Peninsula
Dingle Bay

Kenmare Bay
Bere Island
Bantry Bay
Mizen Head
Old Head of Kinsale
Cork Harbour
Youghal Harbour
Dungarvan Harbour
Waterford Harbour
Hook Head
Bannow Bay
Carnsore Point
Rosslare Harbour
Wexford Harbour
Cahore Point
Wicklow Head

Some islands

Lambay Island
Rathlin Island
Tory Island
North Aran Island
Achill Island
Clare Island
Aran Islands
Blasket Islands
Valentia Island
Bere Island
Great Island
Saltee Islands.

Activities

1. *Ria, fiord, estuary, peninsula*
   Find out the precise meaning of each of these words. Name *one* Irish example of each.
2. **What kind of rock is found at the Giant's Causeway? How was this rock formed?**

Some mountain ranges

In Leinster

Wicklow Mountains
Blackstairs Mountains
Slieve Bloom Mountains

In Ulster

Mourne Mountains
Antrim Plateau
Sperrin Mountains
Donegal Mountains

In Connaught

Ox Mountains
Nephin Mountains
Connemara Mountains

In Munster

Slieve Aughty
Silvermines Mountains
Mullaghareirk Mountains
Macgillicuddy's Reeks
Caha Mountains
Boggeragh Mountains
Galtee Mountains
Knockmealdown Mountains
Comeragh Mountains

Some rivers

Liffey
Boyne
Lagan
Ulster Blackwater
Lower Bann
Upper Bann
Foyle
Erne
Moy
Clare
Shannon
Brosna
Suck
Inny
Feale
Bandon
Lee
Munster Blackwater
Suir
Nore
Barrow
Slaney

Some lakes

Lough Neagh
Lough Derg (Ulster)
Lower Lough Erne
Upper Lough Erne
Lough Conn
Lough Corrib
Lough Mask
Lough Derg (Munster)
Lough Ree
Lough Allen
Lakes of Killarney

Activities

1. In which general direction does each of the following rivers flow: the Upper Bann, the Feale and the Lee?
2. Which river drains from Lough Neagh?

Some cities and towns

Coastal cities and towns

Dublin
Drogheda
Dundalk
Downpatrick
Belfast
Carrickfergus
Larne
Portrush
Derry
Letterkenny
Killybegs
Donegal
Ballyshannon
Sligo
Westport
Clifden
Galway
Limerick
Tarbert
Tralee
Kenmare
Bantry
Kinsale
Cork
Cobh
Youghal
Dungarvan
Waterford
Rosslare
Wexford
Arklow
Wicklow
Bray
Dun Laoghaire

Inland cities and towns

Trim
Navan
Newry
Lisburn
Coleraine
Antrim
Lurgan
Armagh
Strabane
Lifford
Omagh
Enniskillen
Cavan
Ballina
Castlebar
Tuam
Athenry
Ennis
Shannon
Ballinasloe
Mullingar
Carrick-on-Shannon
Athlone
Killarney
Macroom
Fermoy
Mallow
Clonmel
Thurles
New Ross
Kilkenny
Carlow
Athy
Naas

Activities

1. Name *three* Irish ports which have car ferry connections with these overseas ports: Stranraer, Le Havre, Liverpool.
2. Identify *five* popular holiday resorts from the towns listed in this chapter.

Counties

In Leinster

Louth
Meath
Dublin
Wicklow
Wexford
Kilkenny
Carlow
Laois
Offaly
Kildare
Westmeath
Longford

In Ulster

Down
Antrim
Derry
Tyrone
Armagh
Fermanagh
Monaghan
Cavan
Donegal

In Connaught

Leitrim
Sligo
Mayo
Galway
Roscommon

In Munster

Clare
Limerick
Kerry
Cork
Waterford
Tipperary

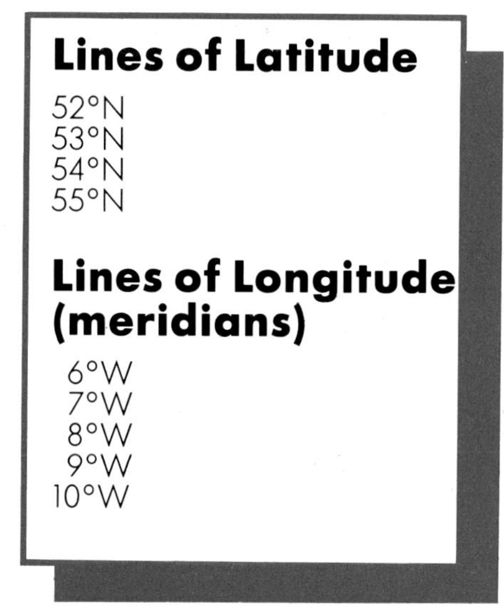

Lines of Latitude

52°N
53°N
54°N
55°N

Lines of Longitude (meridians)

6°W
7°W
8°W
9°W
10°W

Activities

1. **A Giant Test!**

Figure 16.1 shows the locations of 50 features. In your geography notebook, write the names of the features numbered as follows.

1–4 inlets
5–6 headlands
7 peninsula
8–9 islands
10–17 mountains
18–27 rivers
28–32 lakes
33–46 towns
47–50 counties

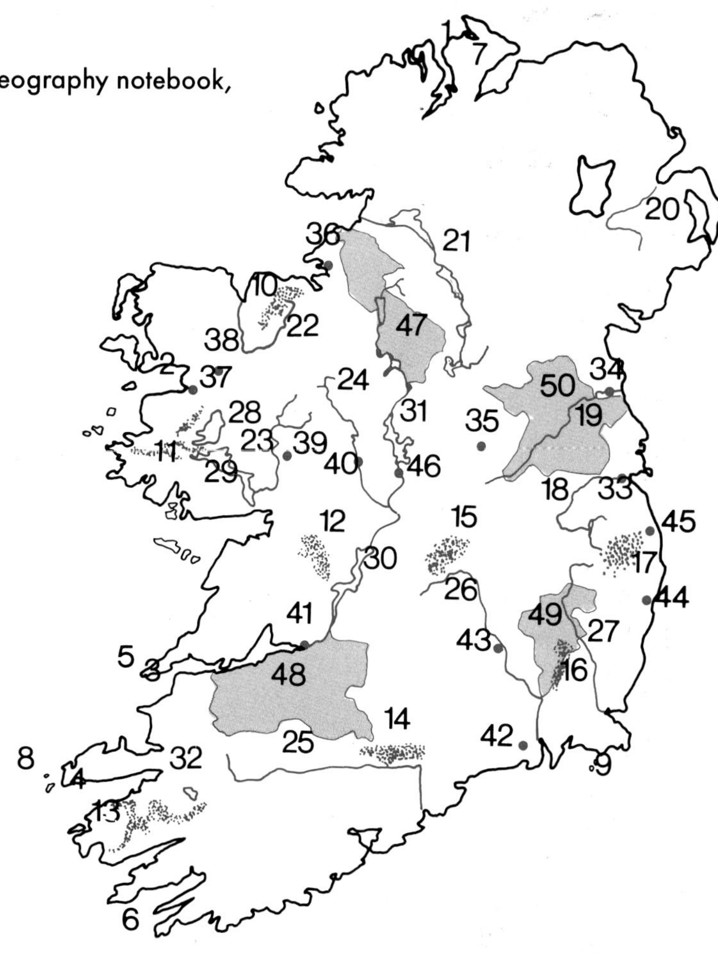

Your grade!

A = 43 to 50 correct
B = 35 to 42 correct
C = 28 to 34 correct
D = 20 to 27 correct
E = 13 to 19 correct
F = under 13 correct

2. For this exercise you will need an accurate outline map of Ireland. You may trace it carefully from your atlas map.
   On the outline map (and without the help of your atlas), show and name a selection of features as listed by your teacher.

17 ECONOMIC ACTIVITIES: AN INTRODUCTION

There are many different ***types of work*** or ***economic activities.***

All economic activities can be grouped into three broad categories. These are:

- ☐ ***Primary*** (first level)
- ☐ ***Secondary*** (second level)
- ☐ ***Tertiary*** (third level)

Category	**Primary**	**Secondary**	**Tertiary**
Description	These activities provide unprocessed ***raw materials*** from the earth's rocks, soils and waters.	These often take raw materials and ***process*** or change them. Secondary industries sometimes take already-processed materials and process them further.	These perform ***useful services.***
Examples	farming, fishing, forestry, mining	manufacturing industries, such as engineering, textiles etc.	shopkeeping, truck driving, teaching, hairdressing, tourism etc.

Activities

1. Into which category of activities would you place each of the following: taxi driver, shipyard worker, TD, potter, farmer, shop assistant, mechanic, forestry worker, chimney sweep, doctor, fisherman, nun, pop singer?
2. List other professions under each of these headings: primary activities, secondary activities, tertiary activities.

18 WATER: A VITAL NATURAL RESOURCE

Of all the good things or ***resources*** provided by nature, clean air and water are the most important. Without water, life on our planet could not exist. The importance of water to our survival and comfort is shown in figure 18.1.

Figure 18.1 The importance of water
▽

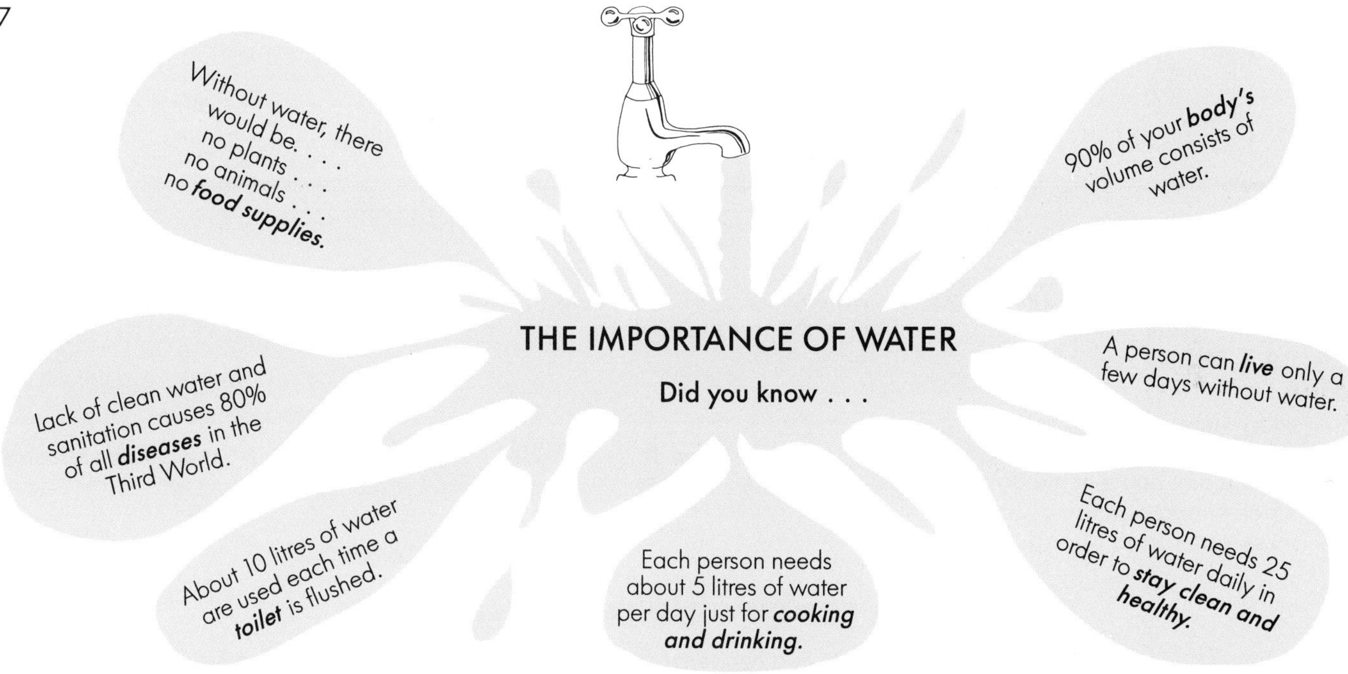

People: the water guzzlers

People use huge amounts of water. With an expanding world population and the increase in manufacturing industries, the demand for fresh water is growing rapidly. In rich countries, such as those in Western Europe, every person uses, on average, a staggering 300 litres of water per day. It is estimated that this rate of usage will have doubled by the year 2000.

Much of Europe's water consumption is accounted for by manufacturing industries. Figure 18.2 shows how much water is needed to make various manufactured products.

Figure 18.2 Water is needed to manufacture many products
▽

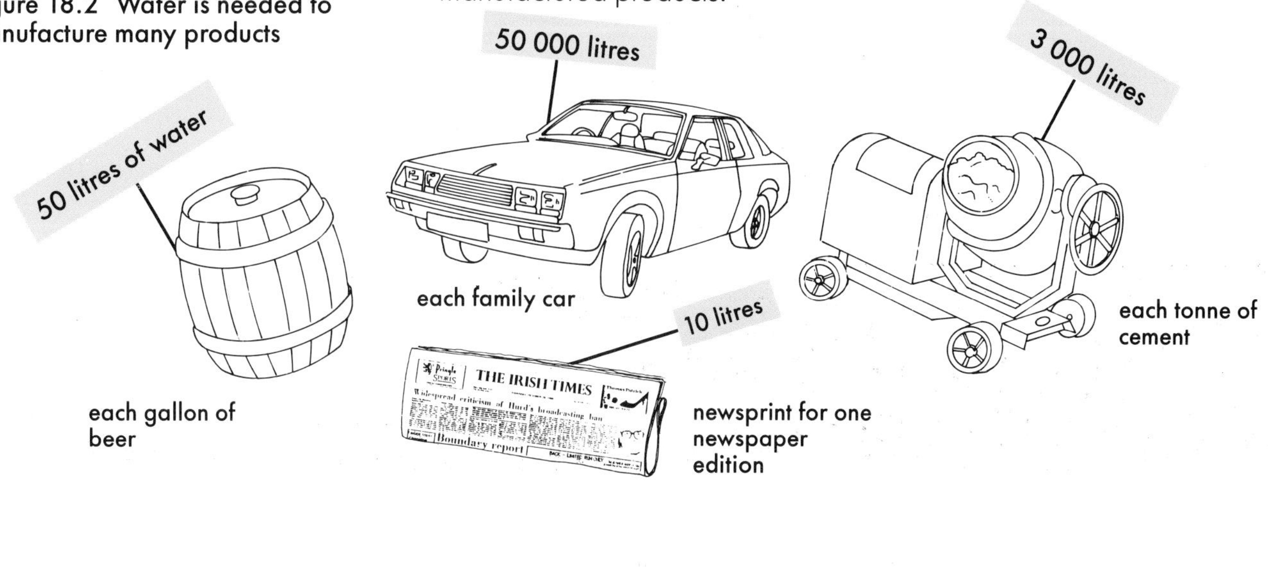

50 litres of water — each gallon of beer

50 000 litres — each family car

10 litres — newsprint for one newspaper edition

3 000 litres — each tonne of cement

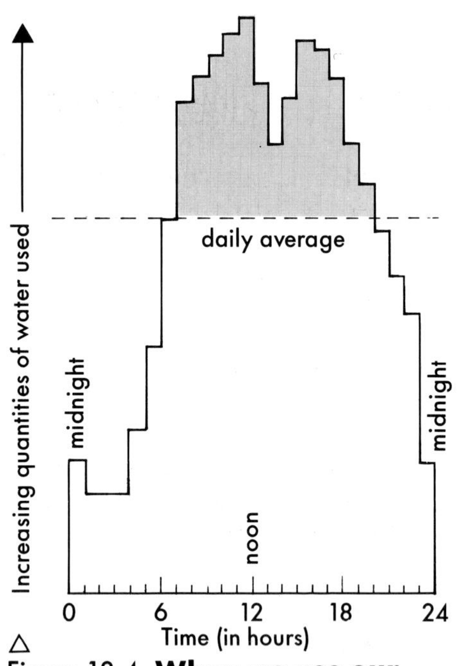

△
Figure 18.4 **When we use our water.**
(a) At which times of day is most water needed? Explain why this is so.
(b) For how many hours in the day is there a below-average use of water? Why is less water used during these hours?

How Ireland uses water

The ways in which fresh water is used in Ireland are shown in figure 18.3.

☐ Draw a ***percentage bar*** to illustrate the uses of water as shown in figure 18.3. Make your bar 10cm long.

☐ Write a few sentences to explain what your percentage bar shows.

☐ ***Trend (line) graph, pie chart, four separate bar graphs.*** Which of these methods would offer the best alternative for showing the information given in figure 18.3? Give reasons for your choice.

☐ List ***specific*** uses of water for each of the general headings given in figure 18.3.

▽ **Figure 18.3**

Percentage uses of water in Ireland	%
A. Domestic	24
B. Industrial (including manufacture of electricity)	58
C. Agriculture	8
D. Other	10

Here is a list of ***natural resources.*** ▶ Separate those which are renewable from those which are non-renewable.

coal	air	iron ore
forests	peat	water
fish	gold	soil
copper	sunshine	oil

Non-renewable resources are those with limited supplies which may eventually be used up.

A renewable resource

Water is a ***renewable resource.*** This means that it can be used again and again, provided it is properly used and conserved.

The same water supply has existed on earth since the beginning of time. This is so because nature acts as a huge distillery, constantly renewing and cleansing the earth's water supply. It does so by a process known as the ***water cycle*** or the ***hydrological cycle*** through which water is drawn from the earth's seas and up into the atmosphere, to fall again as precipitation. Figure 18.5 shows how the water cycle works.

What does the term 'cycle' mean in this diagram?

Figure 18.5 The water cycle ▽

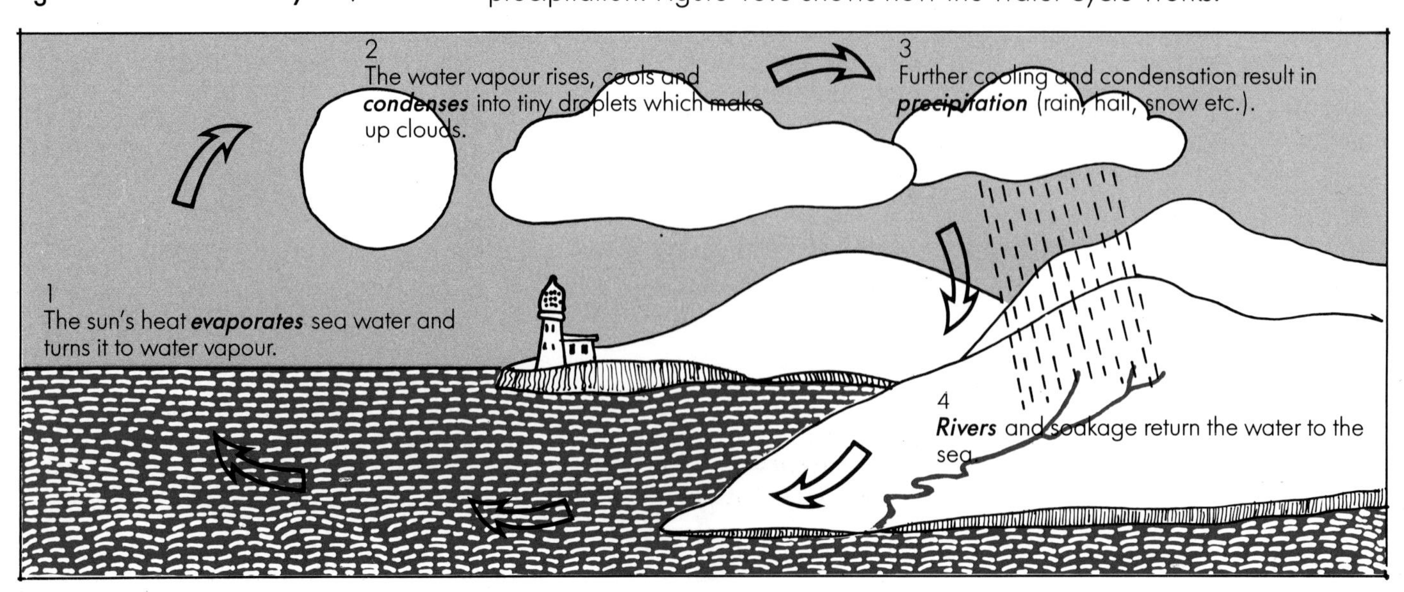

Local water supplies in Ireland

County Councils and City Corporations in Ireland play an important role in supplying fresh water. They collect, store, treat and distribute our water supplies.

The following case study tells the story of South Dublin's main water supply. Read the case study and then do the activities which follow it.

Case Study: Water for South Dublin

From the Liffey to the Tap

South Dublin gets most of its water supply from the ***catchment area*** of the River Liffey. (A catchment is that area drained by a river and its tributaries.) Much of this water is drawn from Pollaphuca in Co. Wicklow.

From Pollaphuca to Consumer

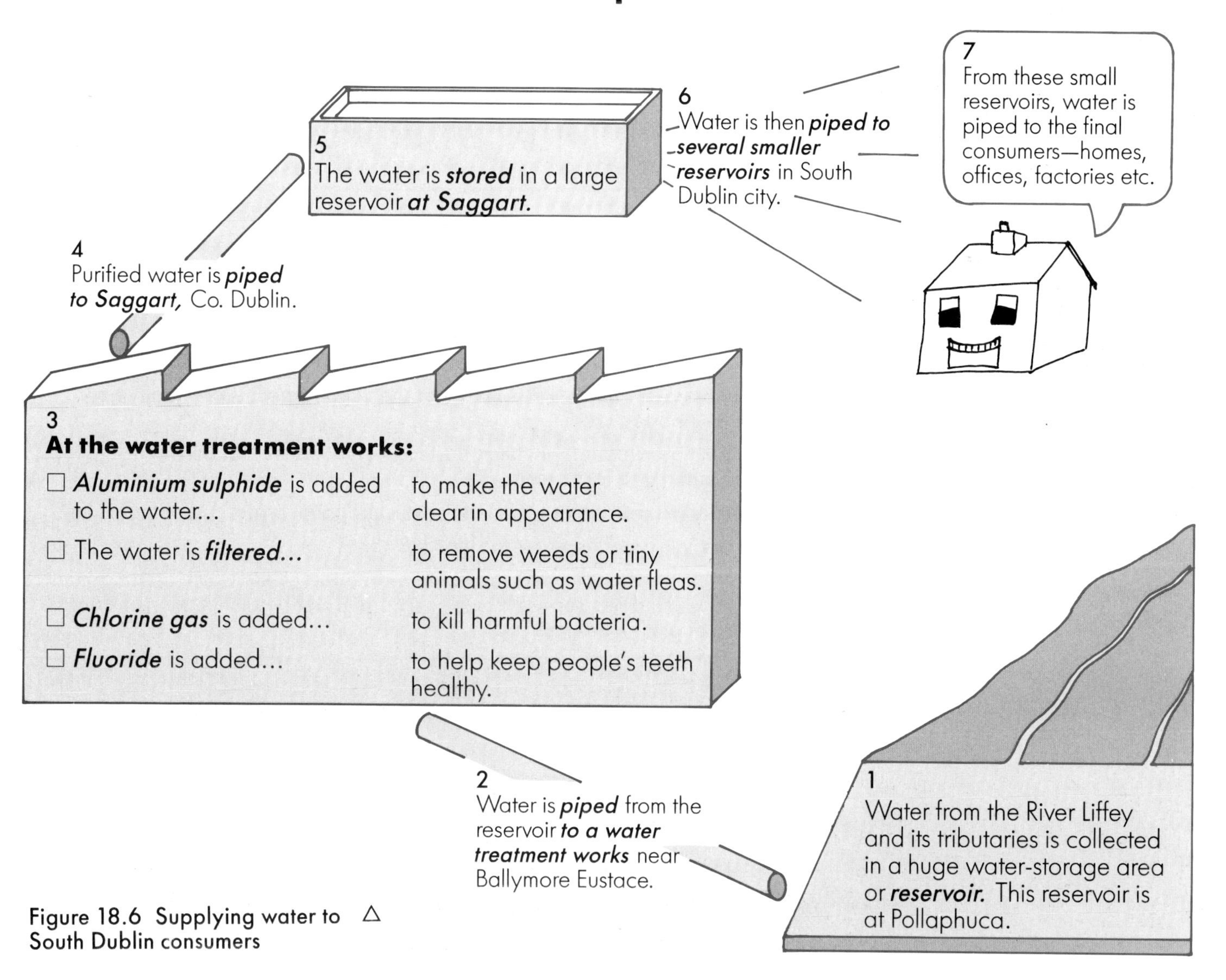

Figure 18.6 Supplying water to △ South Dublin consumers

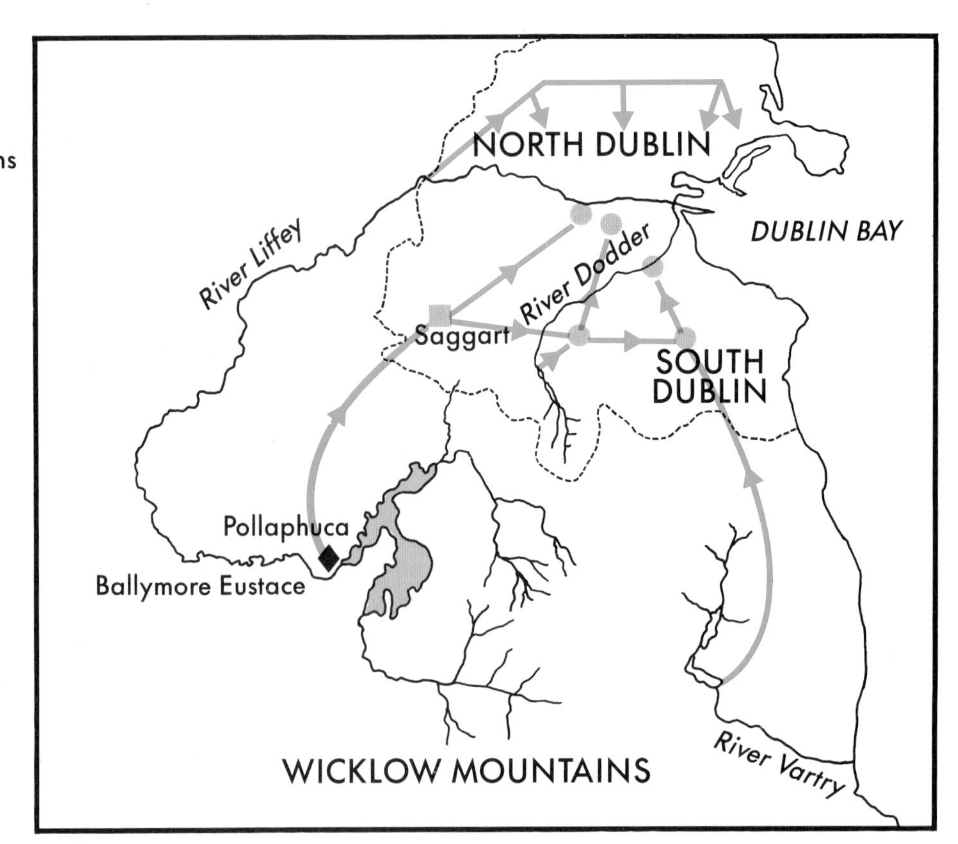

Figure 18.7 Water for Dublin

1. Draw a freehand copy of figure 18.7 to illustrate how South Dublin gets water from the River Liffey. Show and name the following on your map: Dublin city; the Liffey and its tributaries; Pollaphuca reservoir; Ballymore Eustace treatment works; Saggart reservoir; smaller city reservoirs; water pipelines connecting the places listed.
   Using pencil, lightly shade in the catchment area of the River Liffey.
2 Calculate, in kilometres, the length of the water pipeline from Ballymore Eustace to Saggart.
3. Explain how *drainage* in the Pollaphuca area makes that area suitable for the location of a large reservoir.
4. Name a river, other than the Liffey, which supplies water to South Dublin.
5 Which river supplies water for North Dublin?
6. Consult the Dublin District map in the Map Supplement which accompanies this book. Find *two* separate reservoirs and describe the location of each by means of grid references.

A Project

Prepare a report on the local water supply in your area.

☐ In your report, describe:

(a) where your water supply comes from

(b) where, how and why the water is treated before distribution

Your local County Council or City Corporation may be able to help you obtain this information.

☐ Illustrate your report with a map. An OS map of your area could be used to help you complete your report.

Water shortages

Ireland seldom suffers from severe water shortages. Occasionally, however, our precipitation amounts (rain, snow, sleet etc.) may be lower than usual for a prolonged period. When this happens, a water emergency may occur, especially in large urban areas.

1975 was a particularly dry year for Ireland. Figure 18.8 shows the expected precipitation figures for a normal year. It also shows the actual precipitation for 1975.

- ☐ **Make two separate line graphs within a single diagram. One line should show the expected precipitation. The second line should show the actual precipitation for 1975.**
- ☐ **Shade and label your diagram as follows: (a) the periods when precipitation was above average; (b) the periods when precipitation was below average.**
- ☐ **Give your diagram a title.**
- ☐ **During which months was the water supply most likely to cause serious problems? Explain your answer.**

Figure 18.8 Fill in the total figures ▽ for each column

Month	J	F	M	A	M	Jn	Jy	A	S	O	N	D	Total
Expected precipitation (mm)	79	54	53	57	67	54	75	85	82	82	73	90	?
Actual precipitation (mm) in 1975	132	37	27	39	29	7	76	28	124	82	47	43	?

The drought of 1975 affected Ireland in a number of ways. Some of these are shown in figure 18.9.

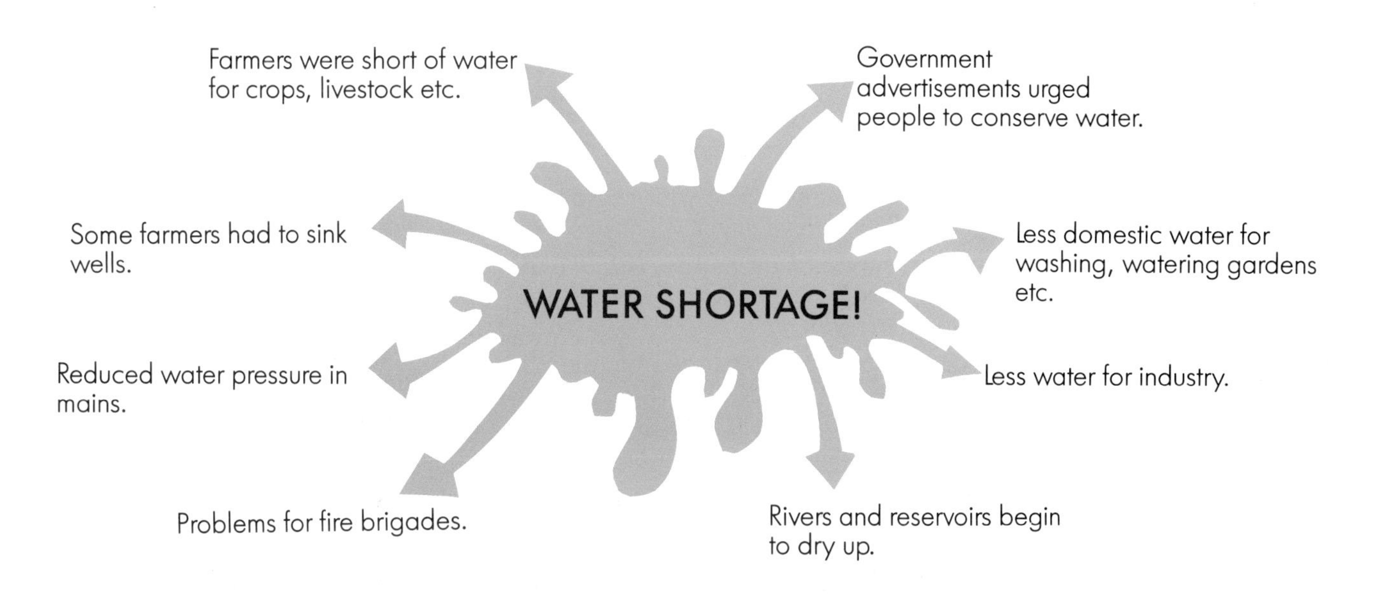

△
Figure 18.9 The results of a water shortage

Where water is really scarce

All living things need water. Yet for millions of people in the Third World, the search for water is a daily problem. The lack of clean, convenient and adequate supplies of fresh water for drinking, sanitation and washing can cause a number of problems.

☐ Each year, millions of people die from ***diseases*** such as diarrhoea, cholera and typhoid. Diarrhoea alone kills 6 million children each year. The spread of these diseases is directly linked to inadequate or polluted supplies of water.

☐ Many people, especially women, are forced to spend a significant part of each day drawing water from distant water supplies. Read 'Nandi's Day in Rural Kenya' and then answer the questions which follow.

Nandi's Day in Rural Kenya

Consider a day in the life of Nandi who lives in a small farming community in the eastern province of Kenya. Just after sunrise, she gets her four children out of bed. The two little girls she sends for water at the village water tap over a kilometre away. The girls carry ten-litre cans back home on their heads. This gives the family enough water for washing and for making breakfast.

Although there is a waterhole just half a kilometre away from Nandi's home, the water there is dirty and has a bad taste. Only the animals use this water now. Since the water tap was installed in the village, there have been fewer deaths from diarrhoea, especially among the babies.

After breakfast, Nandi herself goes to the village. There she joins the queue at the water tap, carrying home the 18 litres of water which she needs for washing clothes and dishes and for preparing the midday meal. She must also carry her 1½ year old son who is too small to be left alone. Because of the heat, Nandi's family will be thirsty and everyone will want a drink.

With temperatures soaring in the afternoon, Nandi must once again make the journey to the water tap to fetch enough water for the evening meal. While at the tap, she washes herself and her baby as do the other women who have come to the standpipe for the precious water. When the children come home from school, they too must go to the tap to wash and drink.

Altogether, Nandi and her daughters spend four hours of every day simply fetching water.

(Adapted from *Dialogue for Development,* Trocaire 1983).

△ Egyptian women on their way to draw water. Across what desert are they walking?

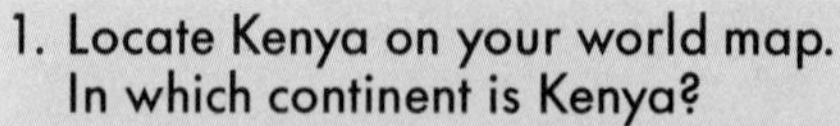

1. Locate Kenya on your world map. In which continent is Kenya?
2. How many trips must Nandi and her family make to the water tap every day?
3. How many trips does Nandi herself make?
4. How many kilometres in all must Nandi travel each day to collect water?
5. Apart from Nandi, which other members of her family must fetch water? What does this suggest about local traditions regarding work? Can you think of any similar traditions which exist in Ireland?
6. Why does Nandi choose to draw water from the water tap rather than from the nearer waterhole?
7. Which of the following words would you use to describe a Third World woman such as Nandi: lazy, hardworking, sophisticated, intelligent, unintelligent? Explain your choice(s).

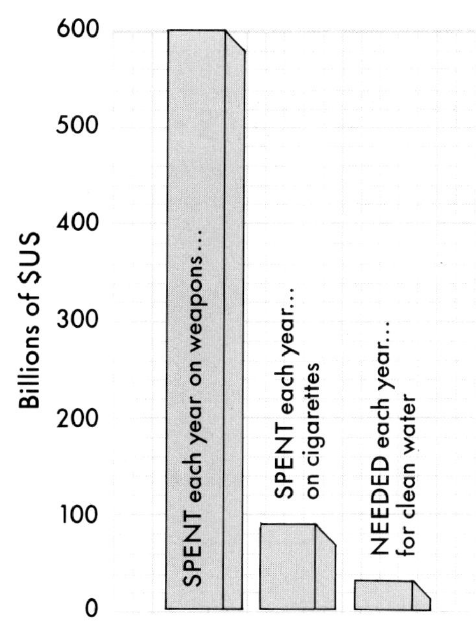

Millions of members of our human family are in desperate need of clean water supplies. It is estimated that this need could be met at a cost of $30 billion per year. This is a great deal of money! Yet we waste many times that amount each year on destructive things like weapons and cigarettes. Study figure 18.10 and then answer the questions which follow it.

◁ **Figure 18.10 Thirty billion US dollars are needed annually to supply the world's people with adequate supplies of clean water. (a) How many times this amount is being spent each year on: (i) cigarettes; (ii) weapons? (b) What could be done by (i) individuals and (ii) governments to change this situation? Give precise answers. Think carefully before answering.**

Pollution: Abusing a precious resource

Water is one of our most precious resources. Yet we continue to endanger our supplies of clean water by polluting our rivers and seas.

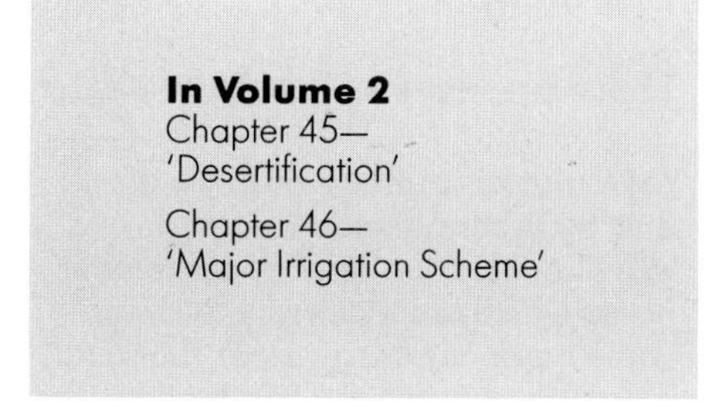

In Volume 2
Chapter 45—
'Desertification'

Chapter 46—
'Major Irrigation Scheme'

Activities

1. 'Water is a much-used yet renewable natural resource which is vital to human survival'.
   (a) Explain the meaning of the term 'renewable natural resource'.
   (b) Explain why so much water is now being used by people in Western Europe.
   (c) Describe *four* ways in which water is important for human comfort and survival.
   (d) Describe *three* difficulties which arise in Ireland when water is temporarily scarce.
   (e) Describe *three* serious problems related to water shortages in Third World countries.
2. Study the cartoon in figure 18.11.
   (a) What message is the cartoon trying to convey?
   (b) How does the cartoon make you feel?
   (c) Do you think the cartoon is giving a true picture of Africa, or is the message too simplified? Explain your answer.

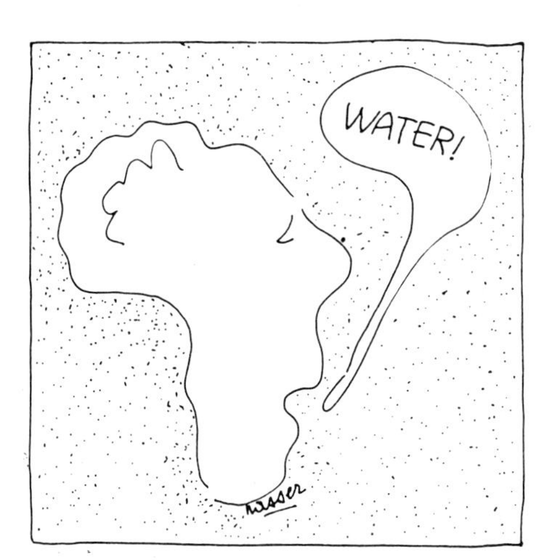

△ Figure 18.11

19 ENERGY SOURCES: FOCUS ON OIL

Renewable and non-renewable energy

The sun is the ***source*** of much of the world's energy. Other sources of energy include wind, tides, waves and falling water. All of these types of energy can be used again and again without being completely used up. They are ***renewable*** or ***infinite resources.***

Some sources of energy are called ***fuels.*** Examples include coal, oil, natural gas and peat. When these fuels have been burned, the energy they contain is released. Once a fuel has been burned, it cannot be used again. It is ***non-renewable.*** Non-renewable fuels will be used up eventually. They are limited or ***finite resources.***

Energy guzzlers

Modern society uses vast amounts of power or energy. This energy is mainly consumed:

☐ in ***domestic*** or household activities

☐ in ***manufacturing*** industry

☐ in ***commercial*** activities

☐ in ***transport***

How energy is used in Ireland

(a) What percentage of Ireland's energy is consumed by domestic activities? List the appliances in your home which use electricity. Name some types of *energy other than electricity* which are consumed in your home.
(b) What percentage of Ireland's energy is consumed by manufacturing industry? Name *four* examples of manufacturing industry in or near your locality.
(c) What percentage of energy is used by transport? Name *four* common types of fuel-using transport.
(d) What percentage of energy is used by commerce? Name *four* commercial activities, apart from commercial transport.

Figure 19.1 The use of energy in Ireland ▽

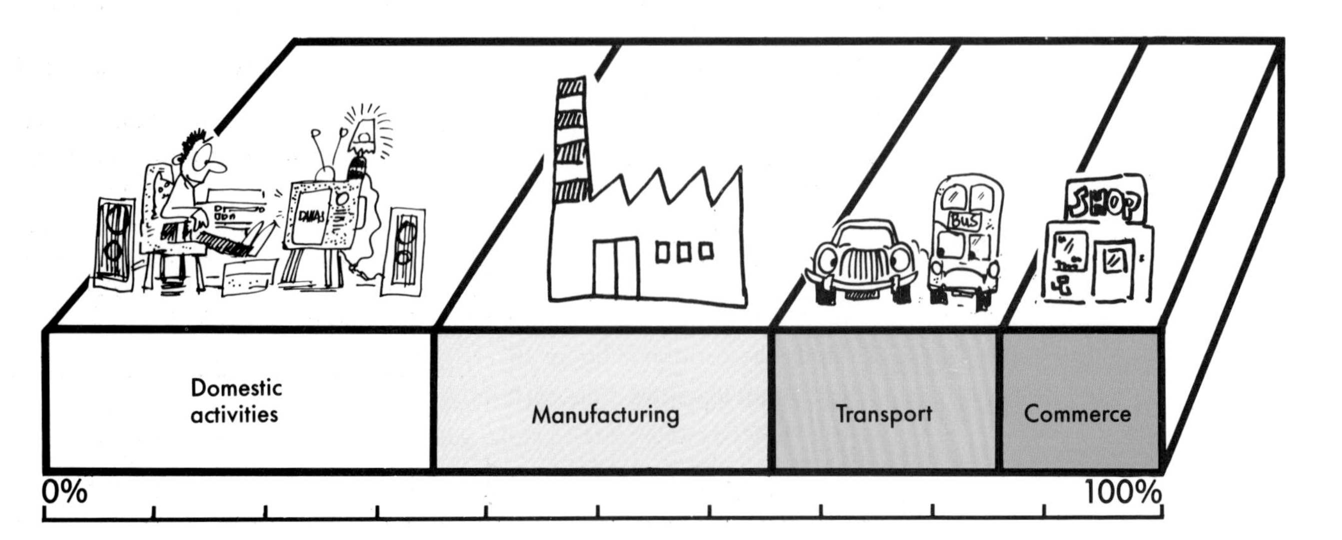

Using energy: It's an unequal world

The use of energy throughout the world is very uneven. People in rich countries use vastly different amounts and types of energy, compared with people in poorer countries.

Figure 19.2 is an example of a ***scatter graph.*** It shows the relationship between the standards of living in several different countries and the amounts of energy used. The ***horizontal axis*** (bottom line) shows the average wealth of people in each country. The ***vertical axis*** (the upright line on the left) shows the average amount of energy used in each country.

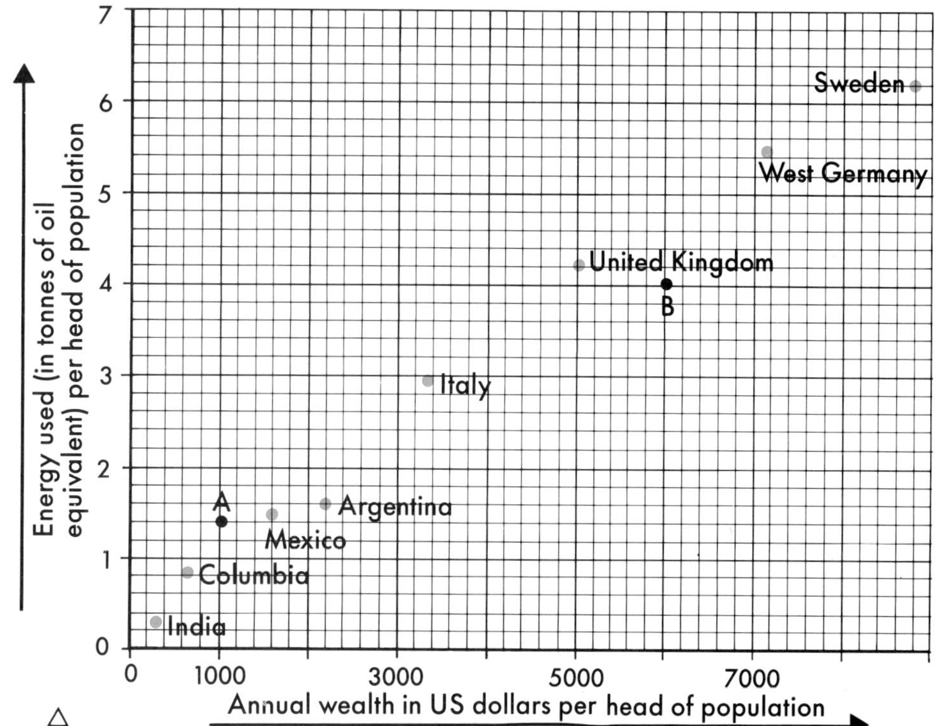

△ Figure 19.2 Standards of living and energy used

(a) Rank each of the eight named countries according to their *wealth.* Begin with the wealthiest and work down to the least wealthy.

(b) Rank the eight countries according to the *amount of energy* consumed per head of population.

(c) Suggest a connection between the wealth of a country and the amount of energy its people consume.

(d) Figure 19.2 shows two unnamed countries marked A and B. These countries are among the four listed in Table 1. Identify countries A and B by matching their wealth and energy-using figures from the statistics given in Table 1.

Table 1 ▷

Country	*Energy used per head of population (in tonnes of oil equivalent per year)*	*Wealth per head of population (in US dollars)*
France	4	6,000
United States	10.5	11,000
Turkey	1.4	1,000
Chile	1.8	1,200

Table 2 ▷

Country	*Energy used per head of population (in tonnes of oil equivalent per year)*	*Wealth per head of population (in US dollars)*
Japan	3.8	5,100
Ireland	3.5	4,000
Brazil	1.2	1,200

(e) Using squared paper, draw a scatter graph similar to figure 19.2. Mark in the correct positions for each of the countries listed in Table 2.

Study these two statements about energy consumption.

☐ ***Statement 1:*** People in rich countries such as those in Western Europe consume much more energy than people in poor countries.

☐ ***Statement 2:*** The amount of energy consumed by people in rich countries is much greater now than it was in the past.

Study the three photographs which show different energy-consuming situations in modern Ireland. How might each of these situations differ:

☐ in a Third World country today?

☐ in Ireland in the past?

How do these differences help to show the truth of each of the two statements about energy consumption?

△ Transport

△ Manufacturing industry

△ Domestic use of energy

Ireland's main energy sources

Ireland's main energy sources are shown in figure 19.3.

Figure 19.3 Ireland's main energy ▷ sources

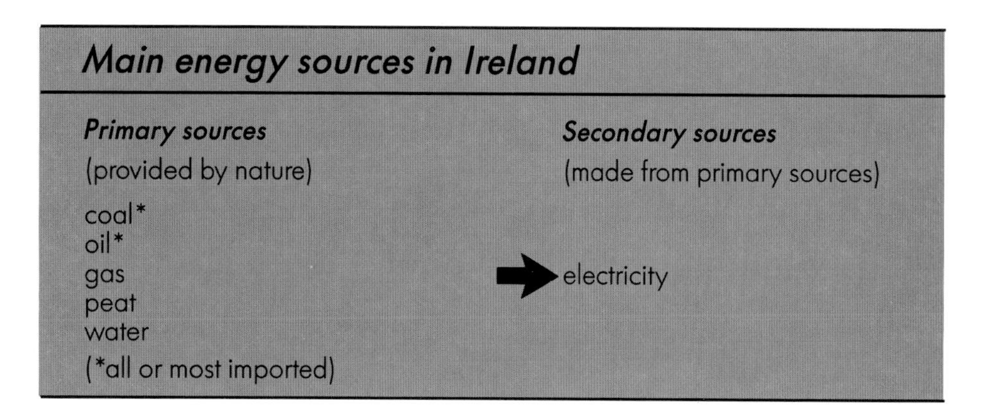

Main energy sources in Ireland

Primary sources (provided by nature)		***Secondary sources*** (made from primary sources)
coal* oil* gas peat water	➡	electricity

(*all or most imported)

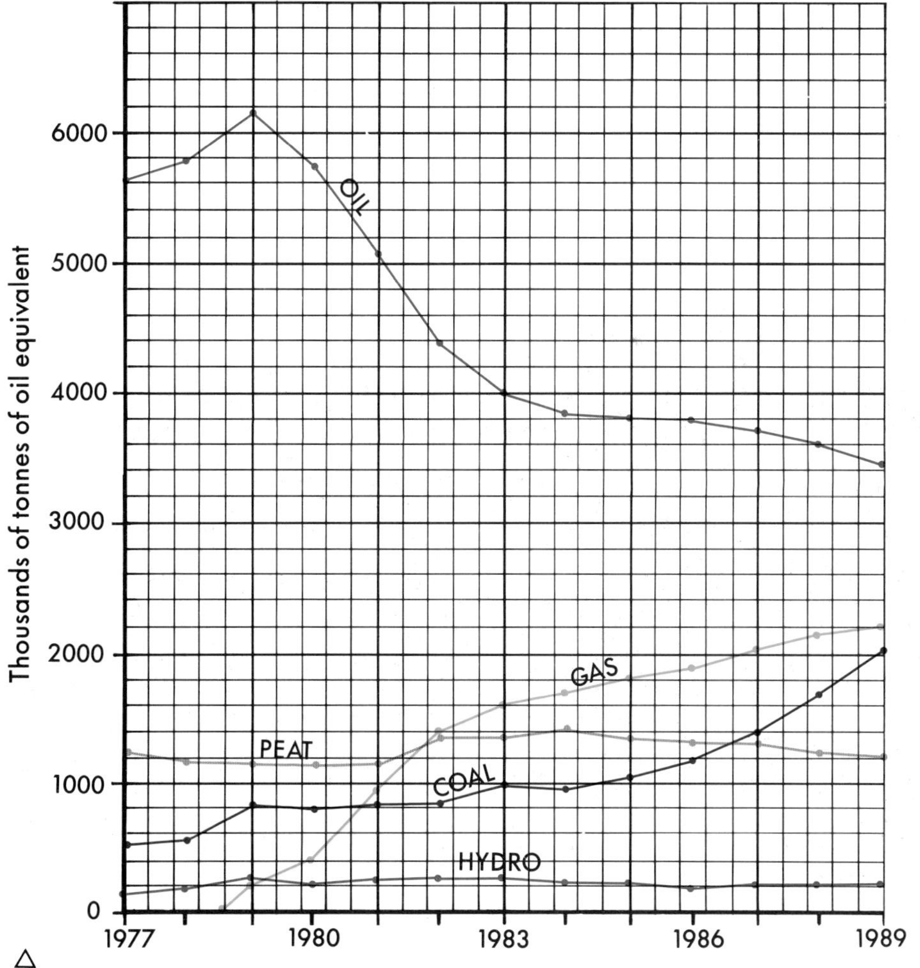

△
Figure 19.4 Energy consumption in Ireland.
(a) List in rank order Ireland's primary energy sources. (b) Name the energy sources which have: (i) shown an overall increase; (ii) shown an overall decline; (iii) not changed significantly between 1977 and 1988. (c) Using evidence from the graph, suggest a reason for the decline in oil consumption since 1979. Suggest another reason which is not apparent from the graph.

How oil was formed

1. Millions of years ago, the bodies of millions of tiny animals collected on the seabed when they died.

2. Layer upon layer of mud and sand covered their bodies.
3. The bodies gradually decayed to form oil and gas. The mud and sand gradually became rock.

Oil: the world's leading fuel

Oil: a valuable but finite resource

Petroleum or oil has many ***advantages*** as a fuel.

- ☐ It gives out a great deal of ***energy*** when burned.
- ☐ It is easily ***transported*** because it is a liquid.
- ☐ It is ***cleaner*** to handle and to burn than solid fuels such as coal or peat.
- ☐ Some of its ***by-products*** (such as petrol or lubricating oil) are needed to run cars, machinery and other forms of transport.

It is no wonder, therefore, that oil is the most popular source of energy in the world today.

But oil is also a ***finite*** resource. Many people are concerned that the world's oil supplies could be used up entirely in the not-too-distant future.

Figure 19.5 shows past trends in world oil production and tries to predict future trends. The graph predicts the almost total exhaustion of the world's oil resources by the end of the next century.

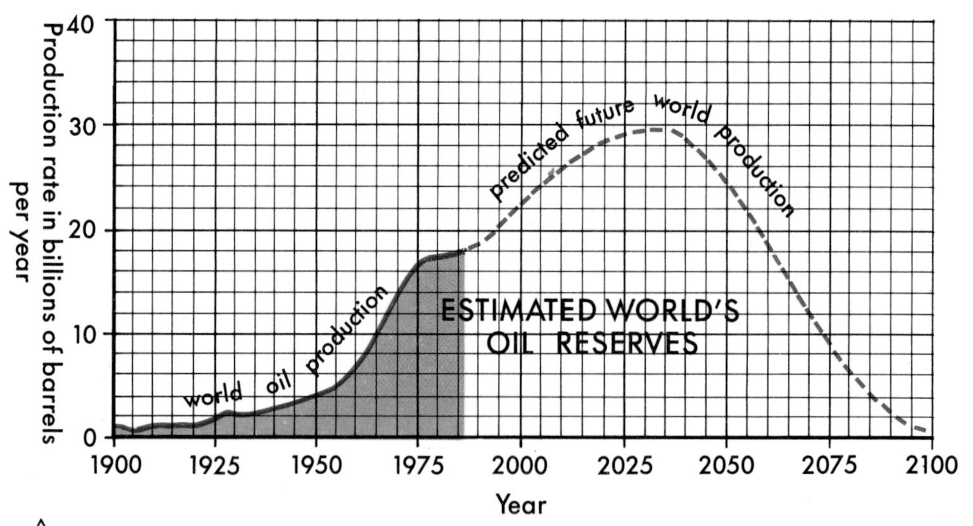

△

Figure 19.5 Use of the world's oil reserves

According to the information in figure 19.5:
(a) When is oil production likely to reach its peak?
(b) When is oil production likely to decline below 10 billion barrels per year?
(c) When is oil production likely to decline below its 1925 level?
(d) Approximately what proportion of the world's oil reserves had already been used by 1985? (The squares will help you to estimate this.)
(e) *Discussion question:* Figure 19.5 predicts the almost total exhaustion of the world's oil reserves by the year 2100. But this prediction may or may not be accurate. Discuss some possible future events which might either prolong or shorten the predicted lifetime of the world's oil reserves.

△
Drilling for oil in the Middle East. How much oil does the Middle East supply each year? To where does the Middle East export most of its oil? (See figures 19.6 and 19.7.)

Oil producing and consuming areas

The world's oil trade

The world's great oil ***supplies*** are unevenly distributed. The ***Middle East*** is the world's main oil-producing region. It produces 25% of the world's oil and contains about half of the world's known reserves.

\+

The greatest ***demand*** for oil products comes from large ***industrialised regions*** such as the USA and Western Europe.

↓

There is a huge ***international trade*** in oil. The pattern of this trade can be seen in the movement of oil ***from producing nations,*** such as those in the Middle East, ***to consuming nations,*** such as those in Western Europe.

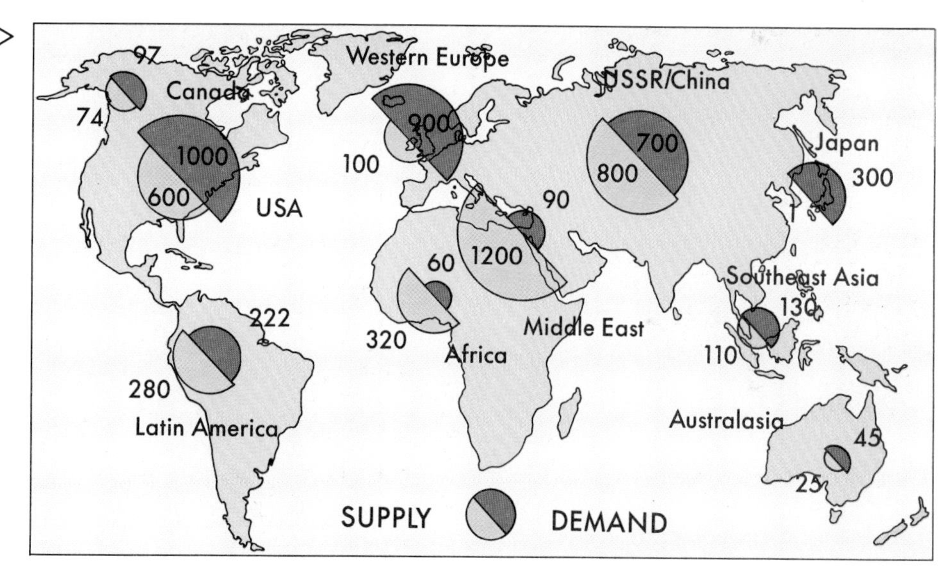

Figure 19.6 Oil supply and demand ▷ in millions of tonnes
(a) List those regions which supply more oil than they use. In which of these regions is the difference between supply and demand greatest?
(b) List the regions which use more oil than they produce. In which of these regions is the difference between supply and demand greatest?
(c) Try to explain why the demand for oil is so great in Western Europe.
(d) Try to explain why the demand for oil is so small in Africa.
(e) Use the information in figure 19.6 to estimate the world's total oil supplies in millions of tonnes.

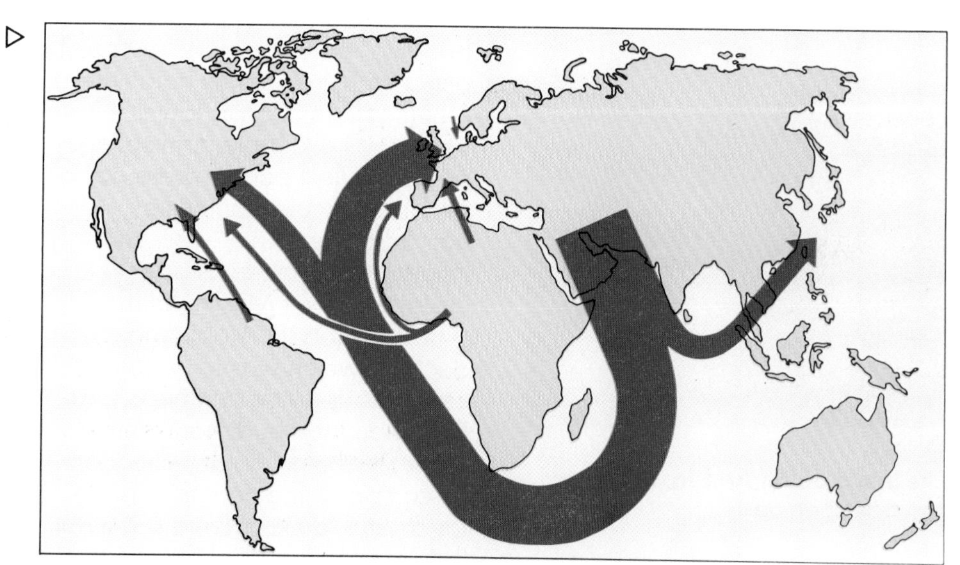

Figure 19.7 The world's oil trade: ▷ main movements by sea. (The thickness of each arrow relates to the amount of oil movements along that route.)
(a) In the case of each of the three largest movements, name the source (supply) area and the destination (demand) area.
(b) Use the information in figure 19.7 to explain why the movements you described in (a) are so large.

Saudi Arabia

Case Study of a great oil-producing state

The area around the Gulf of Iran in the Middle East is the world's greatest oil-producing region. The chief oil-producing country in this region is Saudi Arabia (figure 19.8). About 10% of the world's oil trade is controlled by Saudi Arabia.

Figure 19.8 The Middle East and its oil-producing countries ▷
(a) Which *six* Middle Eastern countries have major oil fields?
(b) Trace an outline map of the Middle East. Show and name Saudi Arabia, its capital city, its main oil fields, its adjoining countries and its adjoining sea areas.

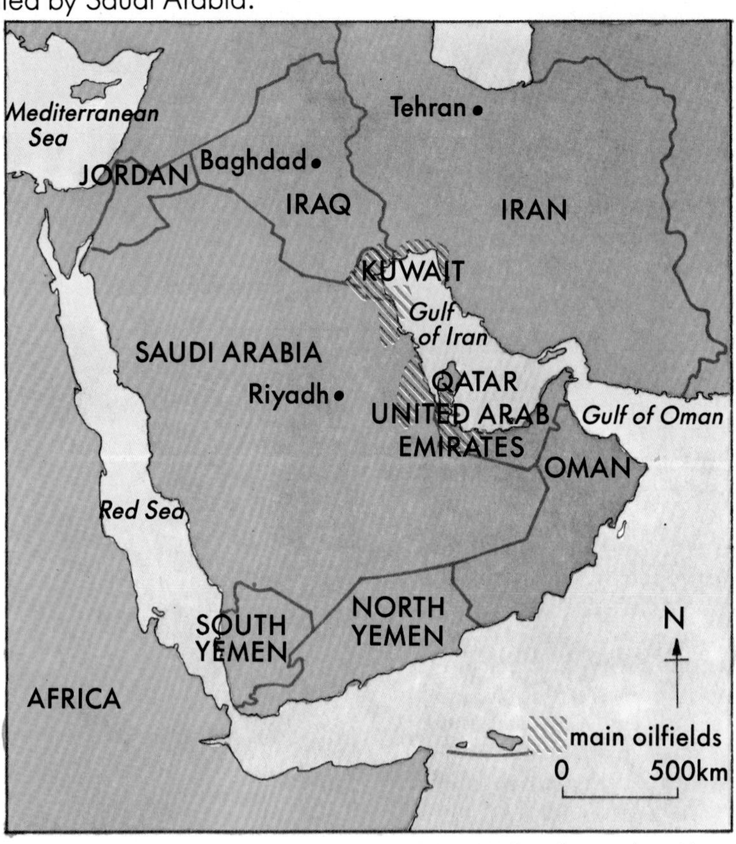

Until 1932, Saudi Arabia was ***one of the world's poorest countries.*** Bedouin Arabs roamed the desert areas with their camel herds while others settled in oasis towns where water was available from springs or wells. Most people were poor and lived simple, sometimes nomadic, lives. They were strongly influenced by the laws and customs of their Muslim faith.

In 1933, Saudi Arabia made a decision which was to change the course of its future completely. In that year, the Saudi government gave Standard Oil, an American oil company, the right to explore for oil and to export and sell what they found. Standard Oil paid only £50,000 for this right. Along with other oil companies, it was soon making vast fortunes from Saudi oil.

It was not long before Saudi Arabia realised that it had made a mistake in signing away vast profits to foreign oil companies. The Saudi government decided to withdraw foreign oil exploration rights and to ***take control*** of its country's vast ***oil production.*** As a result of its oil wealth, Saudi Arabia soon became ***one of the world's richest nations.*** In 1978, for example, Saudi Arabia was producing £150 million worth of oil every day!

△
Life in old Saudi Arabia: traditional nomads at an oasis

△
A supertanker heading towards a port in Saudi Arabia.

When oil comes out of the ground, it is like thick tar. This tar-like substance is called *crude oil.*

Much of Saudi Arabia's crude oil is exported in massive *supertankers* such as the one shown here.

When the crude oil is delivered to its market, it must be *refined* before use. At a refinery, the crude oil is changed into petrol, lubricating oil and other oil-based products.

OPEC – the end of cheap oil

Up to the 1960s, big American and European oil companies controlled oil production in Arabic countries. These companies paid Arab governments very little for their valuable oil.

Arab oil-producing countries soon grew tired of this situation. They joined together to form *OPEC,* the *Organisation of Petroleum Exporting Countries.* Through OPEC, the Arabs were able to act together, forcing oil companies to pay three times more for their oil. The days of cheap oil were over.

This sudden rise in Middle Eastern oil prices caused an *oil crisis* and an economic decline or *recession* in the oil-importing countries of North America and Western Europe. The oil crisis also encouraged Ireland and other European nations to initiate the search for oil in their own territories.

How oil has changed Saudi Arabia

Once one of the world's poorest countries, Saudi Arabia is now oil-rich.

- ☐ Saudi Arabia has become ***one of the world's richest countries*** due to oil earnings.
- ☐ Many ultra-modern towns, roads, hospitals, schools and universities have been built. The overall ***standard of living*** has improved greatly.
- ☐ There is plenty of ***employment.*** Few Saudis now follow the ancient nomadic way of life.
- ☐ The country can employ over one million ***foreign workers.*** Some of these foreigners (including Irish nurses and technicians) do skilled jobs. Others do poorly-paid jobs which the Saudis can now afford to pay others to do.

The lives of Saudi woman

Despite the changes brought about by the oil boom, Saudi Arabia is still a very traditional Muslim state. Even foreign workers who visit the country must adhere to its very strict laws, such as those which forbid the drinking of alcoholic beverages. For many Saudi women, life has changed little. Women are forbidden to drive cars. Few women work outside the home and few will go out alone in the company of men.

Some changes, however, are taking place in the lifestyles of Saudi women. These changes are demonstrated in the photographs on the next page.

◁ New Saudi Arabia: a modern urban area built with oil money. Contrast the scene in this photograph with the picture of the traditional oasis on page 132.

Study these two photographs of Saudi women. Discuss the changes which have taken place for some of them. Why might some Saudi women retain their traditional Muslim customs while others choose a more modern way of life?

The search for oil and gas off Ireland

Large deposits of ***hydrocarbons*** (oil and gas) may exist under the seabed in many areas off the Irish coast. These areas or ***basins*** are shown in figure 19.9.

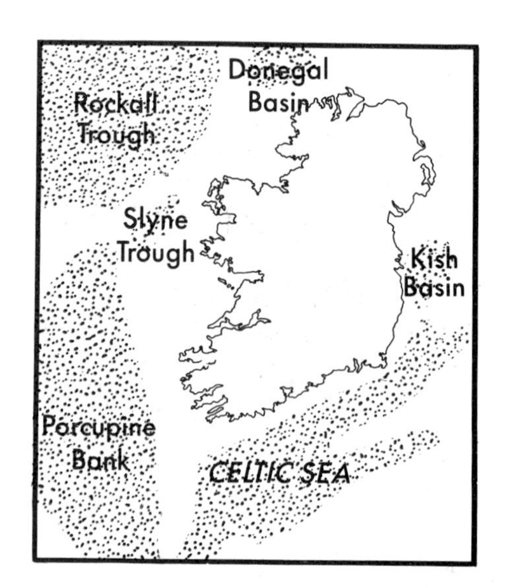

Figure 19.9 Areas of potential oil or gas deposits in Irish waters

Searching and drilling for oil and gas

The Irish government divides the possible oil-producing areas into imaginary rectangles or ***blocks.*** It offers these blocks for rent to ***oil companies.***

▼

A company may first take out an ***option*** on a block for a limited period. It will carry out detailed ***rock studies*** to test for the presence of hydrocarbons (oil or gas) in the area.

▼

If these exploratory tests prove positive, the company may ***lease*** the block. It will then bring in oil rigs and drill deep ***test holes*** to search for the oil or gas. Each of these test holes can cost up to £10 million to drill.

▼

If sufficient hydrocarbons are found, the company may decide ***to develop the oil or gas wells*** and bring the oil or gas ashore.

Most oil companies are ***multinational corporations***—large, wealthy companies which operate in many countries. Most multinationals are American or European in origin. Multinational oil companies include Marathon, Gulf, Esso, Total and BP (British Petroleum).

Can you think of some:
☐ advantages
☐ disadvantages
of allowing foreign multinationals to drill for oil off our coast?

Drilling in the Celtic Sea

The Celtic Sea is the most explored and productive oil and gas area off the Irish coast. A total of 55 drillings have already been made beneath this sea. These drillings have yielded two important results.

- ☐ ***Gas*** has been found and is being exploited in the Kinsale Head gas field. To date, this is Ireland's only commercial oil or gas find.
- ☐ ***Oil*** has been discovered off the coast of Co. Waterford. This oil has not yet been exploited.

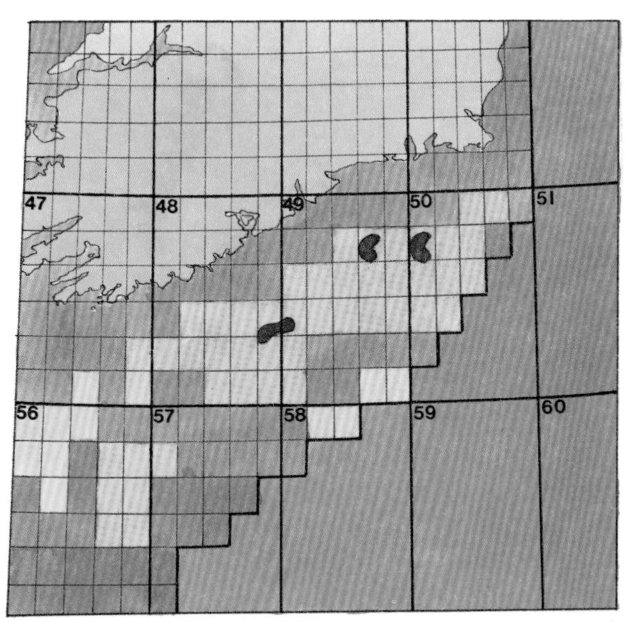

boundary of area under Irish control

block under lease or licence to petroleum company

Kinsale Head gas field

Oil discovered off Co. Waterford

Dividing the Celtic Sea into blocks

- ☐ The area is first divided into *quadrangles* (1° of latitude x 1° of longitude).
- ☐ Each quadrangle is numbered.
- ☐ Each quadrangle is then divided into ***30 blocks*** which are numbered as shown here.

1	2	3	4	5
6	7	8	9	10
11	12	13	14	15
16	17	18	19	20
21	22	23	24	25
26	27	28	29	30

△

Figure 19.10 The precise locations of gas and oil finds in the Celtic Sea. The illustration also explains the system of dividing the area into blocks. Study figure 19.10 and then answer the questions.
(a) In which *quadrangles* is the Kinsale Head gas field found?
(b) The Kinsale gas field exists in three different *blocks.* One of these blocks is numbered 48/20 (quadrangle 48, block 20). Give the numbers of the other two blocks.
(c) Say whether each of the following statements is *true* or *false*.
(i) Quadrangle 50 lies to the south of the Wexford coast. (ii) Most of quadrangle 50 is under Irish control. (iii) None of quadrangle 59 is under Irish control. (iv) 30% of quadrangle 56 is under lease or licence to oil companies. (v) Block 49/29 is under lease or licence. (vi) Oil has been found in block 49/10.

Case Study: Oil off Waterford

Between 1983 and 1985, the Gulf Oil Company discovered oil in two separate wells off the coast of Co. Waterford. (See figure 19.10 for the location of this oil find.)

- ☐ The oil was of ***good quality.***
- ☐ It occurred in ***reasonable quantities.*** Both wells were expected to yield a total of about 14 000 tonnes of oil per day.
- ☐ Good-quality ***gas*** was found along with the oil.
- ☐ The oil was ***conveniently located*** beneath a shallow seabed and less than 70km from the coast.

△
A gas rig in the Celtic Sea. Identify the helicopter pad

It was hoped that these oil finds would prove to be ***commercial*** (worth exploiting) and that Ireland would become an oil-producing nation.

It was around this time, however, that many oil-producing countries began to produce more and more oil. This resulted in a glut or ***surplus*** of oil on the world market. Because of this surplus, the price of oil began to drop rapidly. Between 1983 and 1988, the price of oil on the world market dropped by 50%.

This drop in prices has meant that the Celtic Sea oil wells have ***not been worth exploiting.*** They may yet be exploited, however, if world oil prices should rise again or if more oil is found in this region. If this happens, the south coast of Ireland—and Co. Waterford in particular—might yet experience the advantages (and disadvantages) of an oil boom.

An oil boom in Waterford?

The likely effects

△
Oil spillages can mean death for fish, sea mammals and birds like this guillemot. Find out how oil spillages affect these creatures.

Good effects	*Bad effects*
New ***industries*** would probably be set up. These might include the manufacture or servicing of oil rigs which would provide employment.	New, hastily-built industries might damage the ***natural beauty*** of the area.
Outsiders would move into the area. There would be a boom in the ***building trade.***	As the demand for houses rose, so would house prices. The ***cost of living*** would also rise.
Roads, port facilities and other ***services*** would have to be improved.	Fish and seabirds might face the threat of ***pollution*** from oil spillages.

Activities

1. *World energy consumption*
   (a) **Figure 19.11 shows world energy consumption over a period of thirty years. Use the information to draw a line graph showing this consumption. (Put the years along the bottom on the horizontal axis.)**

Year	1960	1965	1970	1975	1980	1985	1990
Consumption in millions of tonnes (of oil equivalent)	4100	5100	6300	7600	8000	8400	8700 estimated

Figure 19.11 World energy consumption, 1960-1990

   (b) **Now study the line graph you have drawn to answer these questions. (i) During which five-year period did energy consumption increase most sharply? (ii) After which year did the rate of increase in consumption begin to slow down? (iii) Suggest a reason for this slow-down.**

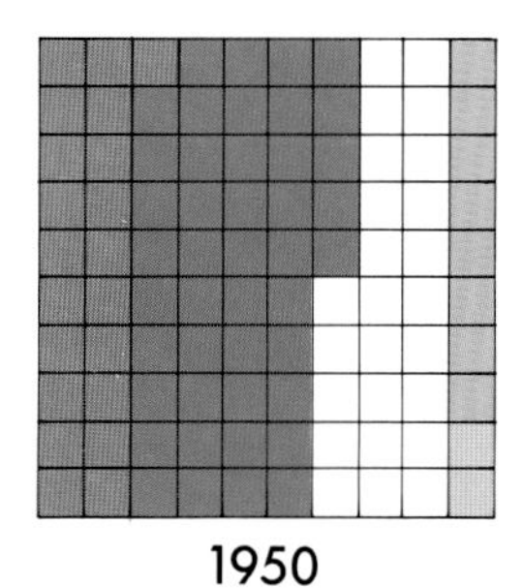

1950

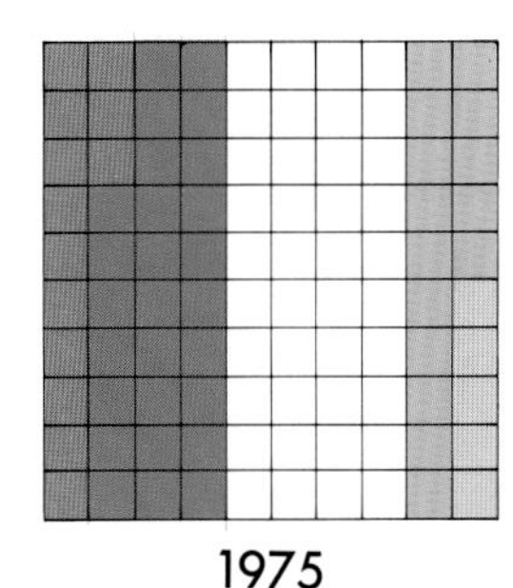

1975

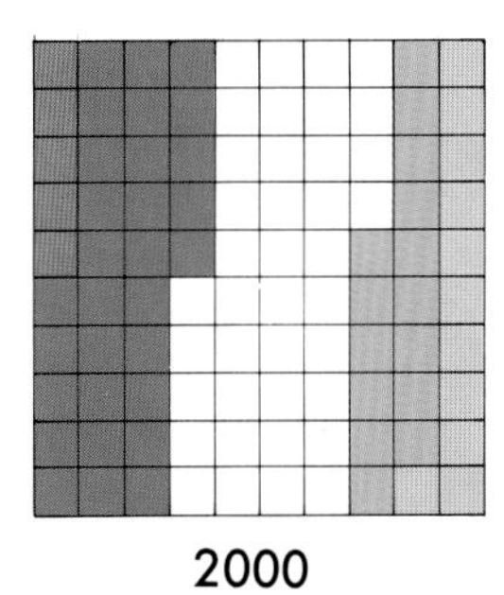

2000

wood
coal
oil
natural gas
other sources

△
Figure 19.12 The consumption of different forms of energy, 1975 and 1985.

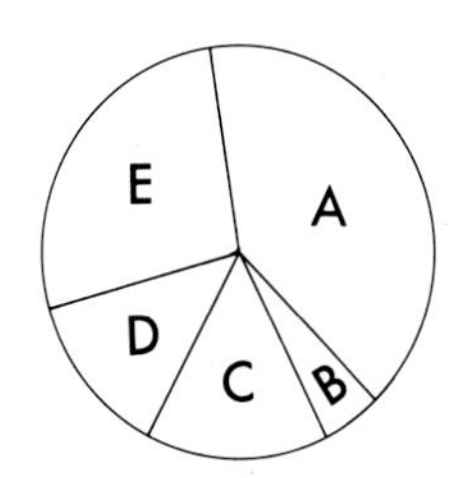

2. *Changing emphasis on the use of world energy sources*

Figure 19.12 shows the relative importance of different forms of energy in 1950, 1975 and 2000.

(a) Name the most important energy source in 1950; in 1975.

(b) The pie chart shows the use of energy sources in 1975. Name the source represented by each segment (A-E).

(c) Total world consumption of energy (oil equivalent) in 1975 was 7600 million tonnes. What *percentage* was oil? How many million *tonnes* of oil were consumed in 1975?

(d) Oil is a *non-renewable* energy source. More oil is being used than is being discovered. What is likely to happen if this trend continues?

3. (a) Identify each of the following features as shown in figure 19.13. The sea areas 1, 2, 3; the natural resource at 4; the city at 5; the countries 6 and 7; the continent 8.

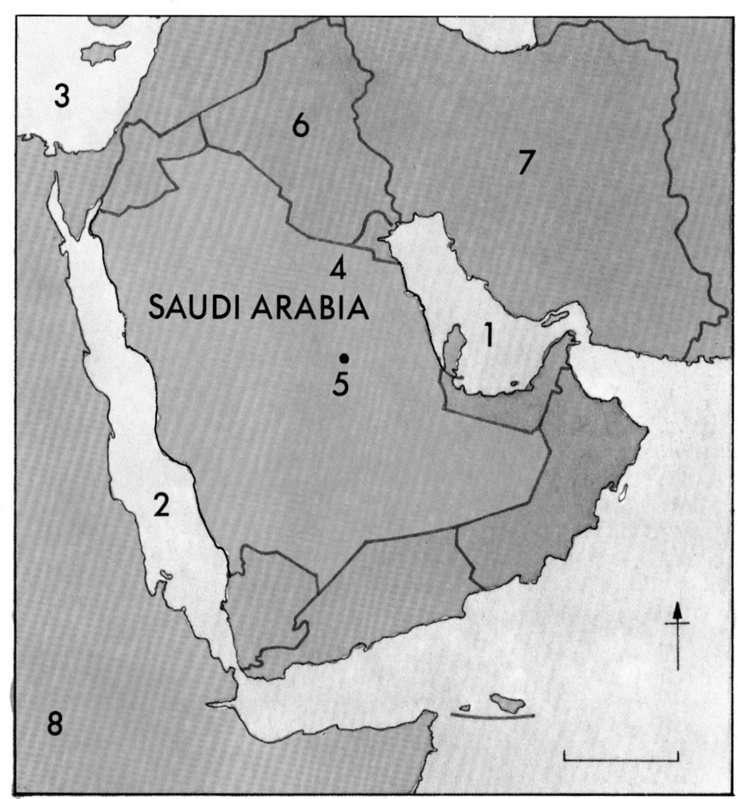

△ Figure 19.13 Saudi Arabia and its surrounding area

(b) What is OPEC? How has the establishment of OPEC: (i) benefited the world's oil-producing states; (ii) affected oil-importing states?

(c) The oil boom has brought about many changes in Saudi Arabia. Describe some of these changes by referring to: (i) the appearance of the countryside; (ii) the people's standard of living; (iii) the people's customs and traditions.

(d) Name one detrimental (bad) effect which the oil boom may have had on life in Saudi Arabia. Explain why you feel this change was detrimental.

(e) Saudi Arabia's finite oil resources will eventually be used up. Suggest ways in which the Saudi government could ensure that its country remains wealthy after the oil has been used up.

20 USING OUR PEAT BOGS: THE ROLE OF TECHNOLOGY

Central idea of this chapter
The rate at which a natural resource is used depends partly on the development of modern machinery and advanced technology.

The development of Ireland's ***Midland peat bogs*** is a special example of this.

Advanced technology and the development of modern machinery have led to a growth in the exploitation (usage) of many different natural resources. Here are two examples.

- ☐ Improved ***bog-draining and turf-harvesting machines*** have led to a faster exploitation of ***peat*** in the Irish Midlands.
- ☐ The development of power saws and heavy transport vehicles has led to a more rapid clearance of the world's ***forests.***

Give two more examples of the role of modern technology based on the information in the photographs.

△ **Photograph A**

△ **Photograph B**

Using Ireland's bogs

What are the advantages of exploiting peat bogs like this one? Can you think of any disadvantages?

About 5% of Ireland's land consists of peat bogs.

For many centuries, Ireland's people used peat for cooking and heating. Up to the present century, however, only ***basic technology*** such as hand tools were used to exploit the peat. As a result, our largest bogs were exploited ***slowly,*** and then only at their edges.

Now our bogs are being exploited ***rapidly.*** Five million tonnes of peat are being harvested each year. One-sixth of all our electricity is being generated from peat. Nearly 800 000 hectares (two million acres) of bogland have already been developed.

The rapid increase in the exploitation of our bogs is due to two main reasons:

- ☐ the establishment of ***Bord na Mona*** in 1946
- ☐ Bord na Mona's use of ***modern technology*** to speed up the rate of peat exploitation.

Bord na Mona was set up by the Irish government for two reasons: to develop the country's peat industry and to increase the amount of peat being extracted from our bogs. The ***two types of peat bogs*** available to Bord na Mona for development are ***raised bogs*** and ***blanket bogs.***

Raised bogs	*Blanket bogs*
☐ have an average peat-depth of *8 metres*	☐ have an average peat-depth of *1½ metres*
☐ occur mainly in *flat* Midland areas	☐ occur mainly in the *mountainous* West
☐ occur in areas where yearly rainfall is approximately *1000mm*	☐ occur in areas where yearly rainfall is approximately *2000mm*

Step 1

The bog is **prepared.** It is ***drained, levelled*** and supplied with ***railway tracks.***

↓

Step 2

The peat is **harvested.**

↓

Step 3

The peat is **transported** from the bog.

↓

Step 4

The peat may be **processed** before being marketed.

△
Figure 20.1 Some stages in peat exploitation

The directors of Bord na Mona knew that the most modern technology would be needed if Ireland's peat resources were to be rapidly exploited. Since the ***relatively flat raised bogs*** of the Midlands were better suited to the use of heavy machinery, Bord na Mona decided to concentrate on developing these raised bogs first.

Engineers from Bord na Mona studied the great advances which had been made in peat exploitation in the Soviet Union and in West Germany. From these countries, they learned about newly-developed machines which could be used to exploit Ireland's bogs. They imported ***foreign technology,*** adapted it to Irish needs and then used it to speed up the exploitation of our peat resources.

Exploiting our peat resources

The role of modern technology

Figure 20.1 outlines four basic stages in the exploitation of Ireland's peat resources. In each of these stages, the use of modern technology has played a vital role in increasing the rate of exploitation.

Step 1
Preparing the bogs

Surplus water must be drained from the bogs before the peat can be harvested.

In the past...

In the past, the drainage of some of our best bogs was almost impossible. Many of these bogs have a ***95% water*** content. They are so ***soft*** that it is difficult for a person to stand on them—let alone drain them—without sinking.

Since the foundation of Bord na Mona...

The problems of bog drainage were solved by changing technology. A special ***drainage machine*** (see photograph) was developed which, though heavy, would not sink into the boggy surface. The machine travels on extremely ***wide tracks,*** distributing its great weight over a large area. This prevents the machine from sinking into the bog.

Drainage machines are used to dig criss-cross networks of drains across the bogs. Surplus water is allowed to run off through these drains for five to seven years. After this time, the bog surface is usually firm enough for the next stages of preparation to proceed.

Large drainage machines have dug 30 000km of drains throughout Ireland's boglands—a distance 60 times the length of Ireland! Note the wide bulldozer-type tracks on which the machine runs. ▷

A variety of heavy machines are used to speed up bog development.

- ☐ ***Peat strippers*** remove surface vegetation.
- ☐ ***Earth movers*** (like bulldozers) level the bog surface. This makes it easier for the big harvesting machines to do their work later on.
- ☐ ***Heavy tractors*** help to lay down railway tracks which will be used to transport the harvested peat.

Step 2
Harvesting the peat

In the past...

In the past, a tool known as a ***slean*** was the principal turf-harvesting implement. Because the slean is worked by hand, exploitation of the bogs by this process is very slow.

△
Handcut turf is piled to dry in the sun and wind

Since the foundation of Bord na Mona...

Bord na Mona now harvests its peat quickly in three main forms.

☐ ***Machine turf*** (sod peat) is used as a domestic fuel and in electricity-generating power stations. ▷

☐ A powdered form known as ***milled peat*** is used in power stations and in the manufacture of peat briquettes. Milled peat is now the main product of Ireland's bogs.

☐ ***Moss peat*** is used as a soil improver and as bedding for poultry. ▷

Each form of peat has its own special harvesting method. All harvesting is now carried out using modern machinery.

Harvesting machine turf

Machine turf is harvested rapidly using huge machines known as ***baggers.*** One bagger can cut 20 000 tonnes of turf in a single four-month harvesting season. The photograph shows how a bagger operates.

The peat sods which are harvested by the bagger are left to dry in the sun and wind for about six weeks. Other machines are then used to gather the sods into piles before they are removed from the bog for packaging.

A bagger and how it works ▷

1. *Mechanical 'buckets'* are used to dig damp peat from the bog to a depth of 4 metres. The peat may vary in quality, according to depth. So it is squashed and blended together into a paste-like substance of uniform quality.

2. The paste-like peat is pushed out along a 50m-long *spreader arm.* The spreader arm drops the peat in long lines on the surface of the bog.

3. *Cutting discs* drawn behind the spreader arm divide each line of peat into approximately 140 sods.

Harvesting milled peat

Modern technology plays a vital role in the harvesting of milled peat. The photographs illustrate these harvesting operations.

△
1. First, a *milling machine* is used to scrape a very thin layer of peat from the surface of the bog. This machine has a rotating drum which contains small spikes that tear up the bog's surface to a depth of 1cm.

△
2. Next a *spoon harrow* turns the peat over to help it dry in the sun and wind.

△
3. After two or three days, the peat is usually dry enough to be gathered into small ridges by a *ridger machine.*

The ridges are gathered by other machines into large storage piles. *Automatic loaders* are used to transfer the peat from the piles into railway wagons which deliver the peat to the power stations or the briquette factories.

Harvesting moss peat

Moss peat is a light peat found near the surface of some bogs.

- ☐ Special ***harvesting machines*** cut the light peat in sod form. The sods are left to dry in the sun and wind.
- ☐ ***Automatic sod collectors*** gather the dried sods into heaps.
- ☐ ***Polythene*** is used to protect the sod heaps from the rain until the moss peat is required at the moss peat factory.

Step 3
Transporting the peat

In the past...

In the past, ***animal-drawn carts*** and large baskets called ***creels*** were used to carry peat from the bogs. Transporting the peat in this way was laborious and slow. This further slowed down the exploitation of our peat resources.

Since the foundation of Bord na Mona...

Bord na Mona uses ***rail transport*** to speed up peat exploitation. Each large bog area has its own railway system, with tracks, level crossings and special turf trains pulled by fast diesel locomotives. These trains carry large quantities of peat to power stations and briquette factories. These bog rail systems are now vital to the rapid exploitation of Ireland's peat resources.

Step 4
Marketing

In the past...

In the past, almost all of Ireland's harvested peat was used:

- ☐ only as a household fuel and...
- ☐ only in areas within easy reach of the bogs.

The ***market*** for peat was therefore limited. This was another factor which slowed the development of our peat resources.

Since the foundation of Bord na Mona...

Since the foundation of Bord na Mona, ***more widespread and varied markets*** have been found for peat. Developments in transport, electricity-making and peat-processing have helped to make this possible.

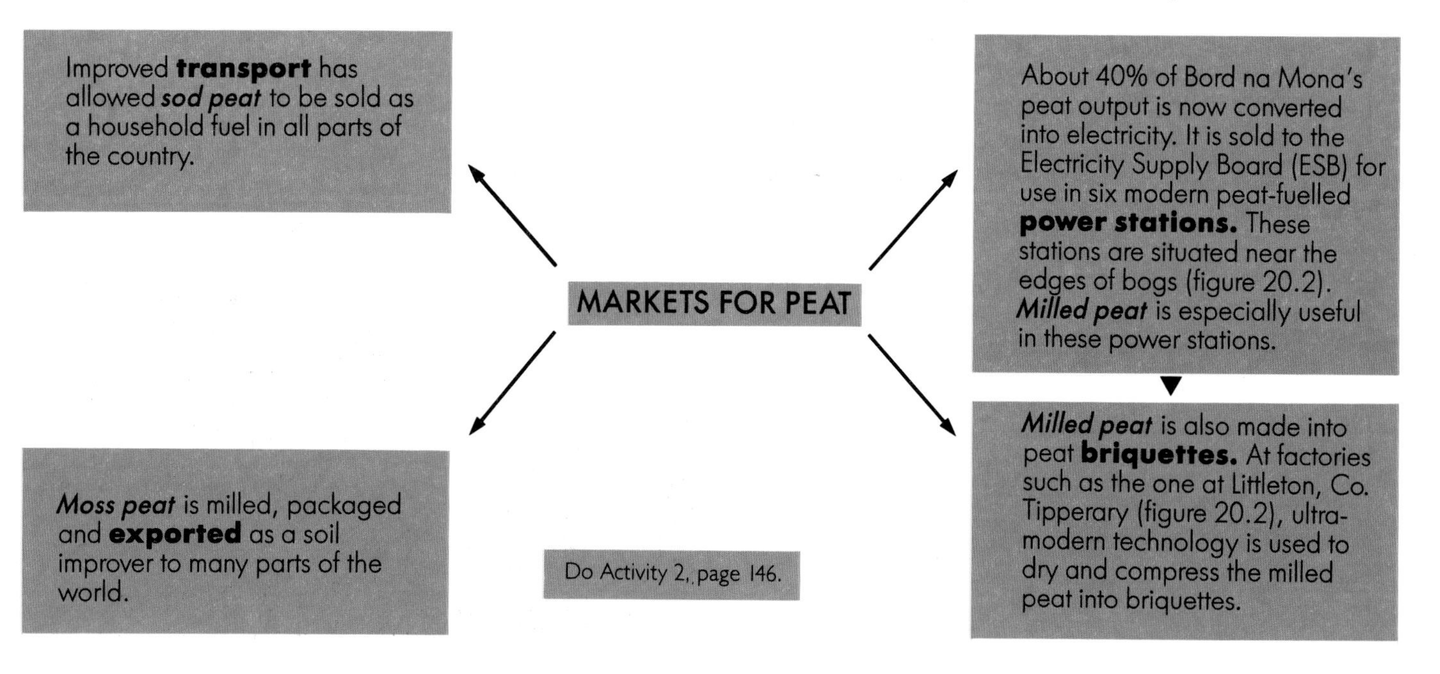

△
A peat-fuelled power station

Some problems

Improved technology has helped to accelerate the exploitation of Ireland's peat resources. But this rapid exploitation has created problems of its own.

☐ Like oil, peat is a ***non-renewable resource.*** Most of our peat will be used up in the next century.

BUT . . .

Bord na Mona has plans to develop the ***cutaway*** (used) bogs as ***grassland*** and ***forest.*** So our present boglands will continue to be useful.

☐ Ireland's unique bogs contain many forms of ***wildlife.*** If our bogs are all exploited, this wildlife habitat will be destroyed.

What should be done about this problem?

☐ Stop peat exploitation?

☐ Ignore the threat to wildlife?

☐ Adopt some other (specific) option?

A cutaway bog developed as grassland
▽

Activities

1. Figure 20.2 shows the locations of some of Bord na Mona's works as well as some ESB peat-fuelled power stations. Study figure 20.2 and then answer the questions.
   (a) In which province are most of Bord na Mona's works located?
   (b) In rank order of power capacity, make a list of the peat-using power stations shown in figure 20.2. After each station you list, name the county in which it is found.
   (c) List the Leinster Midland counties in which there are peat bogs owned by Bord na Mona.
   (d) Explain as fully as you can why Bord na Mona has exploited more peat bogs in the Midlands than anywhere else in Ireland.
   (e) Identify the counties in which each of the briquette factories shown in figure 20.2 are located.

Figure 20.2 ▷

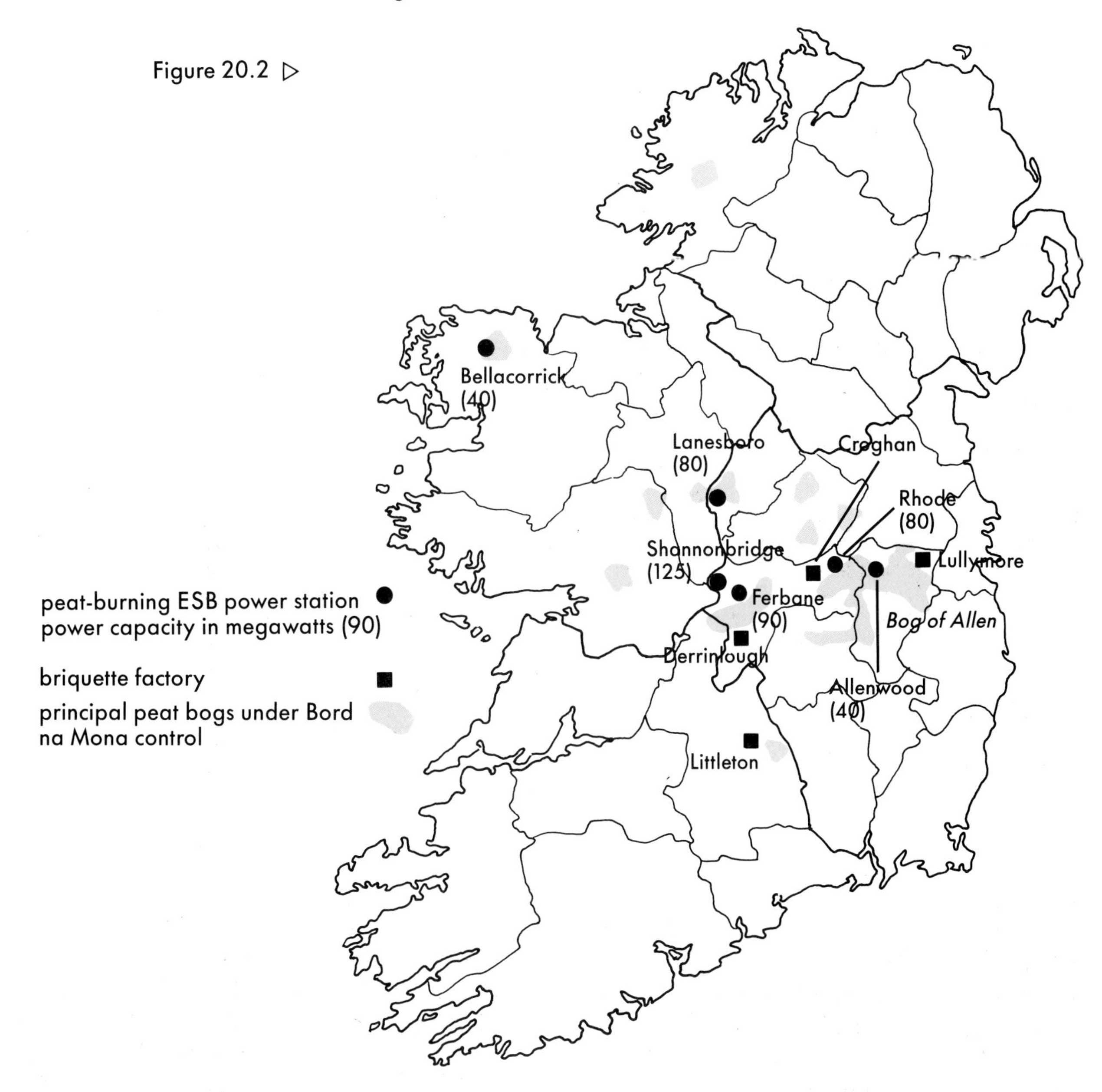

2. The bar graphs in figure 20.3 show the total amounts of electricity generated by the ESB between 1977 and 1985. They also show the energy sources from which the electricity was generated in each year.

   Study the information in the bar graphs. Then rewrite the passage in the box, giving only the correct alternative from the choices given.

☐ oil
☐ gas
☐ peat
☐ hydro and coal

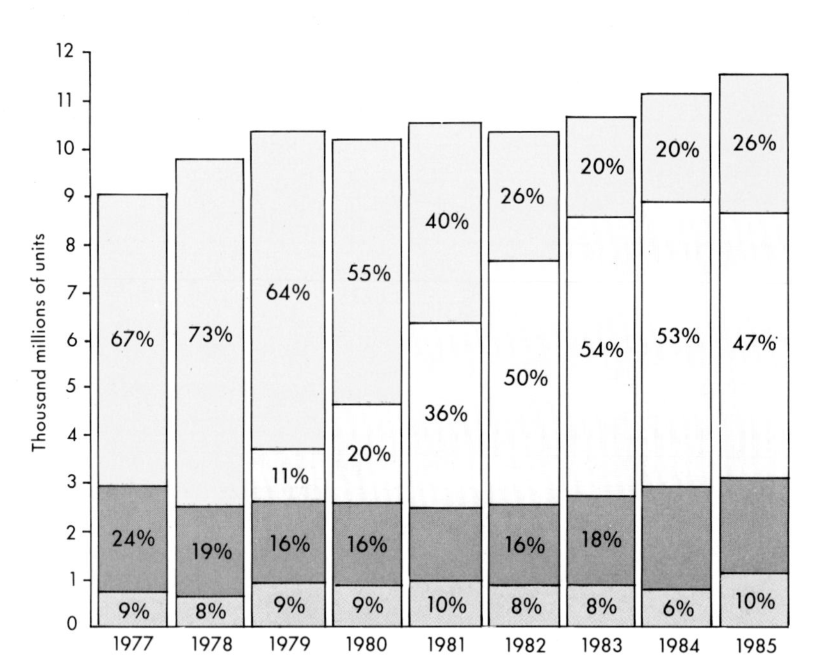

△
Figure 20.3 Electricity generated by the ESB between 1977 and 1985: energy sources and number of units generated.

> The total amount (units) of electricity generated by the ESB showed *(a constant/an overall)* increase between 1977 and 1985.
>
> In 1977 *(9000 million/9 million)* units of electricity were generated. *(67%/24%)* of this was generated from peat. In the same year, peat generated *(just over/just under)* two thousand million units.
>
> Between 1977 and 1981, there was a general *(increase/decrease)* in the annual amounts of energy generated from peat. During the same period, there was a big overall increase in the electricity generated from *(oil/gas)*. By 1981, the percentage of electricity generated from peat accounted for *(16/14%)* of the total.
>
> Between 1981 and *(1984/1985)*, there was a constant yearly increase in the percentage of electricity being generated from peat. In the last-mentioned year, this percentage stood at *(over/under)* 20% of all electricity generated, with peat being *(second/third)* in importance among the sources of electricity.

3. Say whether each of the following statements is *true or false*. Explain precisely why you consider each statement to be true or false.
   (a) Small private companies could have developed Ireland's peat bogs equally as well as Bord na Mona.
   (b) The drainage of many of our peat bogs was made possible by technological developments.
   (c) The ESB has played an important, though indirect, role in Irish peat exploitation.
   (d) Peat exploitation will always play an important role in the Irish economy.
   (e) The rapid exploitation of natural resources is always desirable.

OSPNO RHOWAR
ETPA RPPTISRE
NRIADEAG machine
IMNLILG machine
AGREBG

◀ 4. (a) Unscramble the *italicised* words in the box to spell the names of five machines used in peat exploitation.
   (b) Describe briefly the function (purpose) of each of these machines.

21 OVER-USING OUR NATURAL RESOURCES: FOCUS ON FISHING

Fish are among the world's ***renewable resources.*** Through breeding, fish can renew their numbers again and again. With careful use, the world's fish stocks need never be ***depleted*** (used up).

But some renewable resources, including fish, can be seriously depleted if they are ***over-exploited*** (over-used). In many parts of the world, ***overfishing*** has become a serious problem. When fish stocks are not given a chance to recover through breeding, their numbers will decrease quickly.

Some examples of overfishing

Figure 21.1 Overfishing: a worldwide problem

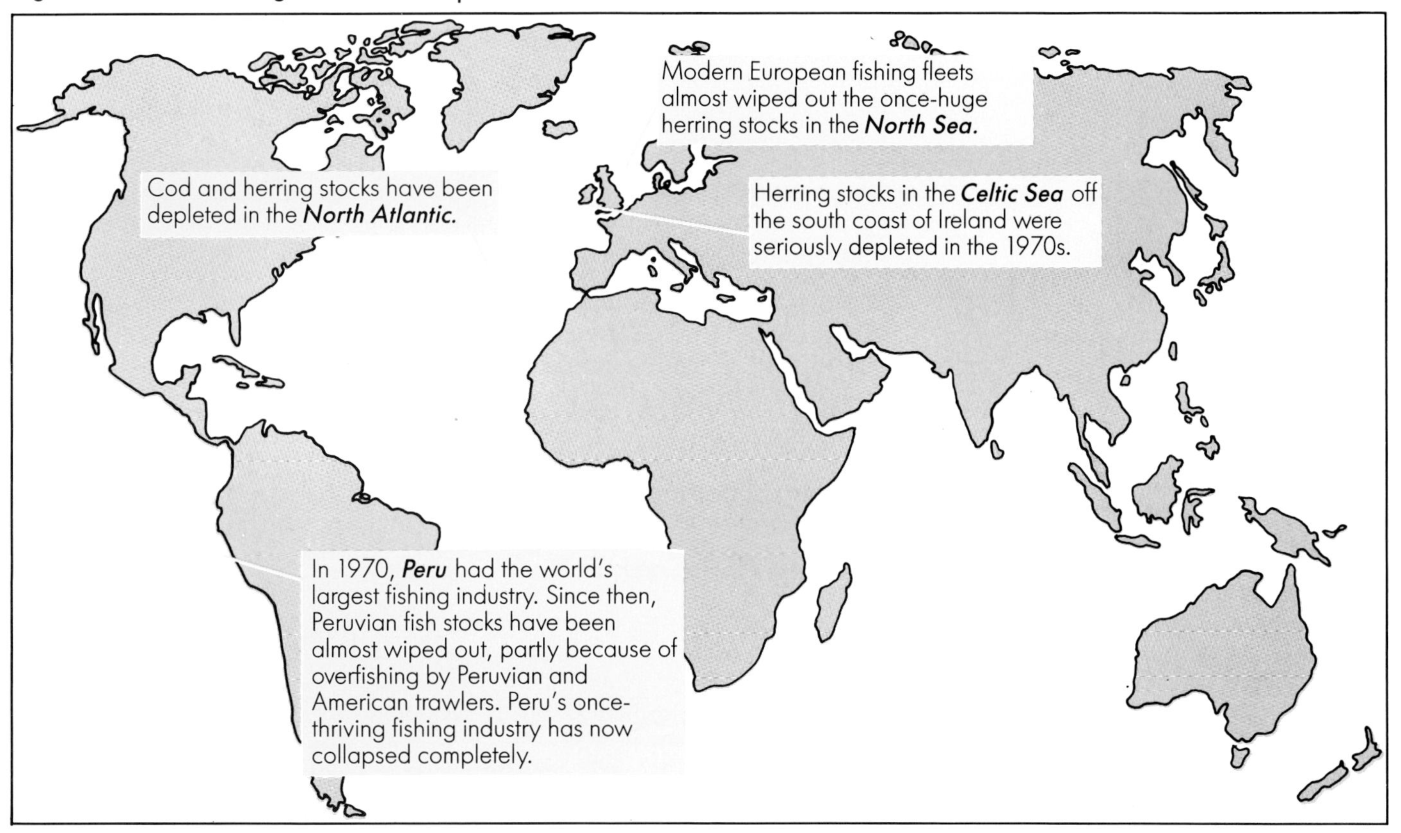

◁ **An abandoned wharf and fish-processing plant in Peru. Overfishing has helped to destroy Peru's once-thriving fishing industry.**

- ☐ Make a list of other natural resources which may be depleted if they are over-exploited.
- ☐ Name two natural resources which do not become depleted through use.

Some reasons for overfishing

Vast amounts of ***money*** and ***research*** have been invested in the world's fishing industries

This has resulted in the use of greatly ***improved equipment.***

This has led to ever-increasing ***fish catches*** . . .

. . . which have resulted in ***overfishing*** and stock ***depletion*** in many areas.

Overfishing is very much a 20th-century problem. Here are some of the factors which have led to large-scale overfishing.

Increased investment and research

☐ Throughout the 20th century, vast amounts of ***capital*** (money) have been invested in the world's fishing industries. Most of this capital has been invested in the fishing industries of rich, developed countries such as Japan, the USSR, the USA, Norway and Britain.

☐ Vast amounts of ***research*** have helped to make fishing more efficient.

Improved technology

Technological advances are made possible through increased investment and research.

☐ Large, modern ***fishing fleets*** have been built. Each fleet consists of many large trawlers. Huge factory ships which process the fish at sea travel with the fleets.

☐ Modern ***navigational aids*** such as echo sounders and sonar equipment are used to locate fish shoals. Radar is used to locate nearby boats or other obstacles.

☐ Improved ***catching equipment*** includes long, strong and lightweight trawls (figure 21.2). ***Derricks*** and automatic ***winches*** are used to haul the heavy catches on board.

☐ ***Storage and transport facilities*** have been greatly improved, especially with better methods of refrigeration.

☐ ***Modern processing methods*** allow much of the world's fish catches to be converted into fishmeal which is widely used as an animal foodstuff.

Increased fish catches

☐ Improved technology has meant bigger and bigger catches. Between 1940 and 1970, for example, the world's fish-catch trebled. Fish catches have been especially large in certain parts of the world. This has led to ***overfishing*** and the ***depletion*** of stocks in these areas (figure 21.1).

Huge factory ships such as this one overfished the Celtic Sea in the 1960s. What is a factory ship? ▷

Figure 21.2 Trawl nets in operation ▷
Modern trawl nets can catch huge amounts of fish. A trawl is a cone-shaped net which is pulled slowly through the water, scooping up the fish in its path. Trawls may be pulled through any depth-level, including along the seabed as shown here. 'Pair trawling' is a very efficient fishing method in which two boats pull a single trawl between them.

A modern trawler

Ireland's largest trawlers now cost about IR£7 million. They can hold several thousand tonnes of fish and are about 65m long—nearly the width of Croke Park!

In this photograph of a modern trawler, identify each of the following features: ▷

- ☐ A *radar* screen;
- ☐ the *bridge* or navigation cabin, which contains many of the trawler's navigational aids.
- ☐ a *derrick,* which is used to help load the fish catch.
- ☐ refrigerated *holds* (inside the trawler) in which fish can be stored for long periods.

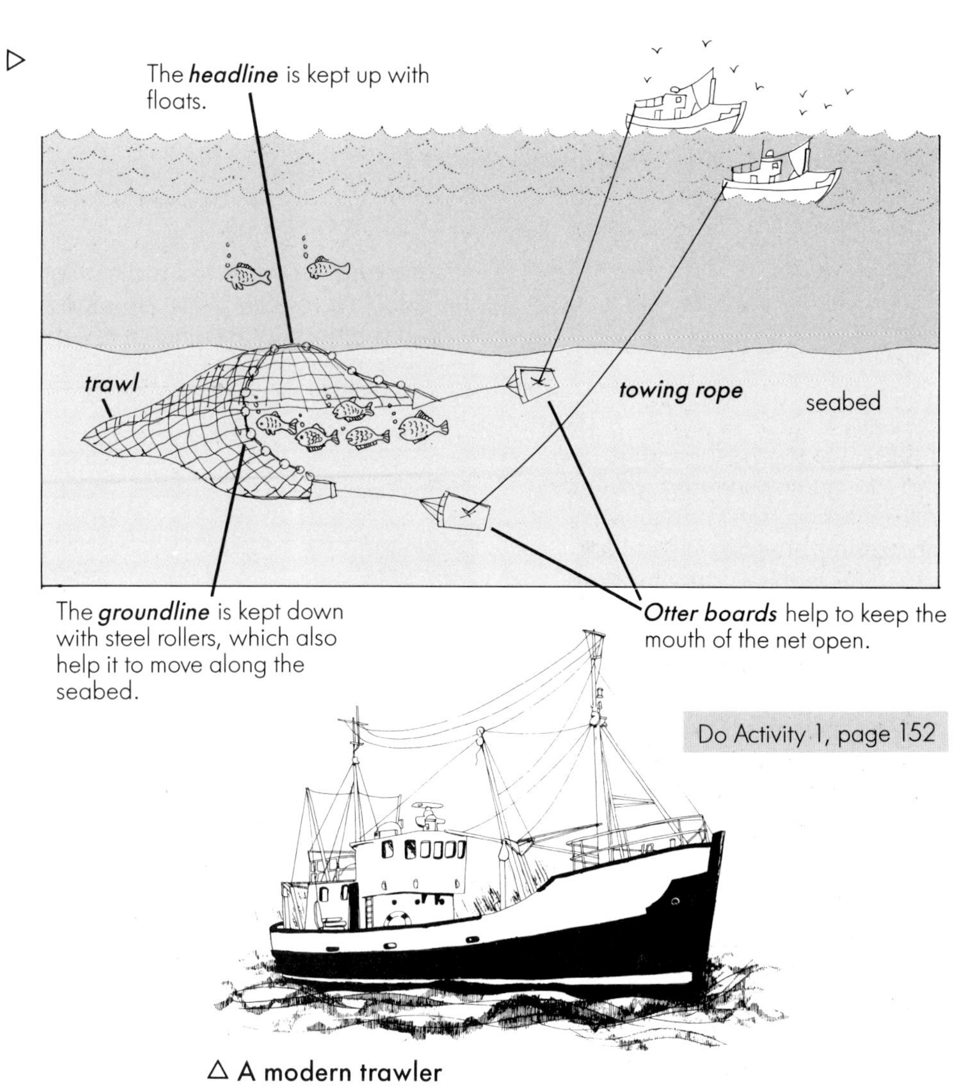

Do Activity 1, page 152

△ A modern trawler

Overfishing in Irish waters

Fishing in the waters around Ireland has increased greatly in recent years. Here are some of the reasons for this.

1 Shallow waters around Ireland contain vast quantities of microscopic life called ***plankton.*** Fish feed on plankton, so great numbers of fish also live in these waters.

2 The species (types) of fish found in these waters are usually good to eat and so have a ***high commercial value.***

3 The fish often occur in ***large, single-species schools.*** This makes it easier to catch large quantities of a chosen species.

4 Fish stocks in some European fishing grounds such as the North Sea have been depleted through overfishing. This has encouraged many ***foreign trawlers*** to come to Irish waters to fish.

5 ***Ireland's own growing fishing fleet*** has taken more and more fish from the seas around Ireland.

Over-exploitation has led to the depletion of fish resources in the seas around Ireland. But this over-exploitation has been very uneven. It has varied greatly:

- ☐ among different fish species
- ☐ in different sea areas

The ***herring*** is one of Ireland's most important fish species. Unfortunately, it is also the species which has suffered most from overfishing.

The depletion of herring stocks through overfishing has been especially severe in the waters off Ireland's south coast—an area called the ***Celtic Sea***.

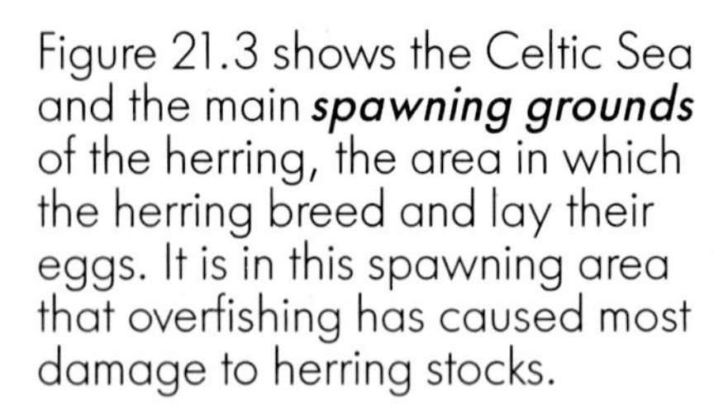

Figure 21.3 shows the Celtic Sea and the main ***spawning grounds*** of the herring, the area in which the herring breed and lay their eggs. It is in this spawning area that overfishing has caused most damage to herring stocks.

Why overfishing has occurred in the Celtic Sea spawning grounds

1. The spawning grounds are in ***shallow and sheltered waters*** very ***near the coast.*** Both large and small trawlers can therefore fish these waters, even in stormy weather.
2. Between October and January each year, the herring return to the same place to spawn. These ***predictable movements*** make the herring easy to catch.
3. Herring ***move slowly*** just before spawning. This also makes them easy to catch.
4. ***Huge numbers*** of herring come together to spawn. This means that they can be caught in large numbers.
5. Herring are in their ***best condition*** just before spawning—and therefore at their highest commercial value.

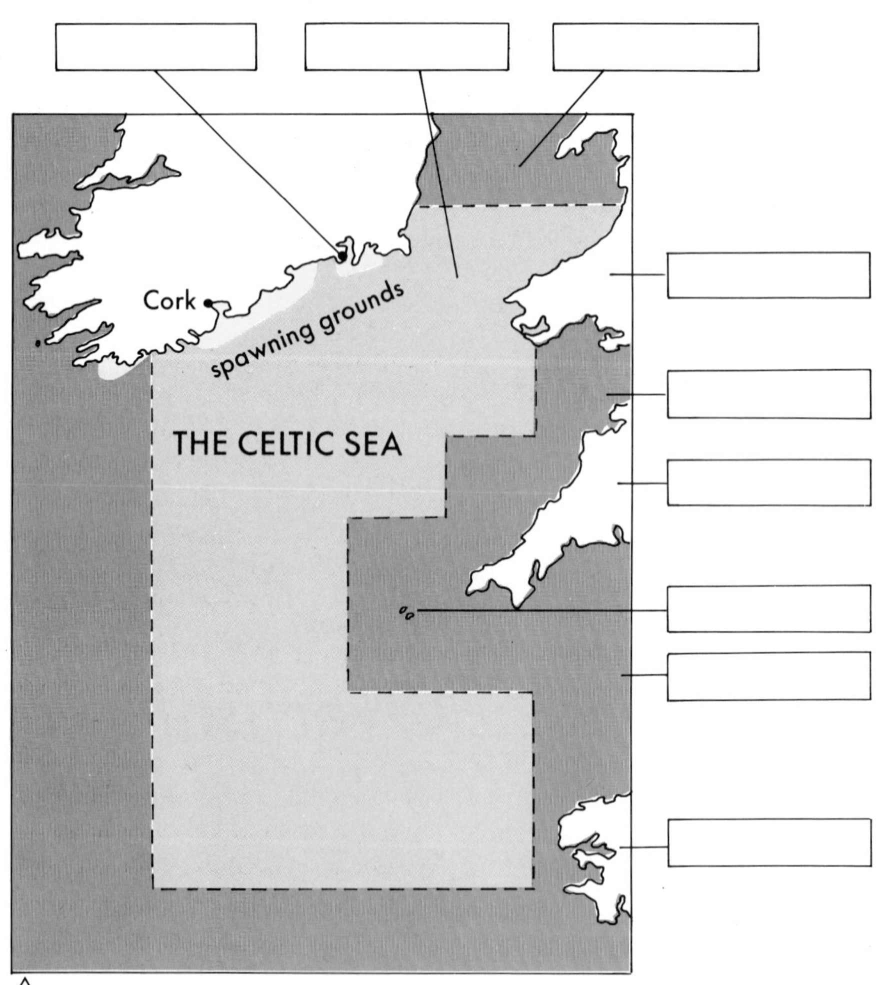

Figure 21.3 The Celtic Sea, with herring spawning grounds
With a pencil, write the correct label in each box: England, Wales, France, Scilly Isles, Irish Sea, English Channel, Bristol Channel, St George's Channel, Dunmore East (a main fishing port).

The Celtic Sea herring...

Overfished, depleted...and finally conserved

Until 1950, the Celtic Sea was fished almost exclusively by fairly small boats from Ireland and Britain. Catches remained small and no overfishing occurred.

Throughout the 1950s, large ***fishing fleets from continental Europe*** were attracted to the Celtic Sea. The Belgians arrived first, then the West Germans, then the Dutch. Increased numbers of Irish trawlers also began to fish the area. Catches increased dramatically. By the end of the 1950s, it was recognised that Celtic Sea herring were being heavily ***overfished.***

In 1961, the Irish government extended the limits of Ireland's ***exclusive fishing grounds.*** Most foreign boats were not allowed to fish inside these limits. As a result, fish ***catches declined*** in the early 1960s and fish ***stocks increased.***

By 1965, herring stocks were nearly ***wiped out*** in the ***North Sea,*** another important fishing area. Those ***continental fleets*** which had once fished the North Sea began to look elsewhere for well-stocked fishing grounds. Many of them ***moved to the edges of the Celtic Sea,*** in spite of government restrictions. These fleets were large and efficient, and made use of the most advanced technology. Celtic Sea herring catches soared. ***Massive overfishing meant seriously depleted stocks.***

Throughout the 1970s, fish ***catches dropped dramatically*** in the Celtic Sea as the results of overfishing became more and more obvious. Herring stocks were nearly wiped out in that area. Something had to be done to conserve the small remaining stocks and to give the species a chance to recover.

In 1977, ***herring fishing was forbidden*** throughout most of the Celtic Sea. When the fishing stopped, ***stocks slowly began to increase.*** The last-minute conservation efforts had saved the Celtic Sea herring.

In 1982, fishing began again in the Celtic Sea. But fish catches are now strictly controlled. Various other ***conservation measures*** (see box) are now being used to prevent the herring stocks from being overfished and depleted again.

Do Activity 2, page 152

1. Account for the fact that Celtic Sea herring: (a) were not overfished up to 1950; (b) were being overfished by 1959.

2. (a) Why did continental fleets arrive in the Celtic Sea after 1965?
(b) Why did the arrival of these fleets cause serious damage to Celtic Sea herring stocks?

3. What conservation measures were taken by the Irish government to conserve fish stocks: (a) in 1961; (b) in 1977?

Conservation measures protect Celtic Sea herring

☐ ***Surveys*** are carried out each year to check the size of Celtic Sea fish stocks.

☐ The EC sets upper limits for the quantities of fish which each of its member states may catch. These limits are called ***quotas.*** Quotas vary according to the size of fish stocks in any given year. The setting of quotas according to the size of fish stocks is called ***fish management.***

☐ Only ***Irish fishing boats*** may fish in waters within 10km (6 miles) of the Irish coast. Only vessels from EC countries may fish within 20km (12 miles) of our coast.

☐ The law states that ***net mesh sizes*** must be large enough to allow young fish to escape.

Activities

1. *More Irish fishing vessels—one aspect of Ireland's developing fishing industry*

The table in figure 21.4 shows the growth in Irish fishing vessel numbers between 1965 and 1985. Study the table and answer the questions which follow.

Figure 21.4 Growth in the number of Irish fishing vessels, 1965-1985

Category	**1965**	**1970**	**1975**	**1980**	**1985**
Under 15m	1626	1772	2006	2698	2691
16-20m	150	160	175	154	183
21-22m	33	57	106	110	105
23-28m	3	26	57	103	97
Over 28m	–	–	2	12	20
TOTALS	?	?	?	?	?

(a) Complete the 'total' columns at the bottom of the table. Calculate the total increases in vessel numbers between 1965 and 1985.

(b) During which five-year period did the total number of vessels grow most rapidly?

(c) Which category of vessel increased most between 1965 and 1985?

(d) Which category of vessel showed the greatest percentage increase between 1975 and 1985? Explain why an increase in this particular category might be especially important to the development of the Irish fishing industry.

(e) On squared paper, draw a set of bar graphs to illustrate the changes in the number of vessels in the 21-22m category between 1965 and 1985. Name another type of graph which would also be suitable for illustrating these figures.

2. Figure 21.5a shows a graph of Celtic Sea herring catches between 1910 and 1988. Match each of the numbered labels in the graph with the appropriate information letter in figure 21.5b. Record your answers in the table as shown in figure 21.5c.

Figure 21.5a Celtic Sea herring catches, 1910-1988

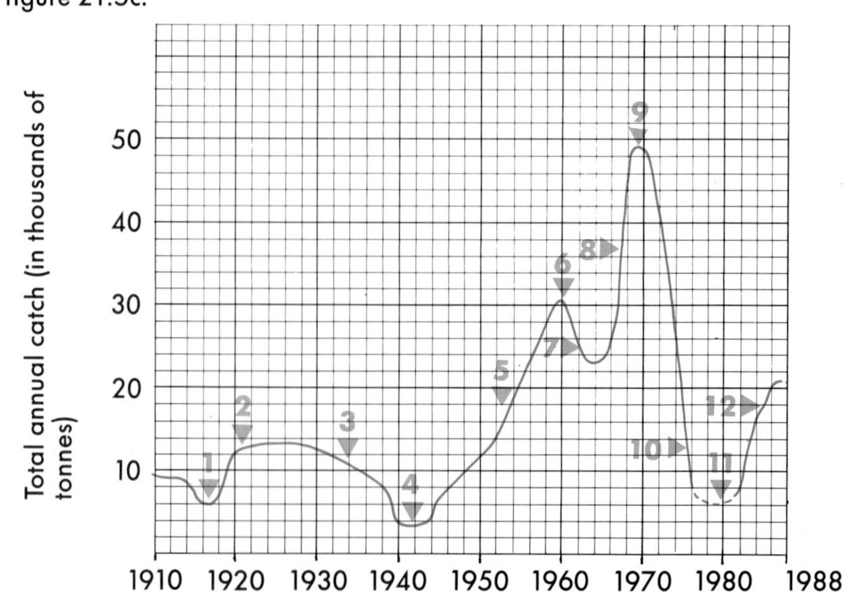

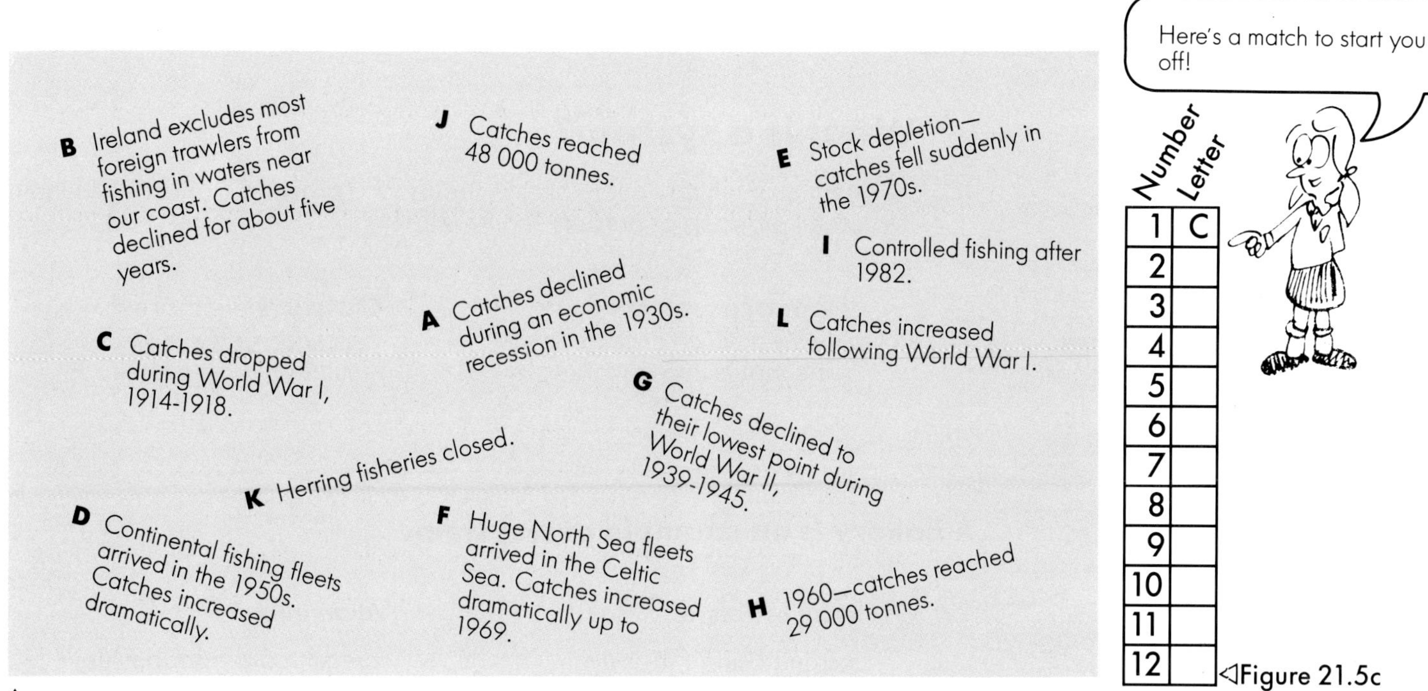

△
Figure 21.5b

◁Figure 21.5c

3. Explain as fully as you can why each of the following statements is true.
   (a) Overfishing in the Celtic Sea has been caused partly by improved technology in the fishing industry.
   (b) Over-exploitation has led to the depletion of fish resources in the Celtic Sea.
   (c) Conservation measures have been employed to reduce the risk of future overfishing in the Celtic Sea.

22 FARMING: AN EXAMPLE OF A SYSTEM

What is a system?

A ***system*** is anything which takes in ***inputs*** and ***processes*** or changes them into useful products or ***outputs*** (figure 22.1).

Inputs are the things which ***enter the system.*** They are either changed by the system's processes or used to help process the work.

The **processes** are those ***activities*** which change some of the inputs into outputs.

Outputs are those things which ***leave the system*** as a result of the processes.

A bakery is an example of a system.

Main inputs

flour, yeast, water, electricity, human labour etc.

Main processes

making dough, baking

Main outputs

bread, cakes, scones etc.

. . . and a lot of heat!

△
Figure 22.1

Which of the following items can be described as a system? Name the main inputs and outputs for each system you identify.

- ☐ a picture
- ☐ a mine
- ☐ a farm
- ☐ a chair
- ☐ an oil refinery
- ☐ a book
- ☐ a hydroelectric power station
- ☐ a milk delivery round
- ☐ an ice cream factory

Farming as a system

A *model* is a diagram which gives a simplified but clear impression of something.

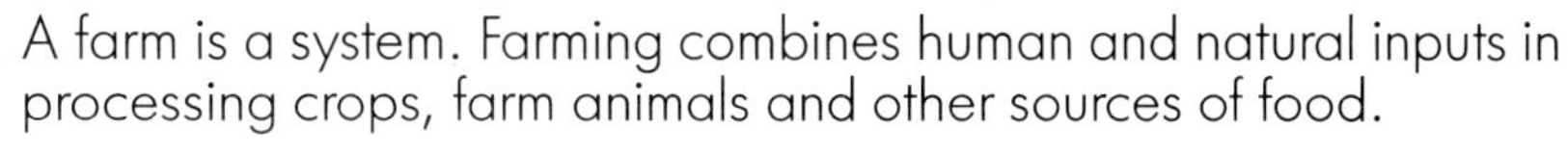

A farm is a system. Farming combines human and natural inputs in processing crops, farm animals and other sources of food.

Examine figure 22.2, which shows a model of farming as a system.

Figure 22.2 Farming as a system
▽

Farm inputs

1 **Climate**
heat
sunshine
rainfall

2 **Land**
soil
slope
drainage

3 **Labour and expertise**
work of farmer, farmer's family and hired farm workers

4 **Farm buildings**

5 **Farm machinery**

6 **Stock**
seeds
animals
etc.

7 **Capital**
money needed to run the farm

8 **Others**
fertilisers
veterinary services
etc.

Farm processes

Crop farming
ploughing
planting
spraying
fertilising
harvesting
etc.

Animal farming
breeding
feeding
milking
etc.

CASH

PROFIT

Farm outputs

Crop outputs
grains
vegetables
fruits
straw
silage
grass
etc.

Animal outputs
meat
milk
wool
manure
etc.

Markets/Outlets
grain mills
co-operative creameries
cattle marts
etc.

Activities

Study figure 22.2. Then do the following:

1. List some specific farm buildings which could be included in input 4.
2. List some farm machines which could be included in input 5.
3. Most of the inputs numbered 1-8 are *variable inputs.* They may differ from farm to farm and from time to time. Explain how land (input 2) could vary between farms. How might this variation affect the processes and outputs of different farms?
4. Name the principal animal species reared on Irish farms. List the outputs associated with each of the animals you have named.
5. Some farm outputs are not usually sold at markets. Instead, they are put back into the farm system and used as inputs to improve the farm processes.
   (a) Which outputs of arable farming are most likely to be used to assist the livestock farming process?
   (b) Which output of livestock farming would be used as an input in the arable farming process?
6. Most of the milk output of a farm is sold to a creamery. Name two other uses to which some of the milk output might be put.
7. Write a paragraph to describe the farm system illustrated in figure 22.2.

Case Study: A Mixed Farm in Ireland

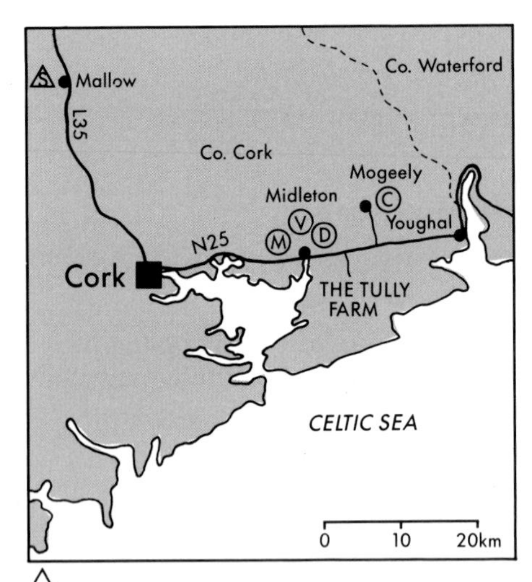

Figure 22.3 The location of the Tully farm

- (V) vegetable processing plant
- (M) co-operative mart
- (D) distillery
- (C) co-operative creamery
- sugar factory
- —— main road
- —— minor road
- - - - county boundary

Most farms in Ireland are ***mixed farms.*** On mixed farms, work is divided fairly evenly between rearing farm animals and growing a variety of arable (tillage) crops.

John and Mary Tully and their family own a mixed farm located between Youghal and Midleton in East Cork (figure 22.3). The Tully farm consists of almost 50 hectares (125 acres). Until the early 1960s, three farm workers were employed on the farm. Now, Mr and Mrs Tully and their eldest son Brian work the farm themselves. (Try to suggest a reason for this change.)

Crops and animals on the Tully farm

Crops	*Animals*
barley	40 milch (milking) cows
sugar beet	2 horses (used for recreation only)
peas and turnips	
ley (grass grown in rotation with other crops)	

Part of the Tully farm. What breed of cows is shown in the photograph? Why have trees been planted as 'fencing' for this field?

Some inputs on the Tully farm

Conditions in East Cork are well suited to mixed farming

The growth of arable crops is favoured by:

- ☐ deep, fertile and well-drained ***soils.***
- ☐ flat or ***gently-sloping land*** which facilitates the use of heavy machinery.
- ☐ a large ***distillery*** at Midleton which uses good-quality barley to make spirits.
- ☐ a ***food processing plant*** at Midleton which provides a nearby market for peas.
- ☐ a ***sugar factory*** at Mallow which provides a market for sugar beet.

The rearing of dairy cattle is favoured by:

- ☐ a mild, damp ***climate*** which encourages the year-round growth of grasses. This reduces the need for expensive stall-feeding of cows.
- ☐ the ***limestone*** beneath the surface which enriches the soil and the grasses with calcium. Calcium contributes to high-quality milk yields in dairy cattle and strengthens the bones of horses and cattle.
- ☐ a co-operative ***creamery*** at Mogeely which provides a nearby market for milk.
- ☐ a co-operative ***mart*** at Midleton where cattle can be bought or sold.

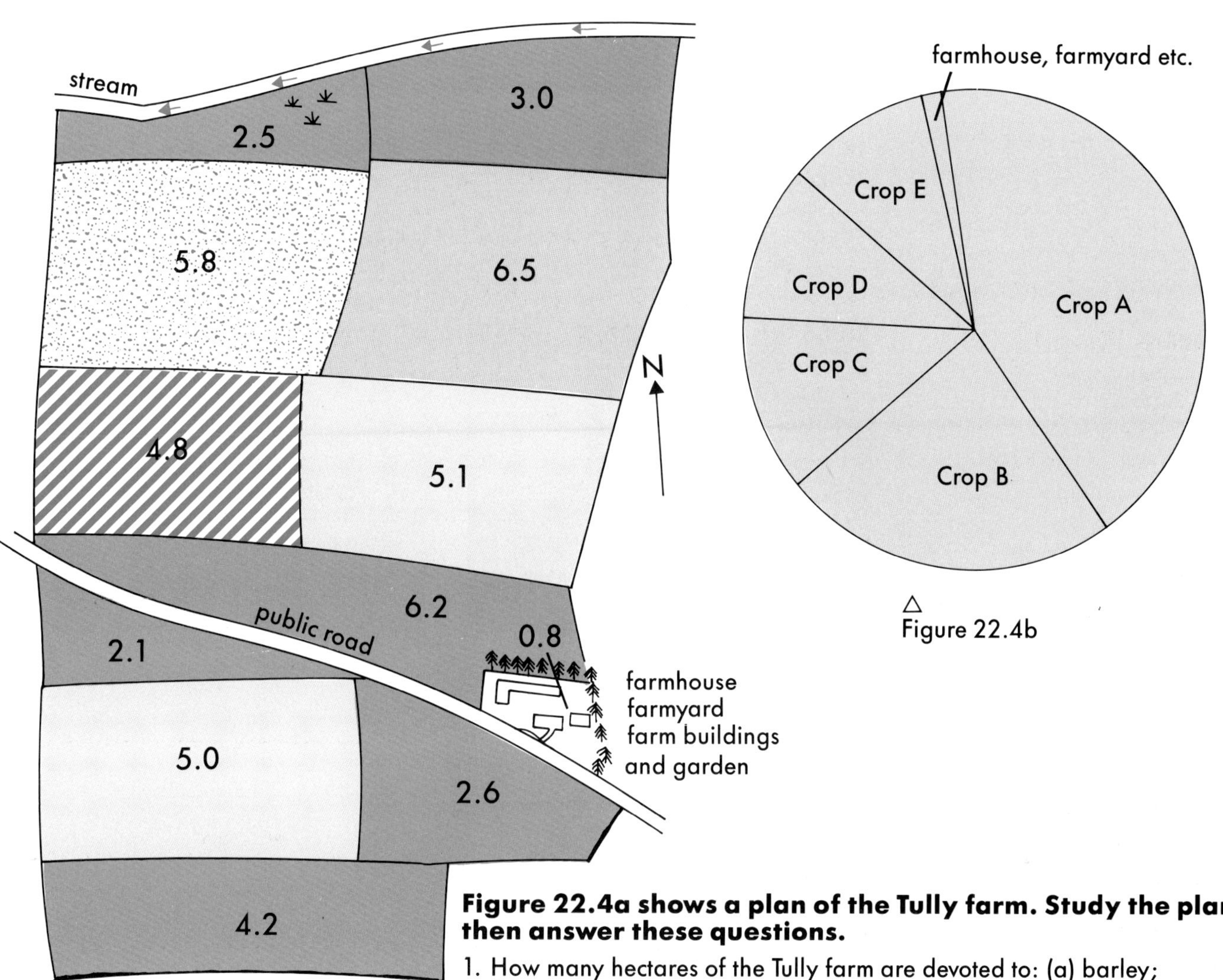

Figure 22.4b

Figure 22.4a A plan of the Tully farm

- permanent grass
- ley
- barley
- sugar beet
- peas and turnips
- coniferous trees
- marsh
- 4.2 size of fields in hectares (1ha = 2.5 acres, approximately)

Figure 22.4a shows a plan of the Tully farm. Study the plan and then answer these questions.

1. How many hectares of the Tully farm are devoted to: (a) barley; (b) permanent grass; (c) tillage crops, including ley?
2. Study the pie chart in figure 22.4b which shows land use on the Tully farm.
   (a) Identify each of the crops labelled A-E.
   (b) Calculate the percentage of the farm devoted to crop A.
3. Suggest a reason why the two most northerly fields are under permanent grass.
4. Ley is harvested as silage or hay. Find out the difference between silage and hay. What are the uses of silage and hay?
5. Are the Tullys' tillage fields: (a) large; (b) small; (c) regular in shape; (d) irregular in shape? Are the size and shape of the fields suited to tillage farming? Explain.
6. Is the public road which runs through the farm an advantage or disadvantage to the farm? Explain.
7. Do you think the farmyard and farm buildings are well situated in relation to the farm? Explain.
8. Why do you think the Tullys have planted a line of coniferous trees bounding the northern and eastern sides of the farmyard?
9. Study 'Crop rotation' and 'Caring for the soil' on page 158. Which crop will be grown next in each of the following fields: (a) the 5.8ha field; (b) the 5.1ha field; (c) the 4.8ha field; (d) the 2.6ha field?

Farm machinery

Many machines are needed on a tillage farm (see box). Some of these machines are very expensive to buy and many are used only for short periods each year. To reduce costs, the Tullys ***share ownership*** of a silage harvester with a neighbouring farmer. They ***hire*** a farm contractor to harvest their barley with his combine harvester. Their peas are cut by special pea harvesters owned by the food processing plant at Midleton.

Some machines on the Tully farm

tractor
rotovator
harrow
trailer
plough
sprayer
fertiliser spreader
silage harvester
(shared)
milking machine
corn drill
manure spreader
bulk milk tank with
cooling equipment

- ☐ **Are more of these machines needed for arable farming or for dairy farming?**
- ☐ **What work is carried out by each of the following: rotovator, sprayer, corn drill, harrow?**

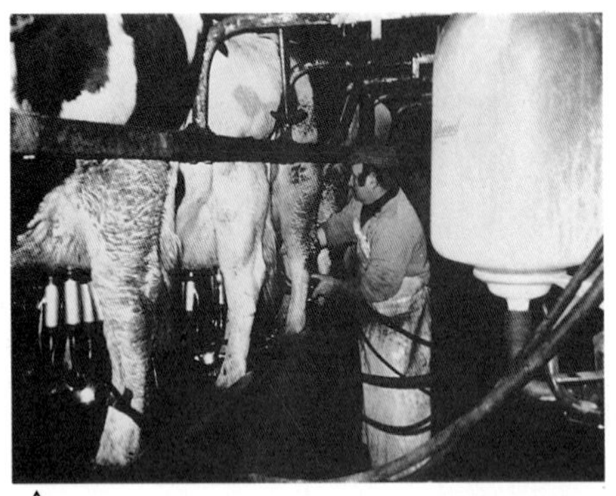

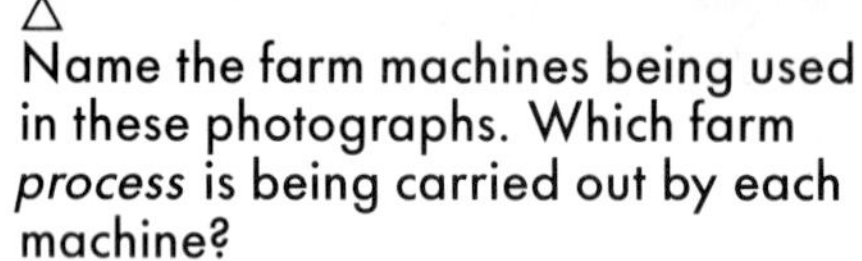

△ ▽
Name the farm machines being used in these photographs. Which farm *process* is being carried out by each machine?

Caring for the soil:

Using crop rotation and fertilisers

Crops remove nutrients or plant food from the soil. These nutrients are vital to healthy plant growth. They can be replaced in two ways: by using ***crop*** rotation and ***fertilisers.***

Crop rotation

Different tillage crops take different nutrients from the soil. To prevent too much of any one nutrient being removed from their tillage fields, mixed farmers use their knowledge of crop rotation. They change their tillage crops from year to year in a planned way.

Tillage crop rotation used on the Tully farm

Year	*Crop*
1	sugar beet
2	barley
3	barley
4	peas and turnips
5	grass (ley)

Fertilisers include:

- ☐ ***manure*** from cattle and horses
- ☐ ***chemical fertilisers*** which contain nitrogen, phosphates and potash.

Pollution warning!
Fertilisers may seep from farmland into rivers and lakes. The pollution which this can cause is discussed on page 160.

Processes

Activities on the Tully farm

Life is very busy on a mixed farm. The Tullys rise early and begin their work by 8.00 a.m. each day. They usually finish work at about 6.30 p.m., although they may work longer hours during busy times of the year such as harvest.

Some farm tasks must be carried out at all times of the year. Other tasks are seasonal and must be done only at certain times (figure 22.5).

Figure 22.5 Tasks throughout the year on a mixed farm

- ☐ Which farm tasks must be carried out at all times of the year? Name three seasonal farm tasks.
- ☐ Which is: (a) the busiest season for dairy activities; (c) the quietest season all round?

All year-round work: Milk cows twice daily. Feed livestock. Clean dairy parlour and its equipment twice daily.

	Spring	Summer	Autumn	Winter
Work relating to dairying	Care for newborn calves.	Fertilise grass. Harvest silage and hay.	Sell surplus calves.	Stall-feed cattle.
Work relating to tillage	Till. Fertilise. Sow barley and peas.	Harvest peas. Sow turnips	Harvest barley and turnips. Fertilise. Plough.	Repair machinery. Mend fencing.

Outputs

The box contains a selection of farm products.

- ☐ Choose nine of these products which would be considered outputs of the Tully farm. (Figure 22.4a will give you some clues.)
- ☐ With the help of information in figure 22.3, identify an appropriate market in which the Tullys could sell ***six*** of the products you have chosen. What is likely to be done with the three remaining outputs?

Some farm products			
milk	*chickens*	*barley*	*turkeys*
surplus calves	*wheat*	*eggs*	*pigs*
barley	*oats*	*sugar beet*	*peas*
straw	*turnips*	*mangolds*	*farmyard manure*

Some problems of modern farming systems

Modern agricultural systems are usually very efficient. A wide variety of farm inputs are processed scientifically, producing large quantities of outputs. But these farming systems also create serious problems. Some of these problems are outlined below.

Sprays, fertilisers and the environment: an input problem

Farmers often spray poisonous chemicals called ***insecticides*** and ***herbicides*** on the land. The insecticides kill insects and the herbicides kill weeds. But these poisons can also kill other things. Beneficial insects are killed by insecticides along with the harmful ones. Birds and other animals which may eat the poisoned insects or weeds are also affected. Many of Ireland's beautiful wildflowers are also killed along with the weeds. And if the poisons seep into the water supply, they can kill fish and other creatures. There is also growing concern about the harmful effects which the heavy use of poisons has on human health.

Large amounts of ***artificial fertilisers*** are also used by farmers, helping their crops to grow more quickly. But the nitrates in artificial fertilisers are washed into rivers by rain. They cause an increase in the level of ***nitrogen*** in the water which in turn causes a rapid growth of ***algae.*** Algae chokes the water and uses up the available oxygen. Fish die as a result, as do the animals which feed on them.

Disappearing hedges: a process problem

Hedgerows have always been one of the pleasant features of the Irish landscape. But in recent years, farmers have been bulldozing their hedgerows, creating bigger fields which are better suited to the use of modern machinery. The destruction of hedgerows has increased the amount of available farmland. But it has also resulted in the destruction of valuable habitat for many birds, mammals and insects as well as traditional tree and hedge species. Crops may suffer with the disappearance of beneficial birds and mammals. Damage to soil and crops may also occur without the hedges which once protected them from the force of the wind.

Study figure 22.6 and answer the questions which follow.

1. Describe the trends suggested in figure 22.6.
2. Draw a trend graph to illustrate the information in figure 22.6. Let the vertical axis represent kilometres of hedgerows and the horizontal axis represent the years.
3. (a) During which period were most hedgerows cleared?
   (b) What percentage of existing hedgerows was cleared during this period?
   (c) Attempt to explain the high loss of hedgerows which occurred during this period.

Year	Kilometres of hedgerows
1800	78
1850	102
1900	99
1960	90
1988	62

△
Figure 22.6 Changes in the length of hedgerows in an area of Co. Cork

Farm surpluses: an output problem

Throughout the European Community, products such as milk are being ***overproduced***—more is being produced than people wish to buy. Much of the milk is converted into butter, so butter is also being overproduced. This overproduction creates a ***danger of price collapse.***

To prevent the price of milk and milk products from falling too low, the EC buys up surplus butter stocks and puts them into storage. This procedure is called ***intervention.*** The amount of butter keeps increasing, creating a 'butter mountain'.

The refrigerated storage of vast quantities of butter is very ***costly.*** Many Europeans now believe that the butter mountain should be reduced drastically.

Class discussion

Here are a few suggestions for reducing the butter mountain. Discuss the merits and/or demerits of each suggestion. Which suggestion do you favour most? Why?

- ☐ Discontinue intervention altogether. Let the price of milk and milk products be decided by demand.
- ☐ Discourage farmers from producing surplus milk. Make farmers pay a special tax or super levy on all surplus milk produced.
- ☐ Pay farmers to reduce their milk yields.
- ☐ Give away surplus milk and butter to the poor and to Third World countries.
- ☐ Convert surplus milk into nutritious milk biscuits. Provide a regular supply of these biscuits to organisations such as the Red Cross which help to combat world famine.

Activities

1. Divide the items in the box into three separate lists: farm inputs; farm processes; farm outputs.

tractor	eggs	combine harvester
buying stock at mart	soil	cattle breeding
seed potatoes	fertilisers	harvesting sugar beet
wool	sunshine	agricultural education
spraying potatoes	newborn lambs	veterinary advice
harvested sugar beet	milking parlour	money borrowed from Agricultural Credit Corporation

2. On a large sheet of paper, draw a detailed diagram to illustrate the farm system which operates on the Tully farm. You may use the diagram in figure 22.2 as a guide, but the inputs, processes and outputs which you illustrate must show specific details of the Tully farm as discussed in this chapter.

3. **Project**

Case Study of a local farm

Arrange a visit to a local farm. During your visit, find out and record the answers to as many of the following questions as possible. Using this information, write a detailed account of your visit

General information:

1. On what date did you visit the farm?
2. Name the farm and its owners.
3. Describe the location of the farm. If possible, show the location of the farm on a sketch map.
4. What type of farm is it: livestock, arable (tillage), mixed?

Farm inputs

5. What rock type underlies the farm? What soil types are found on the farm? Are the soils deep or shallow? Are they fertile or infertile?
6. Is the land surface flat, gently-sloping or steeply sloping?
7. What is the size of the farm? If possible, draw a plan of the farm. Indicate the hectarage and land use of each field. Show the location of the farmyard and any roads or rivers which run near or by the farm.
8. Draw a simple plan of the farmyard. Show and name the main farm buildings and state the use of each building.
9. How many people work on the farm? About how many hours each week does each person work? What agricultural training has each person received?
10. How does the farmer care for the soil? Is there any evidence of the overuse of fertilisers, herbicides and pesticides?
11. What animals are kept on the farm? Are all of them kept for commercial purposes?
12. List the agricultural machinery on the farm and the uses to which each piece of machinery is put. If possible, take photographs or draw sketches of some of these machines.

Processes

13. Describe the farmer's main tasks for each season. Illustrate the farmer's work on a diagram similar to figure 22.5.
14. Make a time chart showing the tasks the farmer does during a typical working day.

Outputs

15. Describe, in rank order of their importance, the outputs of the farm. State one use to which each output is put.

The overall farm system

16. Draw a flow chart to show the main inputs, processes and outputs of the farm.

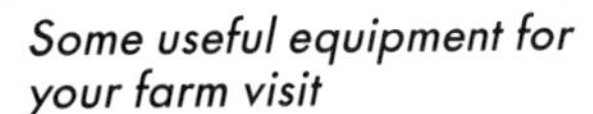

23 POPULATION: A GROWING CONCERN

The growth in world population

The world's population has increased steadily over time. But this rate of growth has been uneven, as shown in the box and in figure 23.1.

Population growth throughout history	*Some reasons why*
In ***early times,*** the planet's population ***grew slowly*** and even ***fluctuated*** (went up and down).	Frequent wars, plagues and famines limited population growth.
From about 1750 AD, world population began to ***grow steadily.***	In ***Europe,*** farming methods and farm machinery improved greatly. So did hygiene and medical knowledge. Europeans were healthier and better fed. They lived longer and had more children. Europe's population began to increase rapidly. This pushed up world population figures.
Throughout the ***20th century,*** world population growth has been more and more rapid. This growth is referred to as the ***population explosion.***	Improved food supplies, better hygiene and increased medical knowledge have spread to ***many parts of the world.*** People in many parts of the world now live longer and have more children.

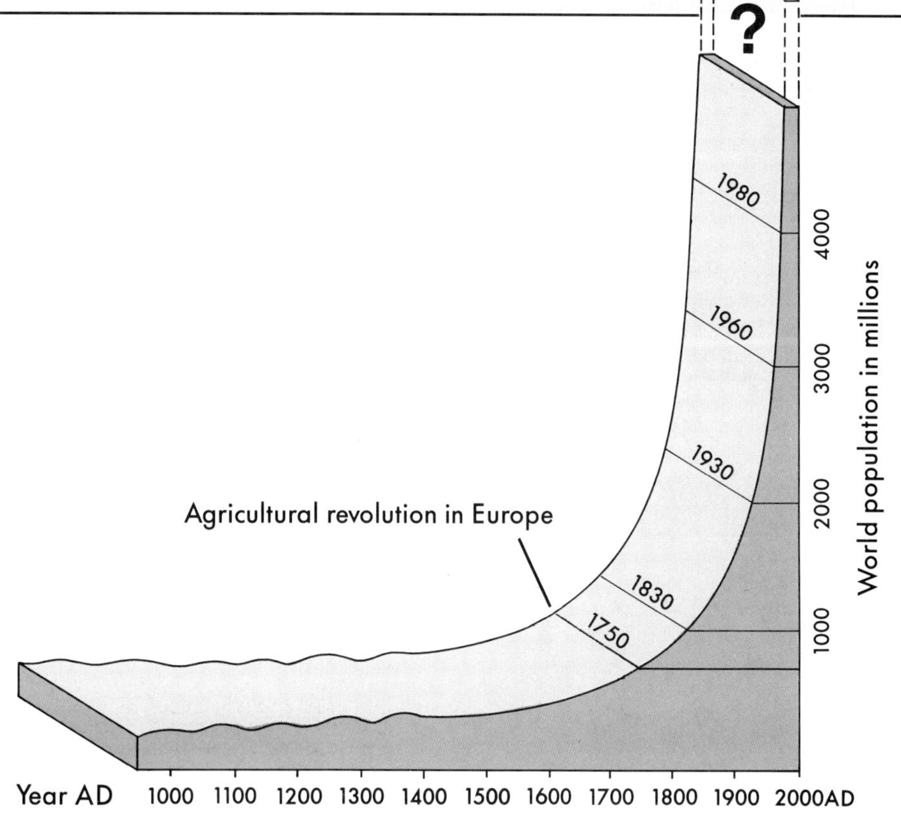

Figure 23.1 World population growth ▷

(a) During which period was population increase slow and irregular?

(b) When did the world's population first reach 1000 million?

(c) How long did it take for the population to: (i) double from 1000 million to 2000 million; (ii) double from 2000 million to 4000 million?
What *trend* is shown by these two facts?

(d) Approximately when did the 'population explosion' begin?

(e) List some of the things which might happen to people if the population explosion continues well into the future.

Population change in any country is the result of two things.

Calculating natural increase

1. The number of ***births*** measured against the number of ***deaths.***
2. The movement or migration of people into and out of the country. ***Immigration*** (movement into a country) increases the population. ***Emigration*** (movement out of a country) reduces population.

Natural increase is calculated by subtracting the death rate from the birth rate.

For example:

Ireland's present birth rate	= 3.4%
Ireland's present death rate	= 1.4%
Ireland's ***natural increase***	= 2.0%

The populations of most countries are growing because more people are born each year than die.

- ☐ The number of people born each year makes up the ***birth rate.*** Birth rates are often given as percentages of the total population.
- ☐ The number of people who die each year makes up the ***death rate.*** Death rates are often given as percentages of the total population.
- ☐ The difference between the birth rate and the death rate is called the ***natural difference***. When the birth rate is greater than the death rate, the difference is called the ***natural increase***. When the death rate is greater than the birth rate, the difference is called the ***natural decrease***.

Country	Birth rate	Death rate
Brazil	3.0	1.2
United Kingdom	1.3	1.2
West Germany	1.2	1.2
India	2.9	1.3

△
Figure 23.2 Birth and death rates in four countries

(a) Calculate the natural increase in each of the countries shown.
(b) Which two countries are rich First World countries? Which two are poorer Third World countries?
(c) What is the difference between the natural increases of the First World countries and those of the Third World countries?
(d) Is this difference caused mainly by differences: (i) in birth rates; (ii) in death rates; (iii) in both?

Changes in population growth

Population growth is very uneven throughout the world.

- ☐ In many ***poor Third World countries,*** the rate of population growth is now very ***rapid.***
- ☐ In ***richer countries,*** the rate of population growth is already beginning to ***slow down.*** In some parts of Europe, population growth has stopped.

These trends have led geographers to the following idea.

'As the economy of a country develops, the population growth of that country goes through a series of predictable changes.'

This idea is called ***demographic transition*** or the ***population cycle.***

Geographers have drawn up a diagram called a ***model*** to show how the population cycle works. Models are not exact representations, but they do give a general idea of how things work. They also present this idea in a clear and simple way.

A model of the population cycle is shown in figure 23.3a. It divides population growth into four ***different stages.*** These stages are:

1. the ***high stationary*** stage
2. the ***early expanding*** stage
3. the ***late expanding*** stage
4. the ***low stationary*** stage

These four stages are connected to the different stages in the prosperity or ***economic development*** of the area concerned.

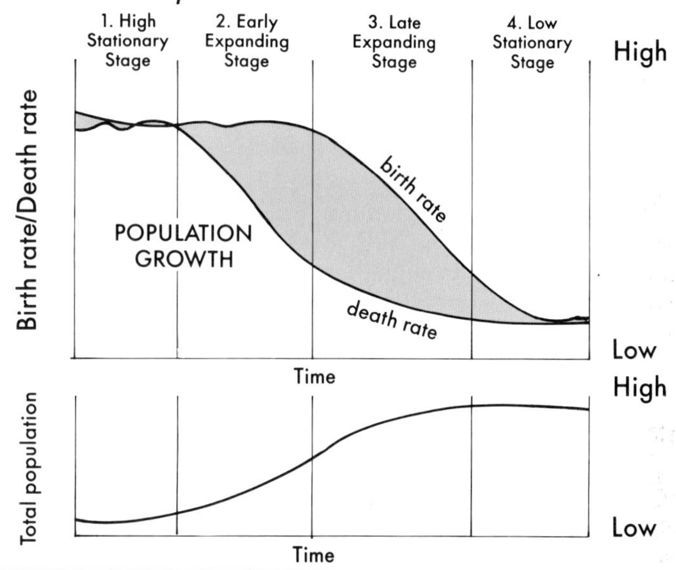

Figure 23.3a A model of the ▷ population cycle

Figure 23.3b Some explanations of the four stages in the population cycle model ▽

Stage in population growth	***High stationary***	***Early expanding***	***Late expanding***	***Low stationary***
Birth rate	high	high	falls rapidly	low
Death rate	high	falls rapidly	low, but levelling out	low
Population growth (due to natural increase)	slow	rapid and increasing	rapid, but slowing down	slow
Some economic conditions	***Economy undeveloped.*** Famines and diseases keep death rate high.	***Economy begins to develop.*** Food and medical supplies improve, so death rates fall.	***Economy continues to develop.*** People decide to have smaller families, so birth rate falls.	***Economy developed.*** Most people are prosperous and prefer small families.
When did these conditions exist in Europe?	before 1750	between 1750 and 1900	early in the 20th century	now
Places where these conditions exist now	parts of the Amazon basin (Brazil)	many Third World countries such as Mali	some Third World countries such as Panama	European countries such as West Germany

1. Study figures 23.3a and 23.3b. Then rewrite the following statements by filling in the blanks and omitting incorrect information.
   (a) The *total* population of a country is at its lowest during the ________ stage.
   (b) A population explosion (very rapid population growth) begins during the ________ stage.
   (c) The population explosion is caused by a rapid fall in the *birth/death* rate, while the *birth/death* rate remains high.
   (d) The population begins to slow down during the ________ stage.
   (e) The population explosion slows down when the ________ rate falls rapidly, while the ________ rate begins to even out.
   (f) At the low expanding stage, the total population is *high/low,* while the rate of population growth is *high/low.*
2. During which stage is the death rate fluctuating (going up and down) and almost even with the birth rate? Suggest a reason why the death rate might fluctuate at this stage.
3. In your opinion, why are stages one and four described as 'stationary' stages?
4. Why are stages two and three referred to as 'expanding' stages?

Some uses of the population cycle model

☐ The model shows that when a country's economy first begins to develop, its population begins to rise rapidly. This helps to ***explain the present population explosion*** in many developing Third World countries.

☐ The model shows that when a country continues to develop, its population growth begins to slow down. This is good news. It suggests that, if prosperity could be improved in the Third World, the ***world's population explosion could be brought to an end.***

Factors affecting changes in population growth

☐ ***'Developing' countries such as Brazil now have rapidly increasing populations.***
This is so because countries such as Brazil are in the expanding stages of population growth. Death rates have fallen each year, but birth rates have remained high. So natural increases of population are very high.

☐ The populations of ***'developed' countries such as West Germany*** increased rapidly during the late 18th and early 19th centuries. But such countries now have ***little or no population increases.***
This is so because these countries are now in the low stationary stage of population growth. Death rates are low, but birth rates are equally low. So natural increases are either very small or non-existent.

Some factors which influence population changes in countries such as Brazil and West Germany are shown in the box.

Some factors which influence changes in population growth

☐ food supplies
☐ health
☐ improved technology (better machines etc.)
☐ war
☐ education
☐ the place of women in society

Factors	*In Brazil*	*In West Germany*

Improved food supplies

Factors: Terrible famines still occur in the poorest countries. In spite of this, world food supplies have shown an overall increase in modern times.

Human health generally improves with better food supplies. People live longer as health improves. Death rates are reduced, especially among children. The population increases.

In general, ***improved food supplies have led to population increases in different parts of the world at different times.***

In Brazil: Food production increased throughout the 20th century. Population has risen as a result.

In West Germany: At the end of the 18th century, improvements in agriculture began to bring about increased food supplies. Population increased rapidly at this time.

◁ **How do improved public health facilities and food supplies help Third World children?**

Health

Factors: ***Improved public health*** in many parts of the world has been due to:

- ☐ improved food supplies
- ☐ improved medicines
- ☐ improved sanitation, especially in the form of piped water and improved sewerage systems.

These things have ***helped to reduce the huge death tolls*** (especially among the young) from easily curable diseases such as measles, whooping cough and severe diarrhoea.

In Brazil: Many Brazilian children still die from hunger-related and easily curable diseases. But improved public health facilities have reduced child mortality by half since 1960. Population has risen as a result.

In West Germany: Medical and sanitary conditions began to improve greatly about a century ago. Population increased at that time.

Improved technology

Factors: New machines and other examples of ***improved technology can have different effects on population change.***

- ☐ Improved farming and food-processing methods have helped to ***increase food supplies.*** This has led to reduced death rates and to an increase in population growth, especially in some Third World countries.
- ☐ Some examples of new technology such as computers can do the work of many people. This often leads to ***increased unemployment.*** In richer countries particularly, increased unemployment has discouraged some people from having children. Lower birth rates have resulted.

△ **Ploughing in Brazil. How might the use of modern agricultural machinery contribute to population growth?**

In Brazil: Improved farm machinery has led to improved food supplies in Brazil.

In West Germany: Computers, factory 'robots' and other scientific advances have led to increased unemployment in recent years. Unemployment has discouraged population growth in West Germany.

Factors	In Brazil	In West Germany
War ***Wars mean increased death rates,*** not only for soldiers but civilians as well. Wars also mean that husbands and wives are separated for long periods of time, resulting in ***lower birth rates.***		Germany lost almost 3 million people during World War I (1914-18). During World War II (1939-45) nearly 4 million people died. These terrible losses reduced West Germany's population growth during the 20th century.
Education ***Better educational facilities often lead to lower birth rates.*** ☐ People who have learned to read are more likely to understand and take part in ***family planning schemes*** aimed at reducing birth rates. ☐ Educated women are more likely to choose working ***careers*** which take them ***outside the home.*** This lifestyle may discourage women from having large families.	Many poorer Brazilians, especially poorer women, receive little formal education. Their families tend to be large.	In West Germany, education is compulsory and of a very high standard. Many West German couples plan small families so that both men and women can pursue careers outside the home.
Women in society In some parts of the world, women have little decision-making powers within the family. They are often expected to marry young and to have large families. In 'developed' countries, women are more likely to choose between having children and pursuing careers outside the home. Birth rates tend to be lower in such countries. In general, ***as the decision-making power of women increases, birth rates tend to decrease.***		

△
A family planning clinic in the Third World. Why would the success of such clinics depend partly on people having a good basic education?

Population change: What about the future?

People who study population trends generally agree that the world's population will continue to increase in the future. They are not in agreement, however, about how rapid this increase is likely to be.

- ☐ ***Some people take a pessimistic view.***
  They believe that the present world population explosion will continue. They say that the world's population will double in the next 35 years. They also believe that the world will become so overcrowded that a terrible world famine will eventually take place.

- ☐ ***Some people take an optimistic view.***
  They remember the lessons of the population cycle graph (figure 23.3a). This graph suggests that population growth declines with economic development. The optimists say that the world's population growth will ease off as the economies of Third World countries continue to develop.

Study figure 23.4 which shows the pessimistic and optimistic views of future population growth.

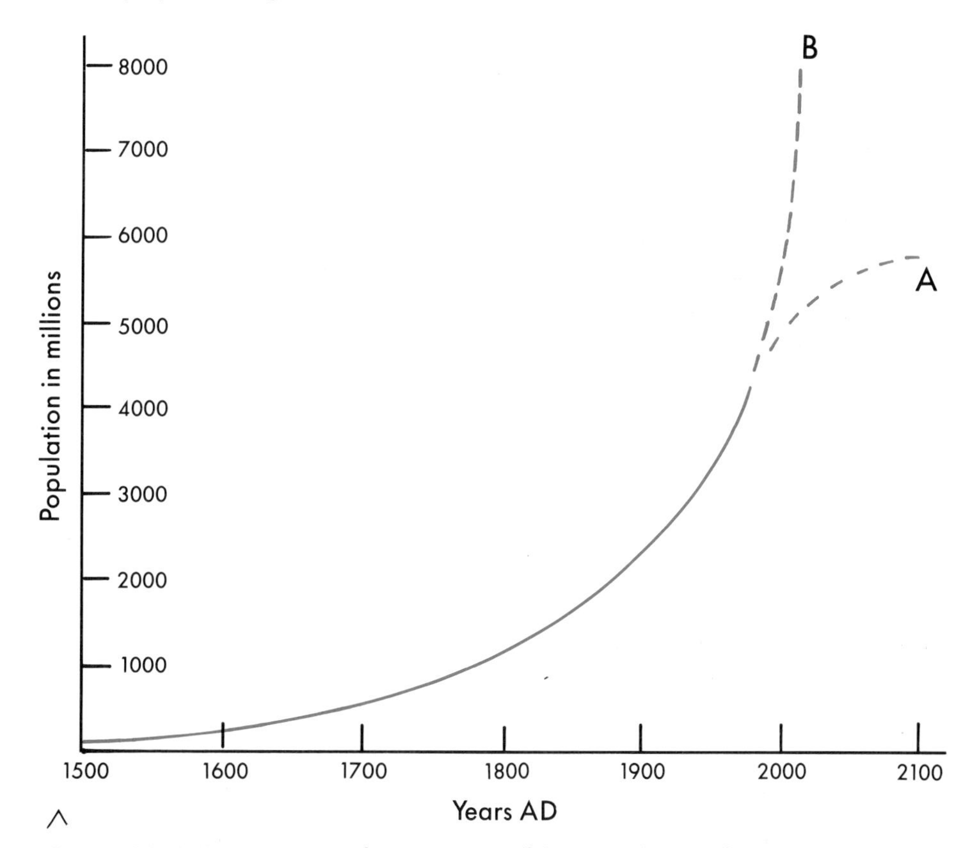

Figure 23.4 Pessimistic and optimistic forecasts of future world population growth

(a) Which broken line, A or B, represents the pessimistic forecast of population growth? Which line represents the optimistic forecast?

(b) According to the pessimistic forecast, approximately when will the world's population reach 8000 million?

(c) According to the optimistic forecast, what will the world's approximate population be by the year 2100?

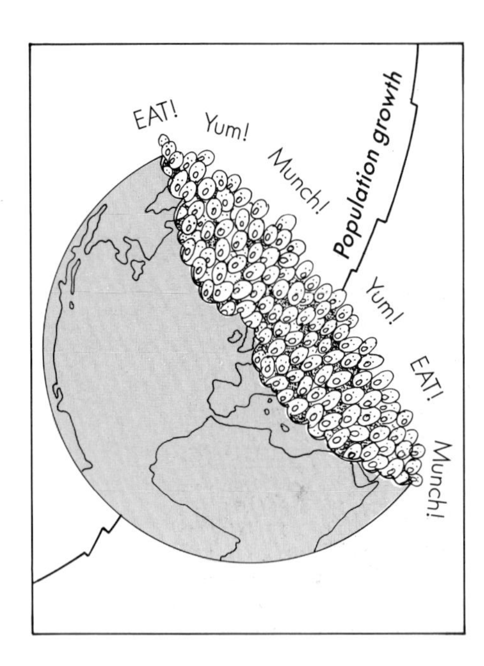

◁ Figure 23.5 Tomorrow's World?
(a) What message do you think this cartoon is trying to convey?
(b) Do you think the cartoon's message is best described as: believable; far-fetched; alarmist? Explain why you have made this choice.

Ireland's population trends since 1800

An assignment

- ☐ Figure 23.6a shows a ***trend graph*** of Ireland's population since 1800.
- ☐ Figure 23.6b shows ***six fact sheets*** describing different stages in Ireland's population since 1800. These fact sheets are not arranged chronologically.
- ☐ Figure 23.6c shows a ***selection box*** of words, terms, figures and dates which have been omitted from the fact sheets.

Study these figures and then do the following.

1. Match labels A – F on the trend graph with the appropriate fact sheet (figure 23.6b).
2. Write an account of population changes in Ireland since 1800. You can do this by:
   - ☐ rewriting the fact sheets in chronological order
   - ☐ filling in the blanks with the correct selection from the box (figure 23.6c)
   - ☐ including only the correct choices from the words in *italics*.

Figure 23.6a Ireland's population trends since 1800 ▷

Figure 23.6b Ireland's population: fact sheets

Between 1920 and 1960, Ireland's population remained steady at slightly under ________ million.
Birth rates were much higher than death rates, so ________ ________ was high.
But the natural increase was almost exactly cancelled out by ***emigration/immigration*** from Ireland. This was particularly noticeable in rural areas where the increased use of modern ________ ________ was beginning to reduce the demand for farm workers.

Matching label on graph —

The Great Famine occurred between 1845 and 1849. One million people died when the potato crop failed. Many more people fled from their homes and migrated to other countries. Ireland's population fell ***gradually/sharply***. By the year ____, it was down to 6.6 million.

Matching label on graph —

In the 1980s, an economic recession resulted in the closure of many factories and in rising ***employment/unemployment***. Greater use of labour-saving machinery such as computers also reduced job opportunities in some areas.
Once again, emigration began to increase. This caused our population growth to ***slow down/increase***. In 1986, a population census showed that there were approximately ____ people in the country.

Matching label on graph —

Throughout the second half of the 19th century, large numbers of Irish people continued to emigrate to countries such as ***America/France*** and ***Italy/England***. Ireland's population continued to ***decline/increase*** quite ***slowly/rapidly***. It had fallen to 4 million by the year ____. By the year ____, it had fallen to 3 million.

Matching label on graph —

Throughout the 1960s and 1970s, manufacturing industry developed rapidly in many parts of Ireland. ***Increased/Decreased*** job opportunities resulted in a rapid ***increase/drop*** in emigration ***into/out of*** the country. By the end of the 1970s, emigration had virtually stopped.
With no emigration to cancel out the continuing high birth rate, population levels began to ________ once more. By 1980, Ireland's population had risen to ***3.4 million/4.6 million*** people.

Matching label on graph —

Between 1800 and 1845, Ireland's population grew rapidly. Population increased from ____ million people in 1800 to ____ million in 1845.
Birth rates were very ***high/low*** during this time. Large families were considered to be of economic ____ to parents. Even young children would help with farm work and household chores. Grown-up sons and daughters looked after their ageing parents.
Meanwhile, death rates were declining rapidly. This was partly because increased ________ crops were producing more food and partly because diseases such as ________ could now be cured.
High birth rates and rapidly-falling death rates led to a population explosion similar to that which is now occurring in ***First/Third World*** countries such as ________ and ***Brazil/West Germany***.

Matching label on graph —

Figure 23.6c Information omitted from the fact sheets

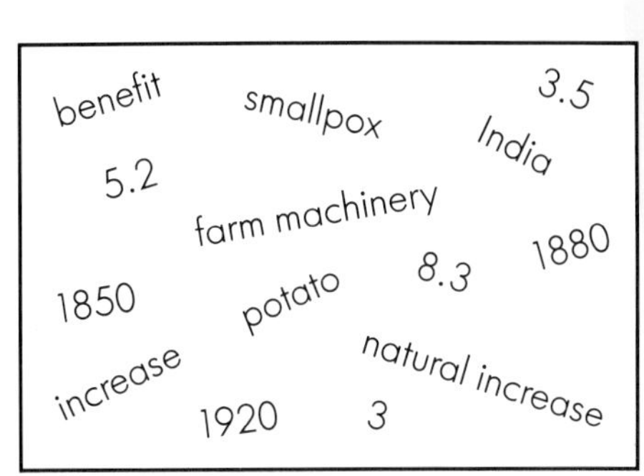

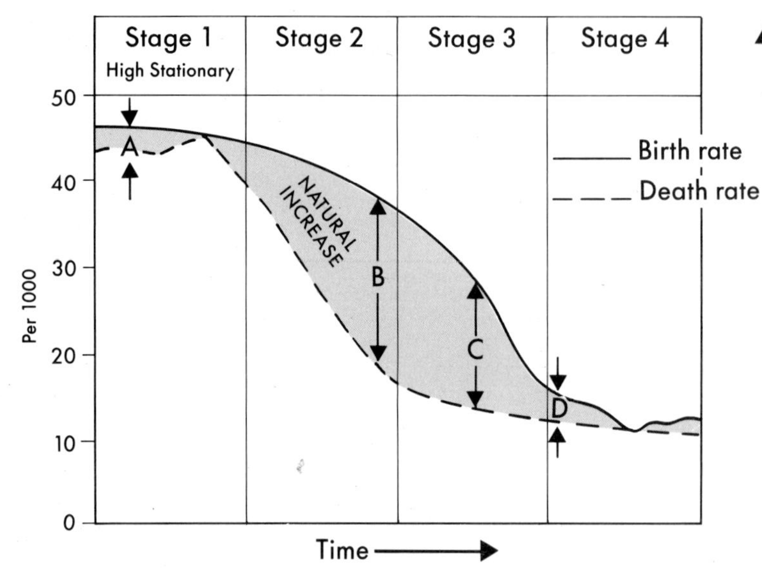

△ Figure 23.7

Activities

1. The model in figure 23.7 illustrates the idea of the population cycle.
   (a) Give another name for the term 'population cycle'.
   (b) Name stages 2, 3 and 4 in the model.
   (c) At which stage does the death rate change most rapidly? At which stage does the birth rate change most rapidly?
   (d) What is meant by the term 'natural increase'? State whether natural increase is greatest at point A, B, C or D in the model.
   (e) Describe the changes in the birth rate, death rate and natural increase throughout the four stages of the model.

2. (a) Name *6 factors* which influence the rate of population change in a country.
   (b) Explain how *any 3* of these factors have influenced population change in Brazil. (ordinary level)

   or

   Explain how *any 3* of these factors have had *contrasting* (differing) influences on population change in Brazil and West Germany. (higher level)

24 WHERE ARE ALL THE PEOPLE?

Here are some important terms which you should learn.

1. *Population density* is the average number of people per square kilometre (km^2).
2. *Population distribution* is the spread of people in any given area.

How to calculate population density

Divide. . . $\frac{\text{the total population}}{\text{the total area}}$ by

Ireland's population is approximately 4 700 000.

Its area is approximately 82 000km^2.

Calculate Ireland's population density.

Figure 24.1 Population distribution ▷

A

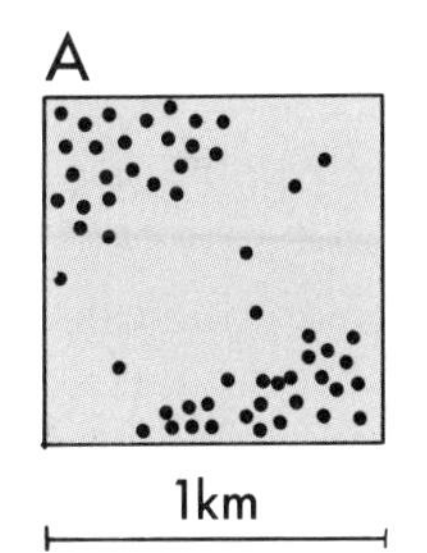

B

C

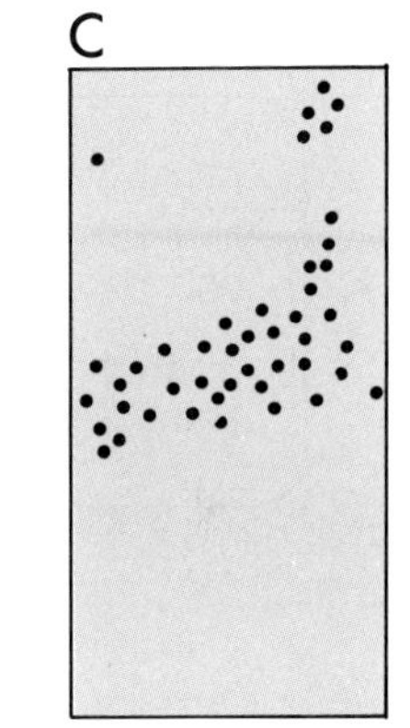

1km

• Each dot represents 1 person

Study maps A, B and C in figure 24.1

(a) Choose suitable words from the list below to describe the population density in each map.
- ☐ density: high
  medium
  sparse (low)
- ☐ distribution: dispersed (scattered)
  concentrated (packed into one or some areas)

(b) Calculate the population density shown on each map.

Do Activity 1, page 185.

An uneven world population

Figure 24.2 shows that the world's population is unevenly distributed. Some areas have very high population densities, while others have very low population densities.

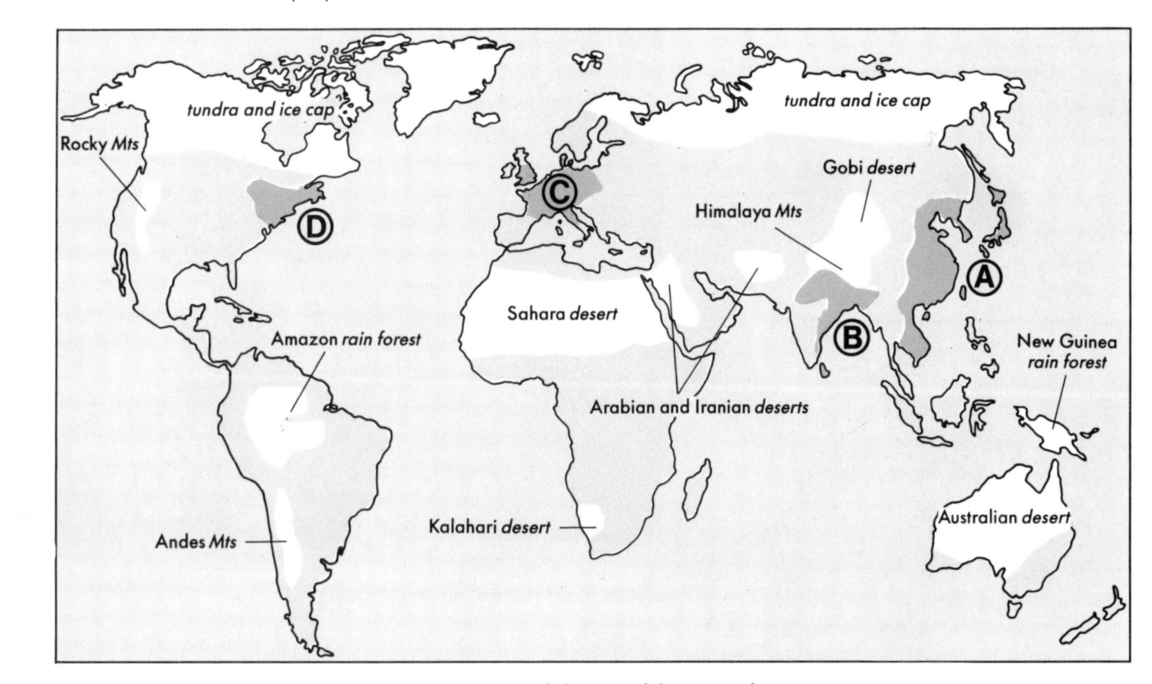

Areas of very high population density

Areas of medium population density

Areas of very low population density

Figure 24.2 Distribution of the world's population

(a) How many main areas of very high population density can you identify?

(b) On the map, *all of Ireland* is shown with a *medium* population density. Is this really so? What does this tell us about the information which can be given on a map of this scale?

People usually live in places where food, water and employment are easy to find. It is these human needs which have caused an uneven distribution of world population.

With the help of figure 24.2 and your atlas . . .

Areas of very high density (See figure 24.2)

In Asia, a high percentage of people earn their living by farming. So population often concentrates in rural (country) areas where more land is available and there are plentiful food supplies. Two very large areas of very high population densities are:

Ⓐ ***Eastern Asia,*** which contains 25% of the world's population. Most people live in rural areas, especially in large and fertile river valleys. Many people also live in highly-industrialised cities such as Tokyo and Bejing (Peking).

Find out the names of the highly populated countries in east Asia.

In which countries are Tokyo and Bejing?

Ⓑ ***Southern Asia,*** which contains 21% of the world's population. Many people live in crowded cities such as Calcutta. But most live in fertile river valleys such as along the River Ganges.

Find out the names of the highly populated countries of south Asia.

In which country would you find Calcutta and the River Ganges?

In Europe and North America, most people live in industrialised cities and towns where there are plenty of jobs. Two very large areas of very high population densities are:

Ⓒ ***Western Europe,*** which contains 20% of the world's population.

Ⓓ ***Northeastern United States and southern Canada,*** which hold 6% of the world's population.

Name some large cities in each of these two areas.

◁ **London. Find out this city's population. Suggest reasons for its high population density.**

Areas of very low density

There are certain types of areas where people cannot live in comfort, where food, water and employment are not easily found. These areas are very sparsely populated. One sparsely-populated type of area is called the ***tundra and ice cap*** (figure 24.2)

Use the information in figure 24.2 to complete Table A. The table gives an account of some of the world's largest sparsely-populated areas.

Sparsely populated areas of the world

Type of area	*Example of area*	*Conditions which discourage population growth in the area*
Tundra and ice cap	☐ Northern Canada	☐ ***Extreme cold*** makes living difficult and farming impossible
Mountains	☐ ________________ ☐ ________________	☐ Some ***high altitude*** areas are too cold and windswept for agriculture or human habitation. ☐ ***Steep slopes*** are often rocky and lack soil. Building houses or roadways in such areas is difficult.
________________	☐ ________________ ☐ ________________ ☐ ________________	☐ Lack of rain causes ***water shortages*** and makes farming impossible in most areas.
________________	☐ ________________ ☐ ________________ ☐ ________________	☐ The climate is ***uncomfortably hot and wet.*** ☐ Heavy rains wash minerals down through the soil and reduce ***soil fertility.*** ☐ ***Dense forests*** make the building of roadways difficult.

Use pencil to fill in the blanks.

Table A △

Population distribution is very uneven. But population densities do not vary on a world scale alone. They can vary within continents, within countries and even within very small areas.

☐ Variations often occur ***between different areas*** at the same time.

☐ Variations often occur ***at different times*** within the same area.

△
Why is this area of the Sahara Desert unpopulated? In which *continent* is the Sahara? Name some *countries* occupied by the Sahara.

△
Why is this area of the Andes Mountains unpopulated? In which *continent* are the Andes? Name some *countries* through which this mountain range runs.

Population variations between areas

Population variations between different areas are usually caused by the varying resources available in these areas.

Case Study 1

Population variations within Sweden

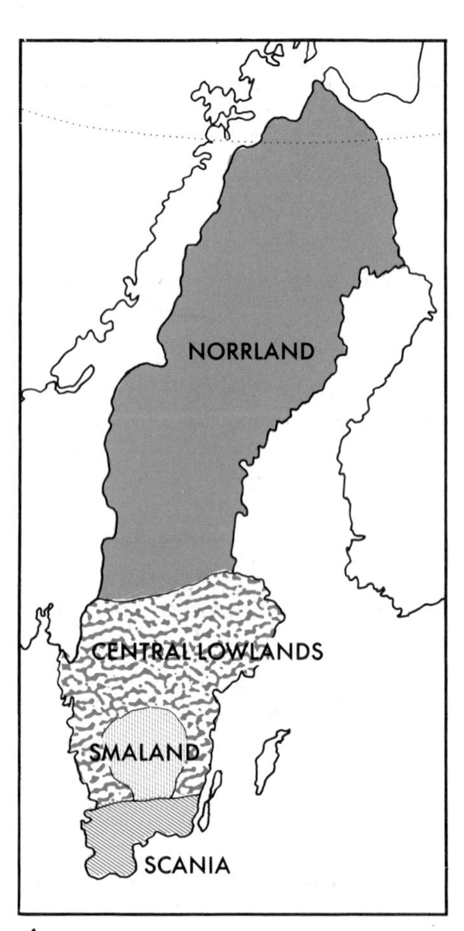

Figure 24.3a Regions of Sweden

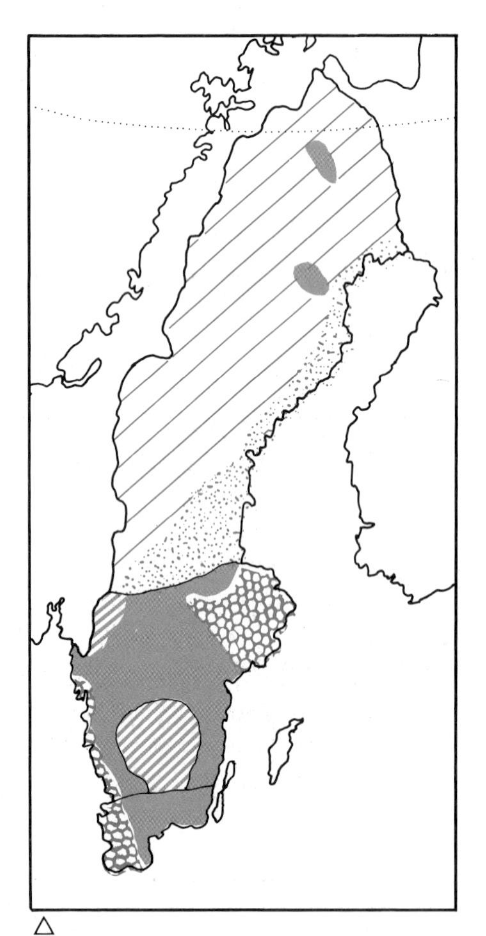

Figure 24.3b Population densities in Sweden

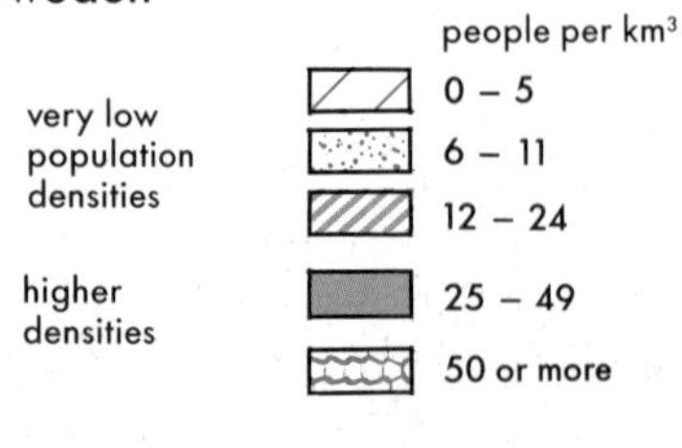

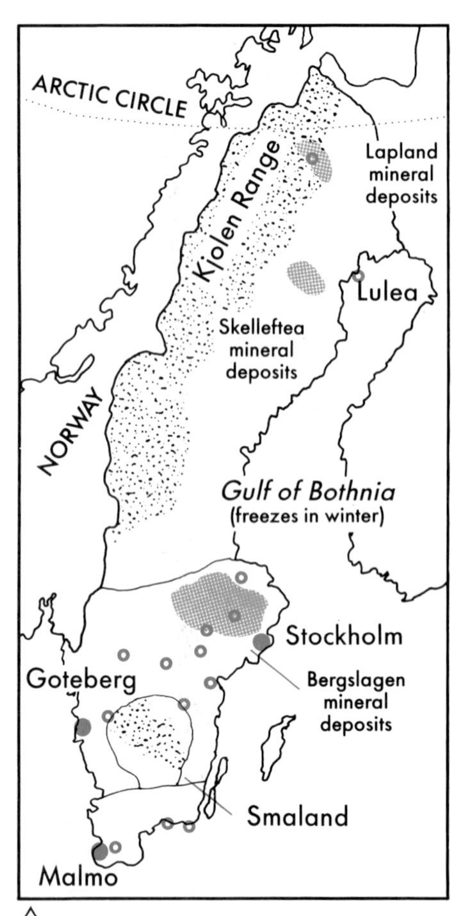

Figure 24.3c Some aspects of Sweden's geography

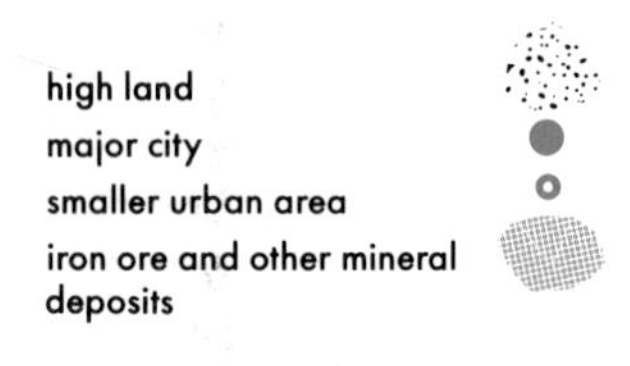

Identify two regions in Sweden with very low population densities and two regions with higher population densities.

- ☐ The map in figure 24.3a shows the four main regions of Sweden.
- ☐ The map in figure 24.3b shows population densities in Sweden.
- ☐ The map in figure 24.3c shows other aspects of Sweden's geography.

Study Table B which explains some of the reasons for high and low population densities in different parts of Sweden.

Table B ▽ **Some factors affecting population densities**

High population densities	*in Sweden*	⬇	*Low population densities*	*in Sweden*
Warmth (favours crop growth)	Scania is the most southerly part of Sweden and therefore the warmest.	**CLIMATE** ← →	***Cold*** (hinders crop growth)	Norrland is on and near the Arctic Circle. It is extremely cold for much of the year.
Lowlands (are more likely to be warmer and more suited to agriculture)	Central Lowlands Scania	**RELIEF** (shape of the land surface) ← →	***Highlands*** (are likely to be cold and windswept and unsuited to farming or human habitation)	Western parts of Norrland Smaland
Flat or gently-sloping land (makes it easier to build roads and houses and to use farm machinery)	Central Lowlands Scania		***Steep slopes*** (make it more difficult to build roads and houses and to use farm machinery)	Western parts of Norrland Parts of Smaland
Deep, rich soils (favour agriculture)	Scania has rich boulder clay.	**SOILS** ← →	***Poor, thin soils*** (hinder agriculture)	Norrland's soils are thin and are frozen in winter. Smaland has poor, peaty soils.
Large-scale mineral deposits (create employment)	The Bergslagen area in the Central Lowlands contains iron, copper & zinc. Iron ore is found in Lapland and at Skelleftea in Norrland.	**MINERAL WEALTH** ← →	***No large-scale mineral deposits***	Smaland
Large, ice-free ports	Goteborg in the Central Lowlands; Malmo in Scania	**COMMUNI-CATIONS** ← →	***No large, ice-free ports***	Norrland's ports are small and ice-bound for much of the year. Smaland has no ports.
Good road & rail networks	Central Lowlands; Scania		***Poor road & rail networks***	Western & Northern Norrland
Large-scale manufacturing (creates employment)	Central Lowlands Malmo in Scania	**MANU-FACTURING** ← →	***Little large-scale manufacturing***	Norrland Smaland
Large cities (are centres of education, commerce & employment)	Stockholm, Goteberg etc. in the Central Lowlands Malmo in Scania	**URBAN AREAS** ← →	***No large cities***	Norrland Smaland

Why Norrland has a very low population density

The population density of almost all of Norrland is between 0 and 11 people per km^2. There are a number of reasons for this low density.

Norrland is on and near the Arctic Circle. It is *extremely cold* for much of the year. This hinders crop growth.

The western part of Norrland is part of the Kjolen Range. This area is *high and windswept* and unsuited to farming or human habitation. The area's *steep slopes* make it difficult to build houses or roads or to use farm machinery.

Soils in Norrland are thin and poor and are frozen in winter. This hinders agriculture.

Communications are generally poor in Norrland. Ports such as Lulea are small and ice-bound for much of the year. Road and rail networks are poor, especially in the north and west of the region.

There is *little manufacturing industry* to provide employment in the region.

There are *no very large cities* serving as centres of employment which would attract people to the region.

Study the information about Norrland. It uses the information in Table B and in figures 24.3a, b and c to draw up the account 'Why Norrland has a very low population density'.
Use this same information to draw up your own accounts about the following:

(a) Why Smaland has a very low population density
(b) Why the Central Lowlands have a relatively high population density
(c) Why Scania has a relatively high population density

Norrland and the Gulf of Bothnia. ▷ At what time of year was the photograph taken? How do you know?

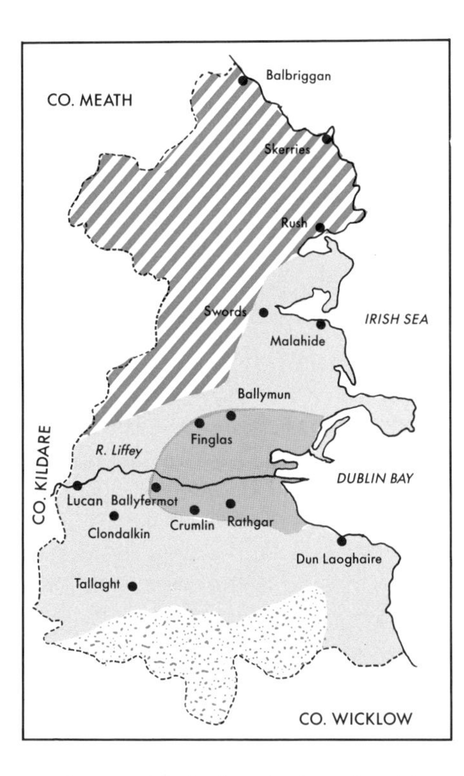

Figure 24.4 Population densities in Co. Dublin

- Very high population density (Dublin city and inner suburbs)
- High and growing population density (outer fringe of Dublin)
- Relatively low population density (North Co. Dublin)
- Very low population density (Dublin Mountains area)

Case Study 2

Population variations within Co. Dublin

Figure 24.4 shows that population densities vary greatly between different areas within Co. Dublin.

Dublin city and inner suburbs

Dublin city and its inner suburbs contain over half a million people. The ***very high overall population density*** of this area has been due to the wide variety of ***employment opportunities and services*** available in Dublin.

- Dublin is Ireland's ***main industrial centre***. It contains many old industries such as brewing and biscuit making, as well as newer industries such as the assembly of computer components.
- Dublin is Ireland's ***chief commercial centre***. Many Irish banks, insurance companies and other financial institutions have their headquarters in Dublin.
- It is Ireland's ***capital city***, which means it is the seat of government. Ireland's civil service, which carries out government operations, has many of its main offices in Dublin.
- It contains Ireland's ***busiest port***. Many exports and imports leave and enter the country through Dublin port.
- Dublin is Ireland's ***largest education centre***. There are many schools and colleges, including Trinity College and University College Dublin.
- It is the country's ***busiest entertainment centre***. It contains numerous theatres, cinemas, sporting facilities and other entertainment possibilities.
- Because of the variety of employment opportunities and services offered in Dublin, the city has long been the ***centre of in-migration*** for people who leave country areas and come to live in the city.

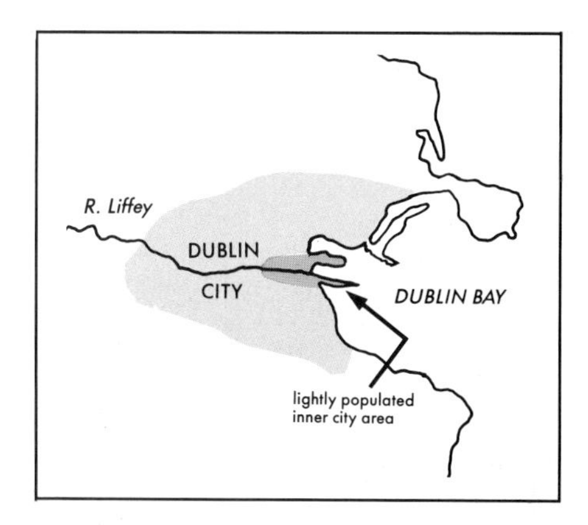

Figure 24.5 Population densities in Dublin city

But remember . . . Not all parts of Dublin city are densely populated. Some inner city areas are now lightly populated because many have left them to live in the suburbs.

Dublin's outer fringe

Immediately outside the city and inner suburbs is ***Dublin's outer fringe***. This area contains suburbs together with some Co. Dublin towns such as Swords and Dun Laoghaire. Its overall ***population density is high and growing rapidly***. This is particularly true in centres such as Tallaght which now contains 75 000 people—a larger population than that of Limerick city.

Two important aspects of growing population densities in Dublin's outer fringe are:

1. The ***migration*** of tens of thousands of people from both rural Ireland and from Dublin's inner city to the outer suburbs. Many of the migrants are young adults whose children further increase the population of the outer suburbs.
2. The growth of ***dormitory towns*** such as Malahide in north Co. Dublin. Dormitory towns develop just beyond the edge of a city. They are inhabited largely by people who work in the city and commute (travel) to and from the city each day.

△
Malahide in North Co. Dublin. Why is Malahide known as a *dormitory town?* Why are such towns experiencing a growth in population?

North Co. Dublin

The overall ***population density*** of North Co. Dublin is ***low and static*** (unchanging). Much of this area consists of ***fertile farmland***, and Dublin County Council wants it to stay that way. The County Council wants to prevent the Dublin suburbs from sprawling too far into the countryside. So it does not generally grant planning permission for new building developments in many parts of North Co. Dublin. It is largely because of this that the population of this area has remained low and static.

The Dublin Mountains

Only a short distance from Dublin's southern suburbs lie the ***sparsely populated*** Dublin Mountains.

△
Study the picture of the Dublin Mountains. Discuss:

(a) how the height and slope of the land might hinder population growth;
(b) why Dublin County Council does not usually grant planning permission for new houses in this mountainous area.

Population variations through time

Population variations through time are often caused by social and historical factors.

Case Study 1 Brazil

Figure 24.6 Population distribution in Brazil ▷

With the help of your atlas, identify the cities which are shown on the map as 3 million clusters and 1 million clusters.

Figure 24.6 shows the population distribution of Brazil. This distribution is very uneven. ***Most people live close to the east coast. Few people live in the interior region of Amazonia.***

Brazil's uneven population distribution is due partly to ***historical factors***. High population densities in the east of the country stem largely from the fact that the Portuguese, who began to colonise (settle in) Brazil four centuries ago, settled mainly in that area.

- ☐ The Portuguese first arrived along the east coast and set up trading towns in that area. They did almost all of their ***trading by sea***. As more and more Portuguese arrived in Brazil, they continued to settle along this coastal trading area. So the population in that area increased rapidly.
- ☐ The Portuguese also set up huge farms called ***plantations***. Most plantations were near the east coast ports through which crops such as sugar and cotton were exported to Europe. Millions of black Africans were kidnapped and brought to work as slaves on the plantations. Most of their descendants still live in eastern Brazil. They make up a large portion of Brazil's population.

- Some of the coastal trading towns eventually grew into ***large industrial cities*** such as Sao Paulo and Rio de Janeiro. As people flocked to these cities in search of employment, the populations continued to grow and grow.

In more ***recent times, further changes*** have been taking place in the population distribution of Brazil. These changes have been caused partly by the movement of people between one region and another (figure 24.7).

Amazonia's population density has increased since the building of ***large roadways*** through the region. One of these is the Trans Amazonia Highway. With the building of these highways, hundreds of thousands of people have moved into the region. They have cleared the rain forests and taken over the land for activities such as cattle ranching and mining.

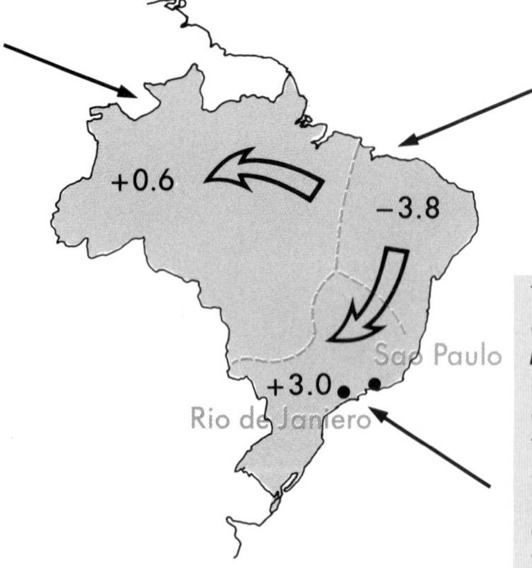

Many people have ***left poor and drought-stricken areas*** of the Northeast region. Most of these have migrated to the South and Southeast. Others have gone to Amazonia.

The population density of the South and Southeast has ***increased rapidly***. This is partly due to the high birth rates. But it has also been caused by the migration of people from the Northeast into cities such as Sao Paulo and Rio de Janeiro. Sao Paulo's population is now increasing at a rate of half a million people per year!

+0.6 General movement of population

Population change in millions since 1950

The building of the Trans Amazonian Highway has increased the population density of this sparsely populated region. But the 'development' brought by the highway has meant the destruction of the rain forests which are the homes of many native forest tribes and countless species of plants and animals. ▷

Case Study 2

The West of Ireland

Population in the West of Ireland has shown two broad trends since population counts began in 1821.

Population grew rapidly up to the time of the Great Famine . . . and has generally declined since that time.

THE GREAT FAMINE (1845-49)

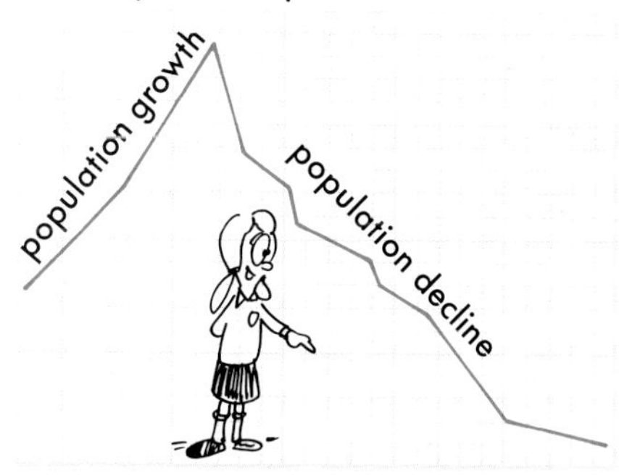

☐ Before the Great Famine

Very high birth rates resulted in rapidly increasing population densities in the West of Ireland. Most people lived on tiny farms and survived on potatoes as their main diet.

☐ The Famine

Between 1845 and 1849, much of Ireland's potato crops were destroyed by 'blight'. Over one million people died of hunger and disease. Another million fled the land and emigrated to foreign countries. Death and emigration reduced population densities, especially in the West of Ireland.

☐ After the Famine

For a century following the famine, population densities continued to decline steadily in the West of Ireland. This was because of a constant stream of migration from the area, both to foreign countries such as the USA and Britain and to Irish urban areas such as Dublin.

☐ In recent years

Overall out-migration has continued from the West of Ireland. But this has been balanced in some places by 'natural increase' in population. As a result, overall population densities have tended to even out. Some western counties experience population growth. Others experience continued population decline.

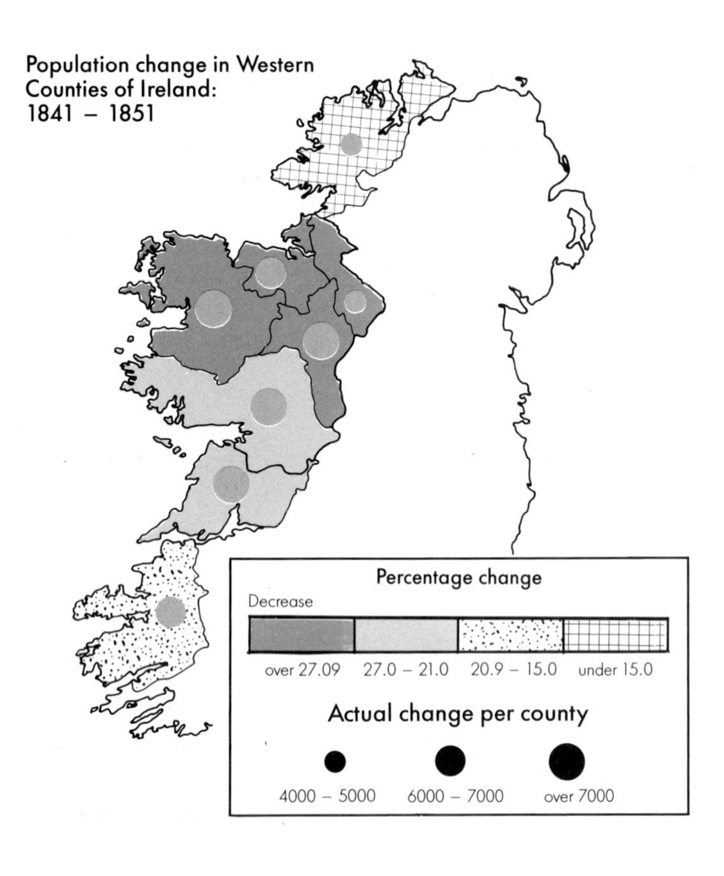

Figure 24.9 Population changes in the western counties of Ireland between 1841 and 1851 (including the period of the Great Famine)

Percentage population change is shown by shading. Actual changes (changes in the numbers of people) are shown by circles of varying sizes.

Study figure 24.9. Then rewrite the following statements, omitting incorrect information and filling in the blanks.

(a) *Some/All* West of Ireland counties had population decreases between 1841 and 1851.

(b) Only Co. ________ had a decrease of less than 15%.

(c) *Four/two/six* counties suffered a population decrease of more than 27%.

(d) Co. Galway had a population decrease of between ________% and ________%.

(e) The 4 counties with the greatest *actual* population decrease were Co. ________, Co. ________, Co. ________ and Co. ________.

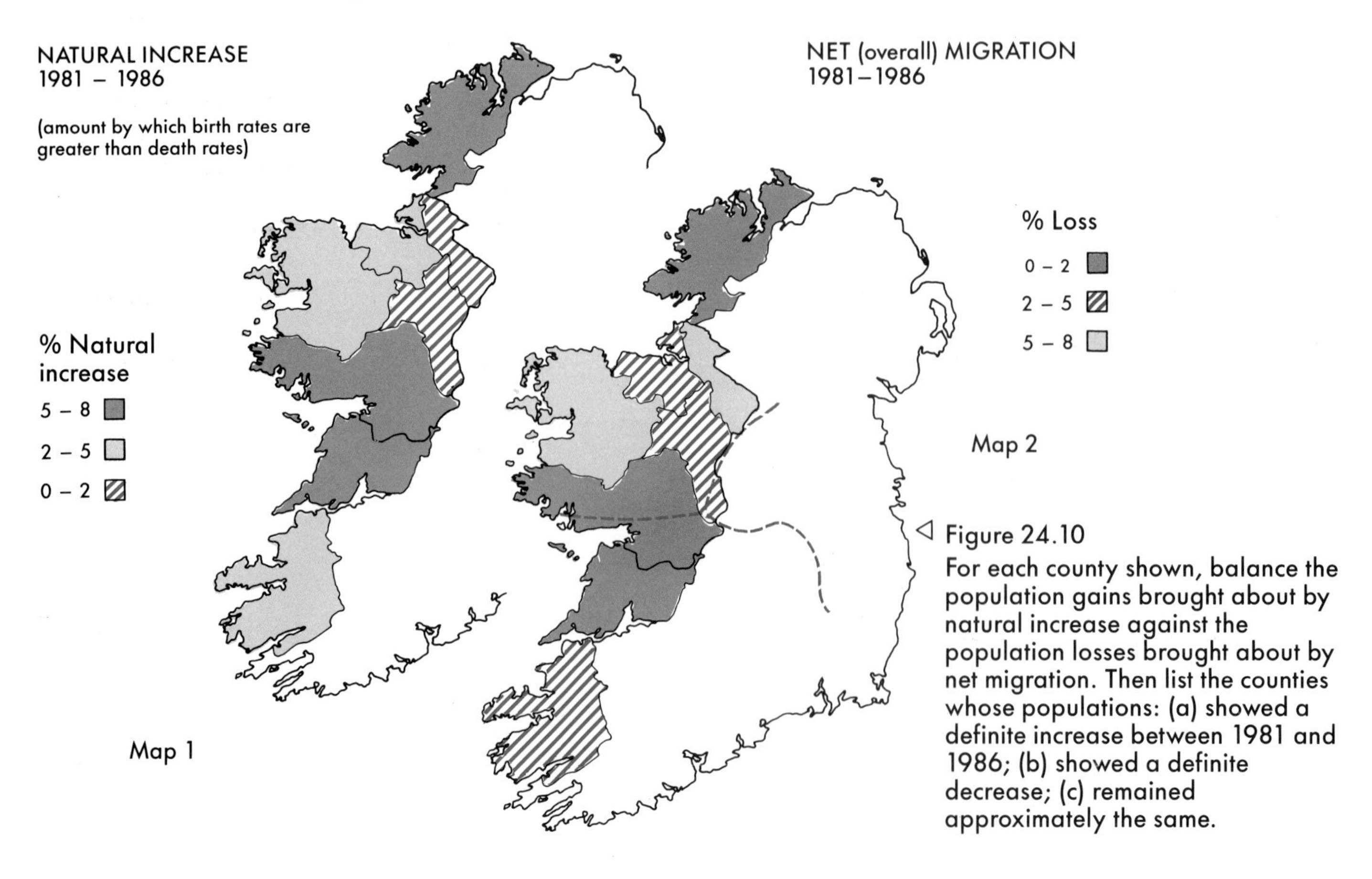

◁ Figure 24.10

For each county shown, balance the population gains brought about by natural increase against the population losses brought about by net migration. Then list the counties whose populations: (a) showed a definite increase between 1981 and 1986; (b) showed a definite decrease; (c) remained approximately the same.

Activities

1. *Map Work*

   Consult the Clew Bay OS map in the Map Supplement which accompanies this book. Study the area which lies east of Easting 00 and south of Northing 70 (near the southeast corner of the map).

   (a) Calculate the population density of this area. To help you calculate the population density, assume the following:

   - ☐ The area covers $50km^2$.
   - ☐ Dwellings in the area contain an average of 3 people.
   - ☐ Schools and churches should be omitted from your building count.

   (b) Describe the distribution of settlement in the area. State briefly the connection between population distribution and each of the following:

   - ☐ altitude (height above sea level)
   - ☐ slope
   - ☐ the presence or absence of roads

2. Study the following illustrations.

   - ☐ Figure 24.11a shows an outline map of the counties of Leinster.
   - ☐ Figure 24.11b gives the percentage rates of population change in each of the counties of Leinster between 1966 and 1971.
   - ☐ Figure 24.11c gives a set of shading codes to be used in symbolising population changes.

   Trace the map in figure 24.11a. On your tracing, show the population change in each county by referring to the information in figure 24.11b and using the shading codes given in figure 24.11c.

3. Explain *why* each of the following statements is true. Write *at least* one paragraph for each of your explanations.

   (a) Climate, relief, soils and communications cause population to be unevenly distributed in Sweden.

   (b) Population density is very high in most parts of Dublin city, but low in part of North Co. Dublin.

   (c) Historical factors have helped to cause uneven population distribution within Brazil.

   (d) The Great Famine had a dramatic effect on the population densities in the West of Ireland.

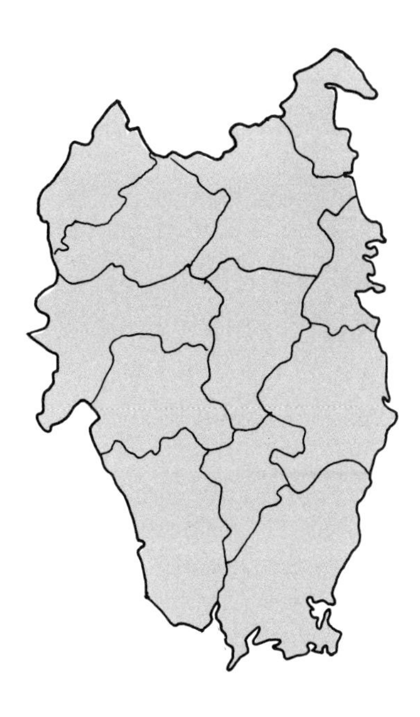

△ Figure 24.11a

County	Change
Louth	+8.2
Meath	+6.5
Dublin	+16.7
Wicklow	+7.5
Wexford	+2.5
Carlow	+2.6
Kilkenny	+1.7
Laois	+0.6
Offaly	+0.1
Kildare	+8.6
Westmeath	+2.2
Longford	−3.5

Figure 24.11b

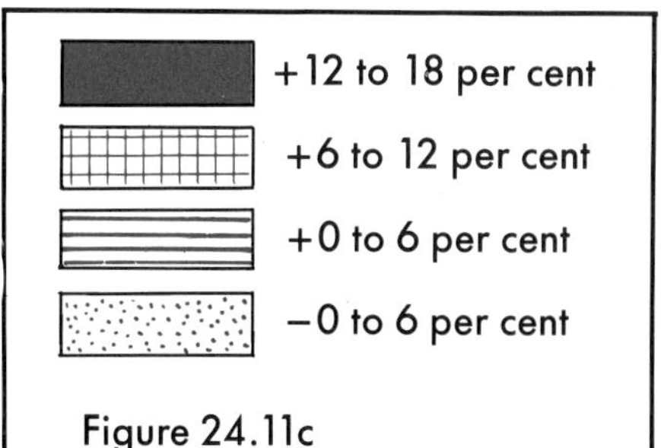

Figure 24.11c

25 POPULATION MAKE-UP

People who study population may need to know certain things about its structure or make-up. They need to know:

- ☐ the ***age structure*** of the population. This is the proportion of people who are in different age groups.
- ☐ the ***sex structure*** of the population. This is the proportion of males to females.

The best way to illustrate age and sex structure is to use a diagram called a ***population pyramid.*** Figure 25.1 shows a population pyramid. Note the following things about it.

- ☐ The pyramid is made up of a series of bar graphs. The bars are laid flat, one on top of the other.
- ☐ The length of each bar shows the percentage of the population (or sometimes the actual number of people) within a certain age group. A scale near the base of the pyramid lets you measure these percentages or numbers.
- ☐ The youngest group (usually 0-4 years) is placed at the base of the pyramid. The next youngest group (usually 5-9 years) is placed on top of this, and so on, until the oldest age group is placed at the top of the pyramid.
- ☐ The bars are divided near the middle by a vertical line. This line separates males from females. Males are always shown on the left of the vertical line and females on the right.

You can make a 'dividers' to help you make these calculations. Use the dividers to measure the length of the population bar required. Then calculate the measurement against the scale near the base of the pyramid.

1. Calculate the percentage of the total population which consists of:

boys between 0–4 years

all children between 0–4 years

girls between 10–14 years

boys between 10–14 years

2. How old are those people in the oldest age group shown in this particular population pyramid?

3. In which population bar would a 13-year-old boy be included? In which bar would the same boy be included fifteen years from now?

4. Why is this diagram called a 'pyramid'?

5. Why does this diagram get narrower towards the top?

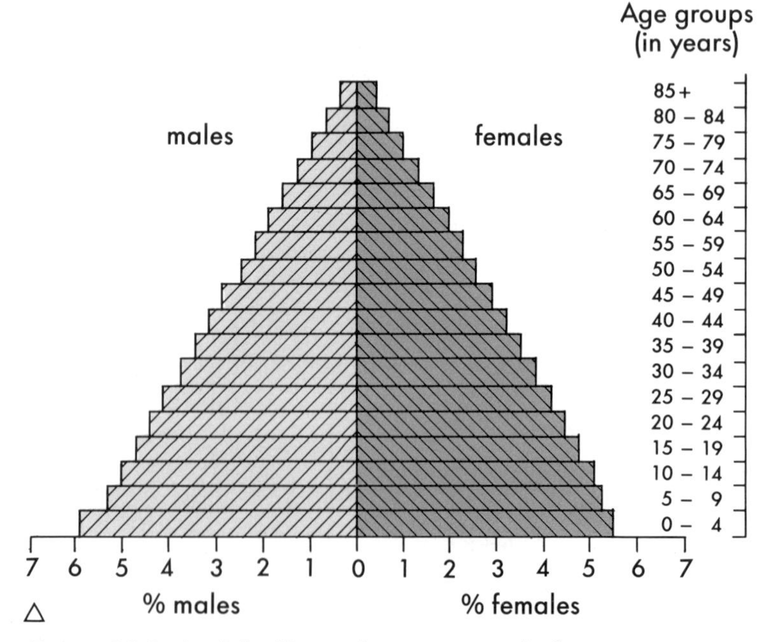

Figure 25.1 An 'ideal' population pyramid of a country

Not all population pyramids are exactly pyramid-shaped, as the one in figure 25.1. Different population pyramids may be of many different shapes.

Not all population pyramids refer to the population of an entire country. Pyramids can also be drawn which show the population structures of a county, a city, a housing estate, a parish or any other population group.

This population pyramid will help you carry out the local population study which begins on page 191.

Look at the population pyramid in figure 25.2. It refers to the population make-up of a street in Midleton, Co. Cork. It was drawn up by a group of local students who carried out a population study of their own area. Notice that this population pyramid differs in several ways from the one in figure 25.1.

- ☐ Each bar shows the *actual number* (rather than the percentage) of people in each age group.
- ☐ Each age group covers a *ten-year age span* (rather than a five-year span).
- ☐ The pyramid has been drawn on *squared paper* so that it can be read more easily.

Study figure 25.2. Then answer these questions.

1. Calculate the total number of people who live in the street referred to in figure 25.2.
2. How many people are under the age of 20?
3. What percentage of the total number of people are under 20 years of age? (see formula in box)

$$\frac{\text{Number of people aged under 20 years} \times 100}{\text{Total number of people}}$$

4. Are there more people between the ages of 10–29 years than there are between the ages of 0–9 years? What does this suggest about birth rate trends in this area?
5. The ratio of males to females in the age group 0–9 years is 18:17 (18 males to 17 females). Calculate the sex ratio (males to females) of the following:
   (a) people in the 10–19 age group
   (b) the entire population of the street
6. On average, Irish women tend to live longer than Irish men. To what extent is this shown in the sex ratio of people aged 70 or over?

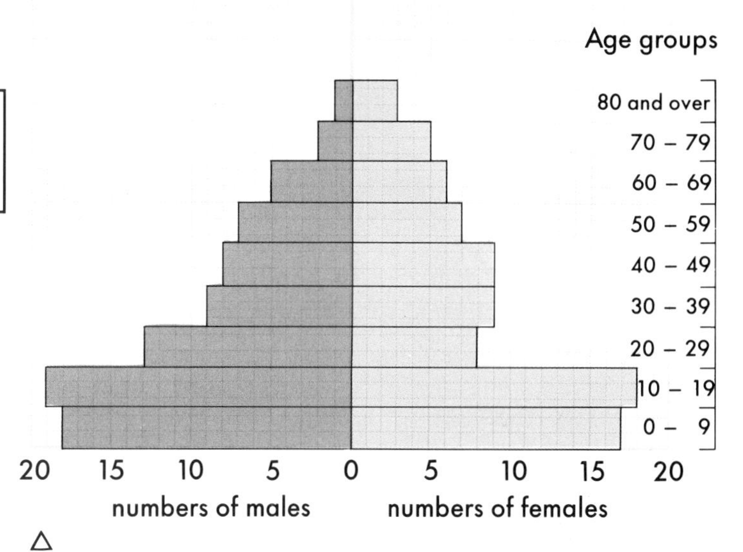

△
Figure 25.2 Pyramid showing age and sex structures of the population of a street in Midleton, Co. Cork

Important idea!

Some places have rapidly growing populations. Others have slowly growing populations.

These two different kinds of places tend to have very different population structures. These are shown using population pyramids with contrasting (differing) shapes.

Case Study 1
Population structures of Brazil and West Germany

Figure 25.3 shows the contrasting population pyramids for Brazil and West Germany.

Brazil is a 'developing' Third World country. It has a rapidly growing population.

West Germany is a 'developed' First World country. Its population has grown slowly in recent years and is now static (neither growing nor declining).

Figure 25.3

Study the differences between the population pyramids of Brazil and West Germany as shown in figure 25.3.

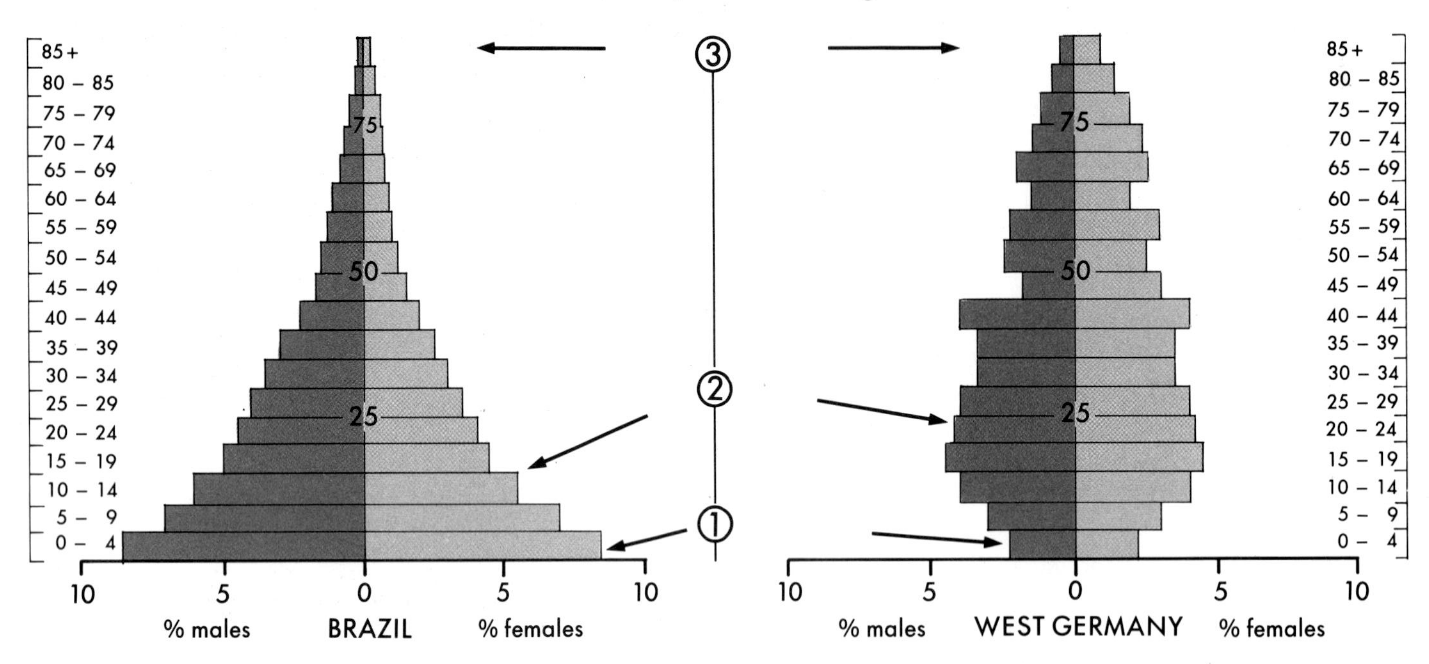

Contrasts between the population structures of Brazil and West Germany

① High birth rates mean that Brazil has a large proportion of children. This causes a ***broad base*** on Brazil's population pyramid.

A broad base is the main sign of a rapidly growing population.

Family planning has led to very low and declining birth rates in West Germany.

Very low birth rates mean that West Germany has small proportions of children. This causes the population base of West Germany to have a ***narrow base***.

A narrow base is the main sign of a slowly growing or static population.

② Death rates among the very young are high in many parts of Brazil. So the ***pyramid narrows rapidly from the bottom up.***

Death rates are low in West Germany. So the ***pyramid narrows slowly***.

③ Relatively few people survive to old age in Brazil. This causes the pyramid to have a ***narrow peak***.

A large proportion of West Germany's population survive to old age. This causes the pyramid to have a ***broader peak***.

Case Study 2

Population structures of housing areas in Tallaght and Inner Dublin

Tallaght	*Inner Dublin*
Tallaght is a 'new town' in Co. Dublin with a rapidly growing population. Most houses are occupied by young couples who have moved into the area from other places. Many of these young couples have young families. Both birth rates and the proportion of young children in the area are very high. Because most housing is so new, very few people in the area have reached old age. There is a small proportion of elderly people in the area.	***Dublin's inner city*** contains several old housing areas which have populations which are slowly-growing, static or even falling. Many houses in these long-established housing areas are occupied by older people. Their families have grown up and left the area, often for new houses in the suburbs. In these inner city areas, the proportion of older people is high. The proportion of young married couples is low. Because of this, birth rates are also low.

◁ This housing estate in *Tallaght* contains a high proportion of young married couples. Birth rates are *high*.

This housing in *Inner Dublin* contains ▷ a high proportion of elderly people. Birth rates are *low*.

Figure 25.4a shows the population pyramid for a new housing estate in Tallaght. Match each of the labels on the pyramid with one of the statements below.

Figure 25.4b shows the population pyramid for an old housing area in Dublin's inner city. Match each of the labels on the pyramid with the appropriate statement below.

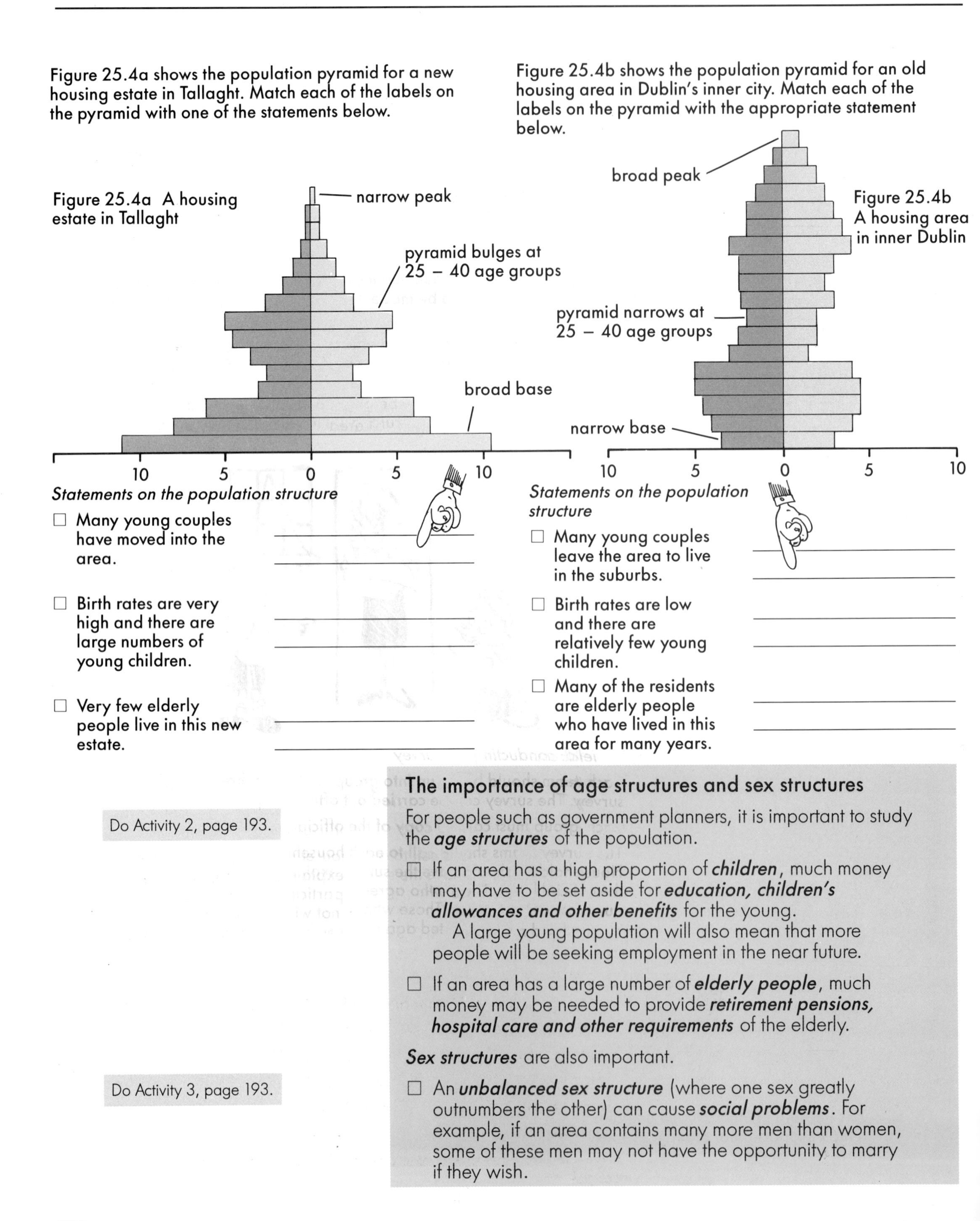

Figure 25.4a A housing estate in Tallaght

Figure 25.4b A housing area in inner Dublin

Statements on the population structure

- ☐ Many young couples have moved into the area. ______________________
- ☐ Birth rates are very high and there are large numbers of young children. ______________________
- ☐ Very few elderly people live in this new estate. ______________________

Statements on the population structure

- ☐ Many young couples leave the area to live in the suburbs. ______________________
- ☐ Birth rates are low and there are relatively few young children. ______________________
- ☐ Many of the residents are elderly people who have lived in this area for many years. ______________________

Do Activity 2, page 193.

Do Activity 3, page 193.

The importance of age structures and sex structures

For people such as government planners, it is important to study the ***age structures*** of the population.

- ☐ If an area has a high proportion of ***children***, much money may have to be set aside for ***education, children's allowances and other benefits*** for the young.
  A large young population will also mean that more people will be seeking employment in the near future.
- ☐ If an area has a large number of ***elderly people***, much money may be needed to provide ***retirement pensions, hospital care and other requirements*** of the elderly.

Sex structures are also important.

- ☐ An ***unbalanced sex structure*** (where one sex greatly outnumbers the other) can cause ***social problems***. For example, if an area contains many more men than women, some of these men may not have the opportunity to marry if they wish.

Activities

1. **Fieldwork – A local population study**

Objectives

- ☐ To make population pyramids based on surveys of population structures in your local area
- ☐ To examine and compare the population structures shown by these pyramids.

In class: preparation

- ☐ The class should be divided into teams, ideally with four to five members each. Each team should be made up of students from the same general area or neighbourhood.
- ☐ Your teacher should draw up an official letter explaining the project and requesting the co-operation of the people surveyed. Each team should have a copy of this letter.
- ☐ Each team should choose (or be given) a local street, small housing estate, small village or small rural area in which to carry out the population survey.

In the field: conducting the survey

- ☐ Each team should break up into groups of two or three to conduct the survey. The survey can be carried out after school or at the weekend.
- ☐ Each group must carry a copy of the official letter.
- ☐ The survey teams should call to each household in their survey area. Householders should have the survey explained to them and be shown the official letter. Those who agree to participate should be asked the questions which follow. Those who do not wish to participate should be thanked and not contacted again. Remember that their privacy must be respected.

Survey questions

1. How many males are in your household within each of these groups?
   - ☐ under 10 years
   - ☐ between 10–20
   - ☐ between 20–30
   - ☐ between 30–40
   - ☐ between 40–50
   - ☐ between 50–60
   - ☐ between 60–70
   - ☐ between 70–80
   - ☐ over 80 years
2. How many females are in your household within each of these same age groups?

Figure 25.5 Survey sheets used to record population structure of a street in Midleton (see figure 25.2) ▷

☐ Record the information gathered on a survey sheet similar to the one in figure 25.5. Record each member of each household by making a tick in the appropriate space on the sheet.

At the end of your survey, calculate the total numbers of males and females in each age group. Record these totals on your survey sheet.

Area surveyed ______________________

Survey team ______________________

Survey dates ______________________

Age group	Males	Females	Age group
80+	I (1)	(3) III	80+
70–80	II (2)	(5) HHT	70–80
60–70	HHT (5)	(6) HHT I	60–70
50–60	HHT II (7)	(7) HHT II	50–60
40–50	HHT III (8)	(9) HHT IIII	40–50
30–40	HHT IIII (9)	(9) HHT IIII	30–40
20–30	HHT HHT III (13)	(8) HHT III	20–30
10–20	HHT HHT HHT IIII (19)	(18) HHT HHT HHT III	10–20
0–9	HHT HHT HHT III (18)	(17) HHT HHT HHT II	0–9

The population pyramid in figure 25.2 can be used as your model.

Back in class: making and examining population pyramids

Making the pyramids

Each team should come together once back in the classroom. On pieces of squared graph paper, each student should make a population pyramid for the area which his/her team surveyed.

Each population pyramid should be given a title naming the area represented by the pyramid.

Studying individual pyramids

Using the information on his/her survey sheet and the population pyramid, each student should calculate the following:

1. the total number of people in the area surveyed.
2. the number of people in the area surveyed who are under 20 years of age.
3. the percentage of people under 20 years of age.
4. the number of people who are over 70 years of age.
5. the percentage of people who are over 70 years of age.
6. the overall ratio of males to females in the area surveyed.

Students should check their calculations with their team mates. The correct answers should then be written beneath each population pyramid.

Comparing pyramids

Copy the population pyramids so that each team now has a copy of all the pyramids drawn. Compare all the pyramids to discover:

1. which place has the highest percentage of young people under 20.
2. which place has the highest percentage of elderly people over 70.
3. which place has the most even sex ratio.
4. which place has the most uneven sex ratio.
5. which place (if any) has a population structure which is very different from the others. How does this particular place differ from the others?

2. Write an account outlining the differences between the population structures of a housing estate with a rapidly growing population and that of a housing area with a static population.

3. Examine the population pyramids in figure 25.4a (a housing estate in Tallaght) and figure 25.4b (a housing area in inner Dublin). To which of the two areas would each of the following statements apply? Explain your choices.
   (a) Large amounts are paid out in this area for retirement pensions.
   (b) Continued rapid population growth can be expected in the near future.
   (c) Some schools may have to close because of falling pupil numbers.
   (d) Larger primary schools may be needed because of growing pupil numbers.
   (e) There would probably be a high demand for baby sitting services in this area.
   (f) Medical services for the elderly are necessary in this area.
   (g) Larger post-primary schools will be needed here in the near future.
   (h) New playschools should do well in this area.
   (i) The population of this area is growing rapidly.

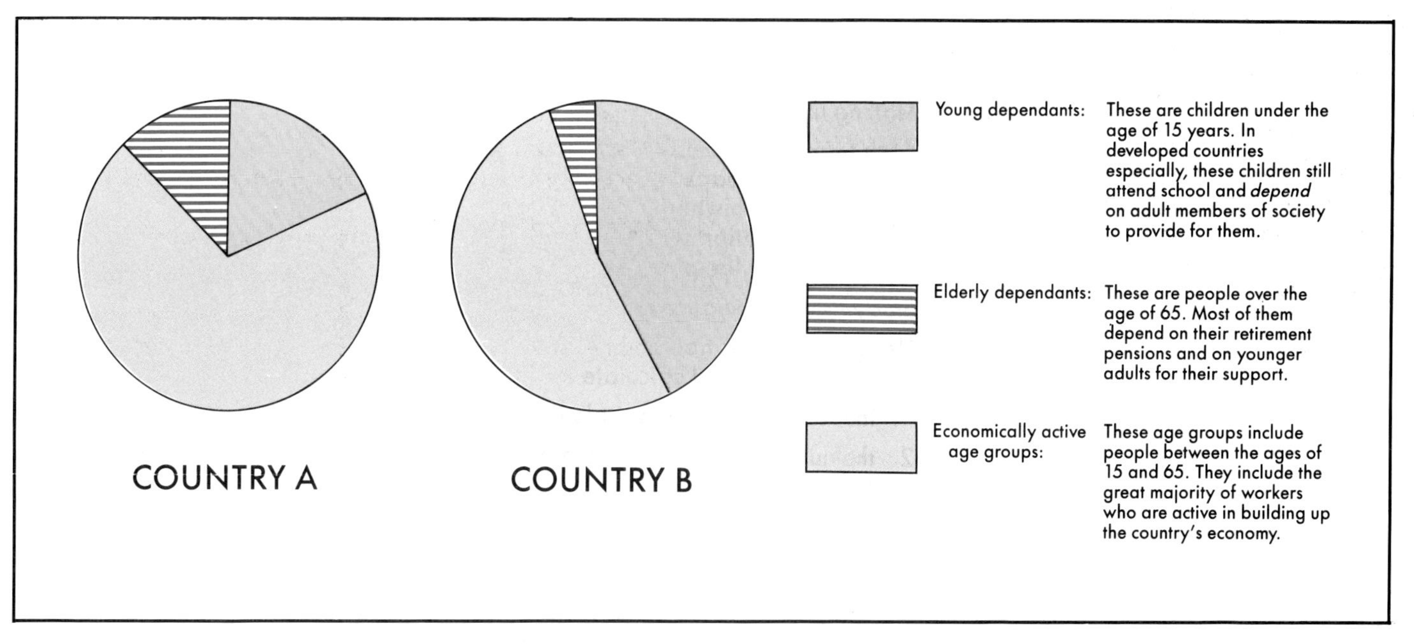

△ Figure 25.6

4. The pie charts in figure 25.6 show the population structures of Country A and Country B. One of these countries is Brazil and the other is West Germany (although they do not necessarily appear in that order). Study the pie charts and answer these questions.
   (a) Which country has the greater proportion of young people?
   (b) Which country has the greater proportion of elderly people?
   (c) Which country is likely to have the higher birth rate?
   (d) Which country is likely to have the higher death rate among young people?
   (e) Which chart represents West Germany?
   (f) Offer some explanations about why one of the countries shown has a higher birth rate than the other country.

5. Study this photograph and then answer the following questions.
   (a) In which of these countries was this photograph taken: Sweden; West Germany; Brazil; Nigeria?
   (b) Name *one economic activity* carried out by the family. Refer to the photograph to support your answer.
   (c) Attempt to explain why many people in Third World countries have large families.
   (d) How are large families such as the one in the photograph likely to affect the following: (i) the shape of the population pyramid representing their area; (ii) the need for social services in the area; (iii) the economic wealth of the family itself.

26 SOME EFFECTS OF HIGH AND LOW POPULATION DENSITIES

The population density of an area can greatly affect that area's economy and social conditions.

These problems are signs that an area is ***over-populated*** – it contains more people than its resources can cope with.

Very high population densities may result in:

- ☐ overcrowding
- ☐ lack of open spaces
- ☐ shortage of clean water
- ☐ pollution

Very low population densities may result in:

- ☐ low marriage rates
- ☐ abandonment of agricultural land
- ☐ political and economic isolation

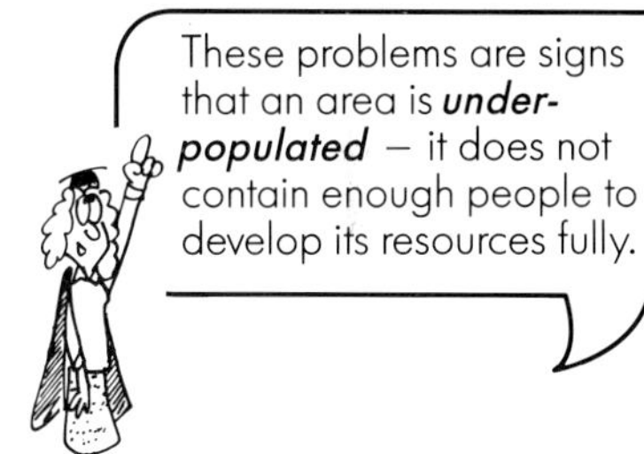

Some effects of very high population densities

Calcutta – A Case Study

Calcutta is India's largest and most densely populated city. It has been described as 'the world's worst city'. Its many problems arise partly from its rapidly growing population, which has doubled in the past 20 years. Calcutta's population is expected to reach 17 million by the year 2000.

A large part of Calcutta's expanding population is made up of poverty-stricken migrants who flock from rural areas into the city in search of employment. These migrants usually crowd into the large urban slums called ***shanty towns*** or ***bustees*** which have grown up around the edges of the city.

The people of Calcutta – and especially those who live in the bustees – face many serious problems. Some of these problems are outlined on page 196.

Figure 26.1 The location of Calcutta ▷

(a) What is a river delta? (see Chapter 7).
(b) On which river delta is Calcutta situated?
(c) What is a distributary? (see Chapter 7).
(d) On which distributary is Calcutta located?

△
The centre of Calcutta. Describe the scene. Why do you think Calcutta is sometimes called 'the nightmare city'?

Overcrowding

Calcutta's population density is almost 35 000 per km^2. It is one of the most overcrowded cities in the world.

- ☐ Many people live in poorly built and ***overcrowded houses***. Fewer than one in four homes in Calcutta has a separate kitchen. Less than half have washrooms.
- ☐ Many bustee dwellers live in ***temporary huts*** made from pieces of timber, canvas and plastic sheeting. Most huts are shared by several families.
- ☐ It is estimated that over half a million people — most of whom have migrated to the city — have no dwellings at all. They are the ***'pavement people'*** who live and sleep in the streets.

Lack of open space

Very high population densities have led to a severe lack of open space in Calcutta. In the poorest areas of the city, almost every available scrap of land has been occupied by people. Even places which were once well outside the city have now been swallowed up by the ever-growing, overcrowded bustees. Calcutta and its bustees now stretch, almost without a break, for over 50km along the banks of the River Hooghly.

△
'Pavement people' bathe on the streetside in Calcutta. The water they are using is unfiltered and meant only for firefighting.

Shortage of clean water

Calcutta's water supplies present many serious problems.

- ☐ The River Hooghly supplies most of the city's water. But this supply becomes ***salty*** when tidal water occasionally flows into it from the Bay of Bengal.
- ☐ The ***pumps*** which help to filter (clean) the water and then pipe it into the city are seriously overworked. These pumps often break down, leaving the city without clean water for a day or two at a time.
- ☐ Up to 30% of Calcutta's people have no clean water at all. These people are forced to drink and cook with ***unfiltered water*** supplies which were originally intended only for street cleaning and fire fighting.

Pollution

The poor of Calcutta must live with many types of pollution.

- ☐ ***Noise pollution*** is almost constant in the overcrowded streets and bustees.
- ☐ ***Visual pollution*** is caused by the many ugly, hastily-built buildings.
- ☐ A shortage of proper ***sewerage*** facilities causes serious pollution. In the many areas without proper sewerage, toilet and kitchen waste flow away through open drains. Heavy rains sometimes flood these drains, flushing the waste onto the streets and into people's homes.
- ☐ ***Garbage and litter*** pile up in the streets. The Indian National Volunteer Force sometimes has to clear paths through this rubbish so people can move around.

1. Discuss the information given in this newspaper report on each of the following aspects of Calcutta.
   - ☐ the streets
   - ☐ pollution
   - ☐ pavement dwellers
   - ☐ the buildings
   - ☐ lack of clean water

2. The article refers to Mother Teresa. Find out what you can about her.

Mother Teresa △

A visit to Calcutta

The following report was published in the *Irish Times* in February 1988.

The Indian Airlines plane which brought us to Calcutta was four hours late. We arrived at one o'clock in the morning. Yet the lighting at the airport was sparse and funereal, and there was the smell — a pungent compound of rotting vegetation and water polluted with the effluent of the millions who make Calcutta possibly the most overcrowded place on earth. . .

The city, when it began to materialise, was even less reassuring — roaming packs of dogs, wattle hovels, shabby buildings, potholes, dimly lit by sodium lamps like slim neon tubes.

Then the 'bodies' began to appear, at first singly, then in large groups. There were people sleeping in doorways, curled up under walls, in rows along the pavements. Some had rectangles of material to cushion them. Many had not. They were refugees from Bangladesh, perhaps, or simply exiles from country villages who came to Calcutta for work. A lucky few, I had been told, might eventually be picked up to die in the care of Mother Teresa. Some might even make it into the congestion of the bustees, the registered slums of the city.

Our hotel was slightly faded, but its comfort would have been palatial to the people I could see, from my window, sleeping rough across the street. Despite the tiredness I could not sleep. I kept thinking about them and wishing guiltily I was thousands of miles away. An official apologised next morning. The best hotels had been booked out for the World Cup cricket matches.

By daylight, Calcutta seemed even shabbier. In the noisy, throbbing streets, people were washing themselves in the muddy gutter water or cooking over fires on the pavements.

Some possible solutions to Calcutta's problems

A. Develop ***farming*** and ***small rural industries***. This will lessen the movement of people from the countryside to the city.

B. Encourage ***family planning***. This will slow down population increases.

C. ***Clear away*** entire shanty towns and ***rehouse*** their inhabitants in large, low-cost apartment blocks. This has been done in Hong Kong, where thousands of people are housed in each block. Each family is given one room, with a toilet and washroom attached.

D. Help and encourage the people of the bustees to ***improve and upgrade*** their own shanty towns. This has been done in African cities such as Lusaka in Zambia. There, the City Council has combined with the dwellers of the shanty towns to improve dwellings and toilet facilities and to provide street taps with clean water.

Discuss!

Study solutions C and D. Which of these two solutions do you think would be more appropriate in solving the problems of Calcutta's bustees? Before answering, consider which of the two solutions would be more likely to:

- ☐ provide more sturdy housing
- ☐ encourage better community spirit and self-help programmes within the bustees
- ☐ provide better toilet facilities and water supplies
- ☐ be less expensive
- ☐ satisfy the bustee dwellers themselves

Some effects of very low population densities

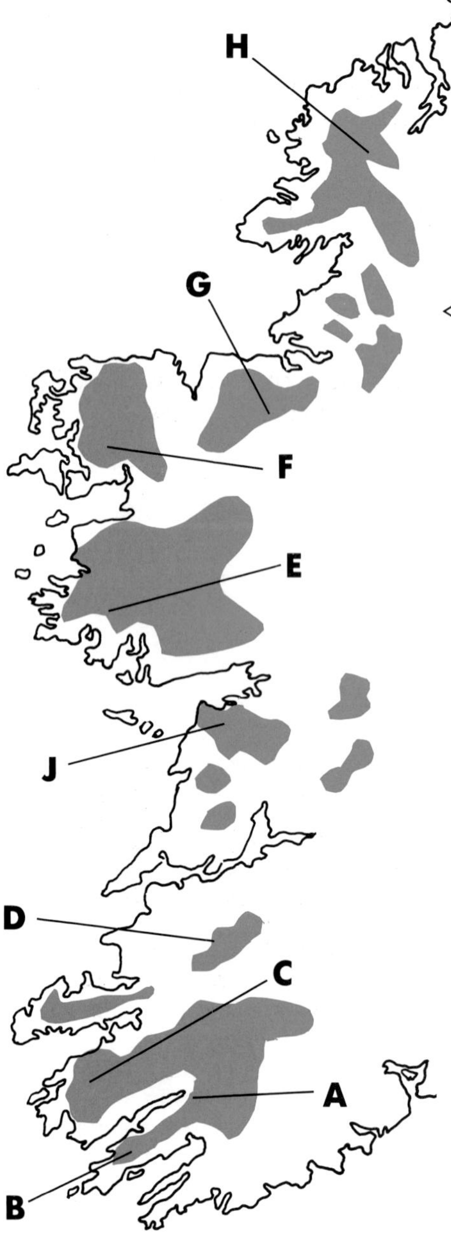

Case Study 1

Rural areas in the West of Ireland

Many rural areas in the West of Ireland have very low population densities (figure 26.2). They have been affected by this in a number of ways.

◁ Figure 26.2 Some areas of very low population densities in the West of Ireland.

Most of these areas are mountainous, while some others contain large areas of bog or exposed limestone.

Name the highland areas labelled A – H. Name the karst (exposed limestone) region labelled J. Name the county in which each of the areas A–J is located.

Low marriage rates

Many ***lightly populated rural areas*** offer few job opportunities. There are also fewer discos, cinemas and other social facilities for young people. These things cause many young, single adults to ***migrate*** from such areas to towns and cities in search of employment and enjoyment.

The loss of many marriageable young people results in ***low marriage rates*** in rural areas. This, in turn, leads to low birth rates and a further decline in population.

A continuous cycle then develops in which low population densities and low marriage rates both cause, and are caused by, each other (figure 26.3).

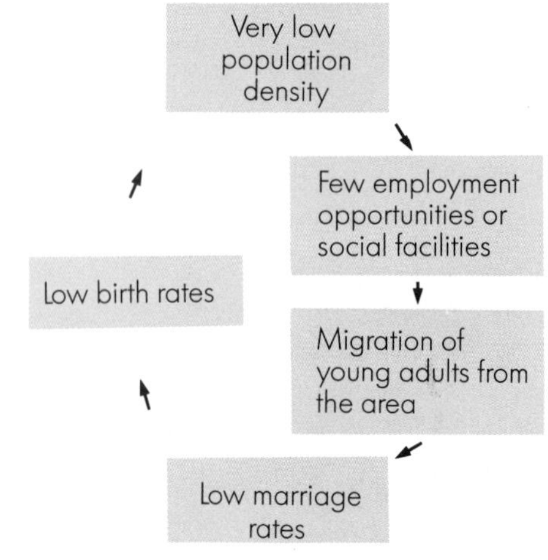

Figure 26.3 A low population density/low marriage rate cycle

Why do you think a situation such as this is called a cycle?

See Activity 3, page 204.

Abandoning agricultural land

Agricultural land has been either neglected or completely abandoned in some lightly populated areas in the West. Here are a few reasons for this.

- Out-migration and low marriage rates have caused a decrease in the population densities of these areas. As this happens, ***fewer people*** are available to work the often difficult and hilly land. Some farmland then becomes neglected or abandoned.
- The people who migrate from these areas are often the energetic, the ambitious and the young. Many farms are left in the care of ***older people*** who do not always have the ability to work them fully.
- European Community (EC) ***farm policies*** favour the reduction of farming in poor, ***marginal land*** and the increased use of that land for recreational or other purposes.

Political and economic isolation

Geographers sometimes divide countries into the prosperous areas called ***core regions***, and the less prosperous, more isolated areas called ***peripheral regions***.

The lightly populated areas of the ***West of Ireland*** do not always share in the wealth and power of the Dublin area. Partly because of their very low population densities and partly because of their distance from Dubin, these areas tend to be ***isolated from the economic and political core of the country.***

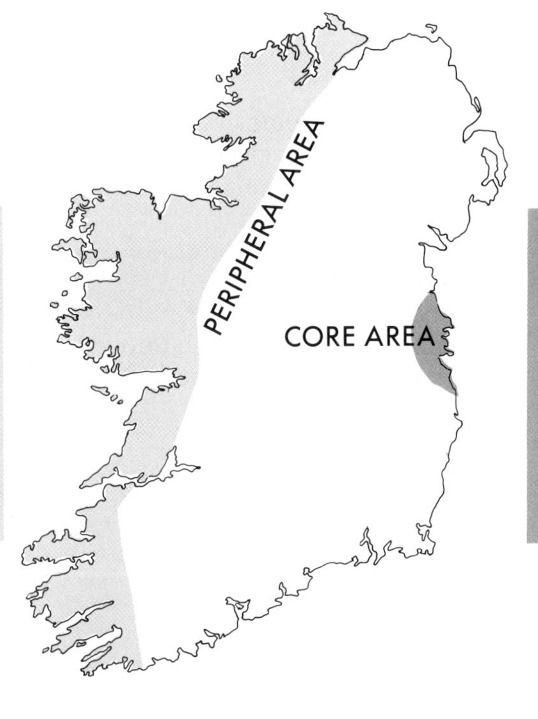

The densely populated ***Dublin area*** can be regarded as the political and economic ***core*** or centre of Ireland. It is the place where our government meets and where many industries, banks etc. have their headquarters. It is the place where important decisions are made regarding the economy and the politics of the country. It is therefore a prosperous and powerful area.

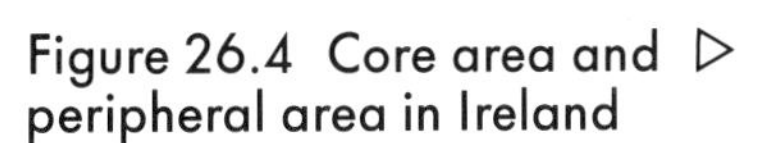
Figure 26.4 Core area and ▷ peripheral area in Ireland

An old man from the West

Read the following interview with a man who lives in a rural parish in the West of Ireland. Then answer the questions.

1. What evidence does this passage present of declining agriculture in Loughfadda? What reasons(s) are suggested for this decline?
2. According to the passage, why does Loughfadda have a low marriage rate?
3. 'Politicians and industrialists don't even know we exist.'

   What reason is given in the passage for the political and economic isolation of Loughfadda?

   What are the results of this political and economic isolation?
4. Suggest a reason for the placename, Loughfadda.

Sean O'Sullivan, approaching his seventieth birthday, looked out over the beautiful but partly abandoned hill slopes of his native Loughfadda. Ancient fields clung to the lower levels of the rocky hill. But many of these fields had clearly been abandoned by farm animals and crops. A few had been planted with young conifers; many more were being conquered by furze and heather. Three former farmhouses occupied the now quiet hill slopes. Two had fallen into ruin. The third, just above the shore of the ribbon-shaped lake, was being converted into a holiday home by a family from Galway city.

'What we really need in Loughfadda is more young people,' said Sean. 'Old fellows like me,' he added wistfully, 'no longer have the energy or strength to work these hill farms.'

Sean explained that many young people are leaving the area. Some go to work in Dublin, while others emigrate to England or America. The girls especially are inclined to go, and the few young men who stay find it difficult to find wives. There had not been a wedding in the little local church for almost two years.

'The trouble is,' said Sean, 'that lightly populated areas like this one get no new industries or social amenities. The politicians in Dublin and Brussels don't pay any attention to the likes of us. Politicians and industrialists don't even know we exist. And no wonder. After all, there are only 300 people left in the entire parish.'

An abandoned farmhouse in the West of Ireland. Why has some land in the West of Ireland been abandoned by its former owners? To what new use has the farmland shown in this picture been put?

Case Study 2

Mali, a country in West Africa

Figure 26.5 Mali and its location ▷

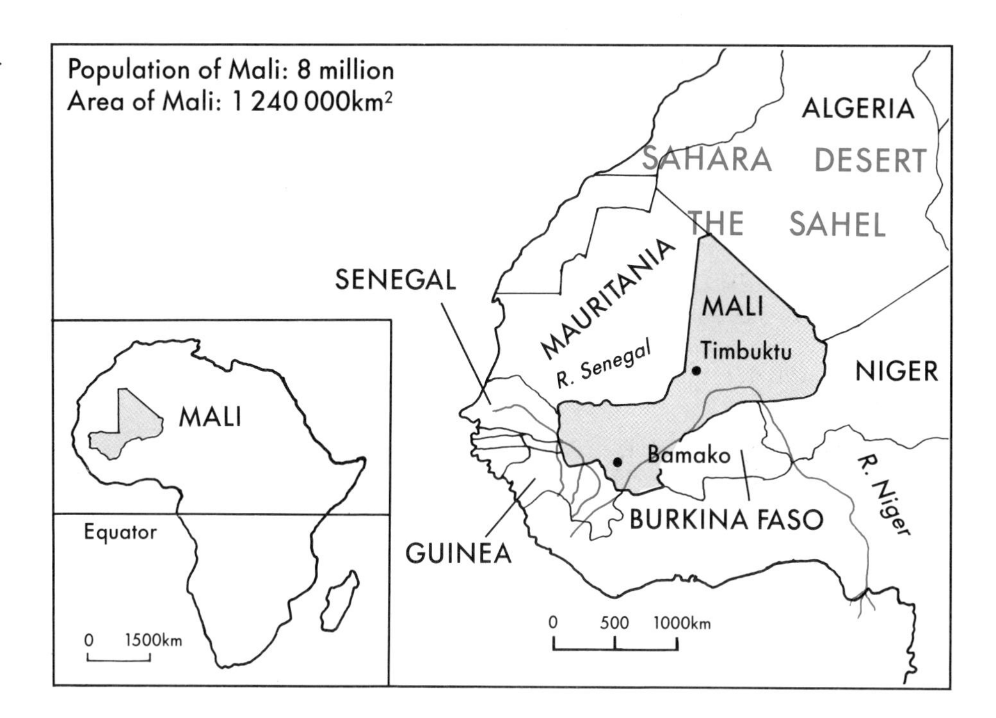

An introduction to Mali

The Republic of Mali is a landlocked (inland) country in West Africa. To the north of Mali lies the country of A_____. To the east lies N_____. Other countries bordering Mali include M_____, S_____ and G_____.

Mali is a hot and generally dry country. The northern part is especially arid (dry). This makes up part of the _______ desert and part of the area known as the S_____ which borders the desert. More rainfall occurs in the south and southwest, where the rivers _____ and _____ dominate drainage.

Mali was once a colony of (was ruled by) France. It became an independent republic in 1960. It continues, however, to be one of the world's poorest nations.

Many of Mali's people are nomadic tribes such as the Fulani and the Taureg. Nomads move from place to place in search of water and pasture for their herds of cattle, goats and sheep. Other groups of people such as the Bambara live mostly as settled farmers.

Mali is a huge country covering about ____km². But its population is a mere _____. This means that the country has a very low average population density of just over ____ people per km². Most people live in sparsely populated rural areas. A small number live in cities such as the capital, B_____, and T_____.

Marriage rates in Mali

Marriage rates in Mali could not be called low. But they have been reduced by the following factors.

1
Many young adults are no longer content to live in remote and very lightly-populated rural areas.

2
Some migrate to the cities in search of work, education and other advantages. As in many parts of Africa, it is mainly the young men who migrate.

3
This results in a 'surplus' of young women in the countryside and a 'surplus' of young men in the cities. . .

. . . and reduces marriage opportunities and marriage rates in both areas.

Abandonment of agricultural land

Over the past ten years, thousands of country people have abandoned very lightly populated land in northern and central Mali. Almost all of them have come to live in or near cities such as Timbuktu. There are two main reasons for the abandonment of agricultural land in Mali.

☐ Almost all health and education facilities, as well as political power, are concentrated in the cities. Because of this, city people are often ***seen as being better off*** than the peasants who are spread out over a vast, dry countryside. Believing this to be so, many country people abandon their isolated lands and move to the cities in search of better lives.

☐ Since the 1970s, ***terrible droughts*** have occurred in that part of Mali known as the Sahel. Vast areas of former grazing land have become desert. Entire animal herds have been wiped out by hunger and thirst, forcing the herd owners to leave their lands. Some have moved to the edge of Timbuktu where they hope to cultivate crops with the help of aid from overseas. (See 'Trocaire in Timbuktu', page 203).

For more about the Sahel, see Book 2, 'Losing to the Deserts: Desertification'.

1. (a) Nomads are mentioned in the first paragraph. What are nomads? Name two nomadic peoples in Mali.
   (b) Which animals are herded by the nomads?
   (c) Why did these nomads traditionally visit Timbuktu from time to time?
   (d) Timbuktu is described as an 'oasis city'. Find out what this means.

2. (a) How has the long-term drought affected the traditional lives of the nomads?
   (b) Why have some former nomads decided to settle near Timbuktu?
   (c) What has Trocaire been doing to help these former nomads?
   (d) Suggest a reason why 'large numbers of trees' have been planted by the former nomads.

▷ **Trocaire in Timbuktu**

Trocaire, the Irish Development Agency, has been helping some Malian country people who have been forced to abandon their land. Read the following passage and then answer the questions.

For centuries, nomadic tribes, with their vast herds of goats and sheep, have roamed through the Sahel and along the edge of the Sahara. From time to time they came to Timbuktu and other oasis towns to sell their livestock and to buy food and other supplies.

But the tens of thousands of nomads camped around Timbuktu today are not there to trade — they have nothing left to trade with. A decade of severe drought has decimated their herds. Families who only a few years ago owned six or seven hundred goats and sheep have seen every single one of their animals die.

Most of these people will never return to their nomadic way of life. And that is why Trocaire is helping them to settle down near Timbuktu on the banks of the River Niger.

At the moment there are nearly 2000 families — divided into 47 groups — who hope to be able to support themselves by cultivating vegetables and maize around the water points.

Already they have cultivated some small areas and planted large numbers of trees. Now they are ready — given a good rainy season — to farm more extensively. Trocaire is helping to supply the water pumps, agricultural tools and seeds they need to take advantage of the rains due around now.

(from *One World*, October 1985)

Political and economic isolation

- ☐ Eight hundred years ago, Mali was the centre of a large and thriving empire. Its merchants traded in gold throughout West Africa. Its ambassadors could be found as far away as Egypt. But Mali's population was not large enough to control an empire. Largely because of this, the empire collapsed in the 15th century. Then, in the 19th century, French armies conquered Mali and made it part of their empire in Africa. Mali's scattered population could do little to resist the French. Once a political power, Mali then became an ***isolated colony*** in the French empire.
- ☐ Because Mali is very poor and lightly populated, its transport networks (roads and railways) are poorly developed. The northern part of the country has no railway lines and only one major road.

Shortages of communications and people keep Mali ***economically isolated***. For example, the country contains many minerals, including iron ore and gold. But few of these minerals are mined because they are located in isolated places which lack people and transport routes.

Activities

1. Imagine you are a newspaper reporter who has just returned from Calcutta. Write an article for your newspaper describing, *in your own words*, the effects of high population density on the lives of Calcutta's poor. Mention overcrowding, shortage of clean water and pollution.

2. (a) *List* three effects of very low population densities on some rural areas in the West of Ireland.

   (b) *Describe* in detail any two of these effects.

3. *Marriage rates in Ireland*

   The map in figure 26.6 shows marriage rates in Irish counties. Study the map and then answer the questions which follow.

   (a) This passage refers to information given in figure 26.6. Rewrite the passage, omitting the incorrect alternatives.

   Marriage rates in Ireland tend to be *higher/lower* in western counties than in eastern ones. Co. Dublin, for instance, has a marriage rate which is *4.5%/over 4.5%* higher than that of Co. Galway.

   The lowest marriage rates occur in the *west and northwest/west and southwest* of the country. This area includes counties *Galway, Clare and Kerry/Galway, Mayo and Donegal.*

   Connaught is the *district/province* with the lowest marriage rates. *Most/All* counties in Connaught have marriage rates of less than 50%.

   No county in the west coast has a marriage rate of *over 50%/over 51.5%*. No county on the east coast has a marriage rate of *less than 53%/less than 54.5%.*

   (b) Make lists of: (i) the counties with the lowest marriage rates which touch the west coast; (ii) the counties with the lowest marriage rates which do not touch the west coast.

   (c) Lightly populated western counties tend generally to have lower marriage rates than those which occur in the more heavily populated eastern counties. Suggest the connections between low population densities and low marriage rates.

Percentages of people of 15 years of age and older who are married

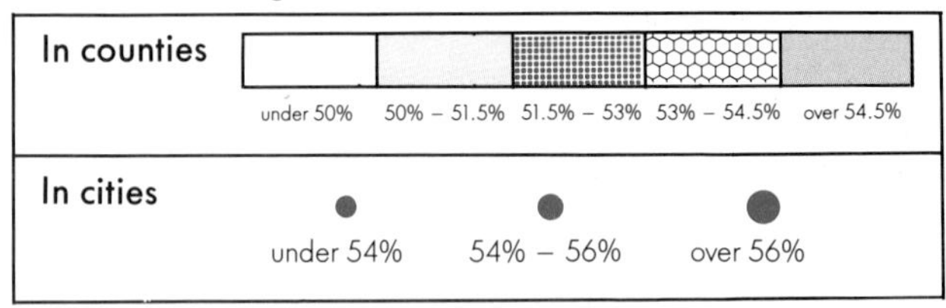

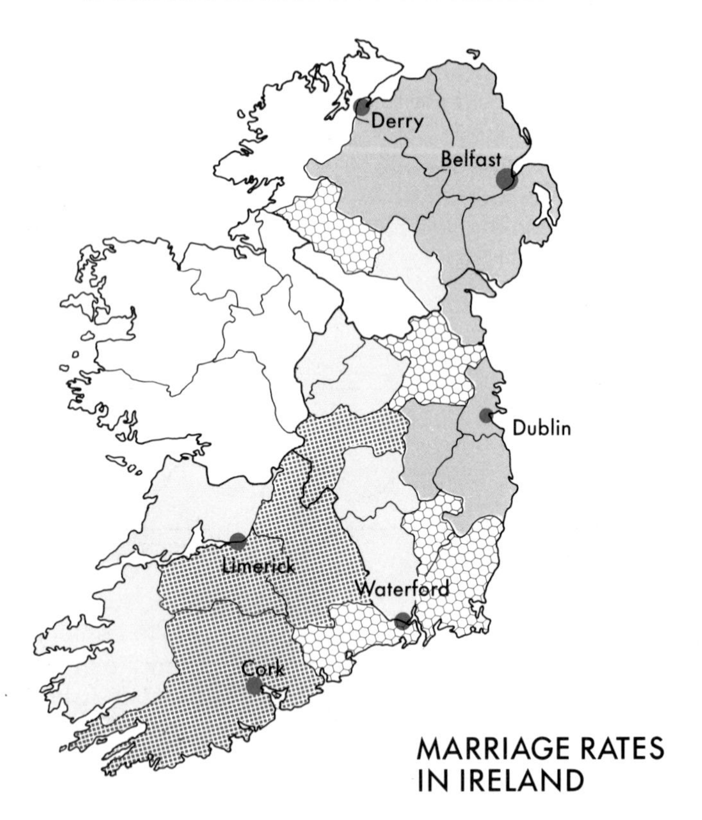

△
Figure 26.6 Marriage rates in Ireland, by county

27 LIFE AND DEATH IN AN UNEQUAL WORLD

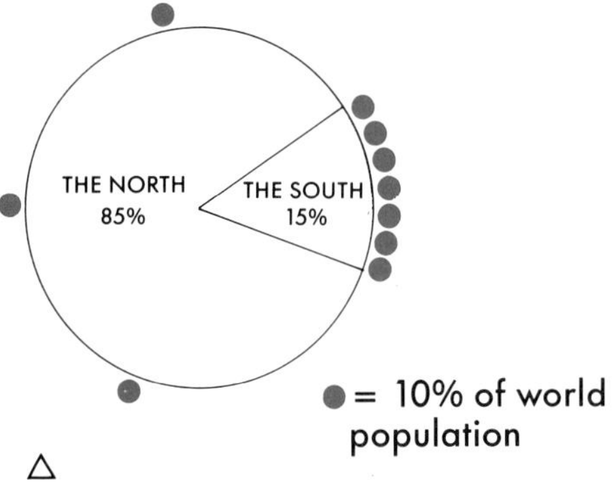

△
Figure 27.1 How the North and South share the world's wealth

What does this pie chart show about the divisions of wealth and population between the North and the South?

In which region is a person more likely to be poor? Why?

An unequal world

One of the greatest problems facing our planet is the terrible difference which exists between the wealth of some parts of the world and the extreme poverty of others. It can be said that the earth is divided into ***two broad regions*** — the ***North*** and the ***South***.

- ☐ The ***wealthy region*** of the world is called ***the North***. It consists mainly of the rich countries of North America and Western Europe (including Ireland), Australia and Japan, along with the better-off communist countries of Eastern Europe.
- ☐ The ***poorest countries*** are referred to as ***the South*** or the ***Third World***. Most of our human family lives in the South.

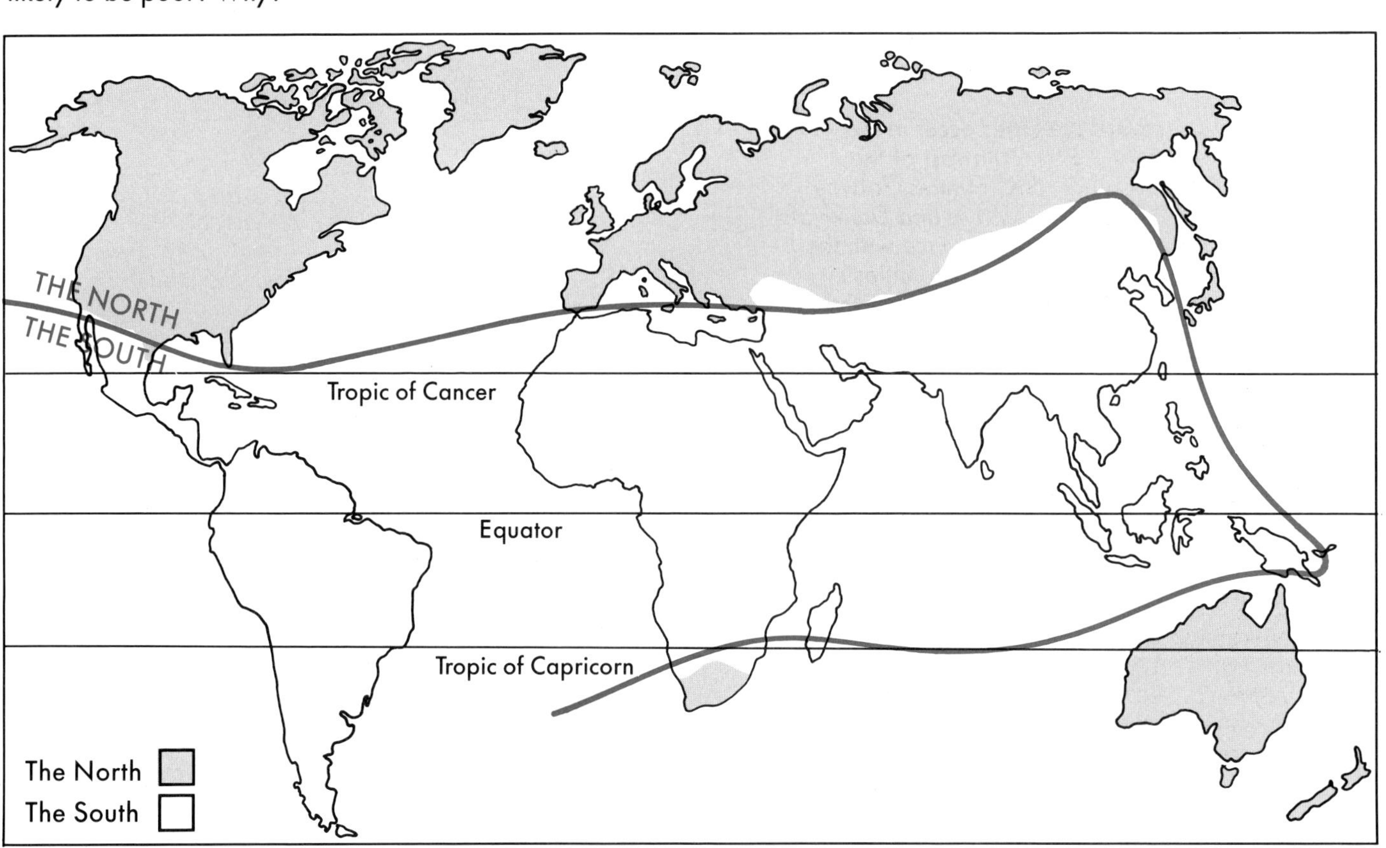

△
Figure 27.2 The world divided into the generally rich North and the generally poor South

Most of the world's richer 'developed' countries lie to the north of the Tropic of Cancer. This is why this richer part of the world is called the 'North'. Most of the poorer 'developing' countries lie to the south of the Tropic of Cancer. So this part of the world is referred to as the 'South'.

Some differences between North and South

	The North	The South
Past imperial rule	Many northern countries such as Britain and France were once ***imperial powers***. They ruled over southern countries and used the wealth of these places to make themselves more rich and powerful.	Many southern countries, such as Mali and Nigeria, were once the ***colonies*** of rich northern countries. Most colonies remained weak and poor while their northern rulers grew strong and rich.
Control of world trade	Northern countries still ***control world trade*** and use it to increase their own wealth.	Southern countries are often ***treated unfairly*** in world trade. This tends to keep the South poor.
Food supplies and diet	Most people in the North are ***well fed***. In Western Europe and other rich countries, unwanted food is stored or destroyed to keep the prices up.	Many people in the South suffer from ***hunger*** caused by chronic food shortages.
Health	Health care is ***adequate*** in most northern countries. Few people suffer from hunger-related diseases.	Health care is ***inadequate*** in many countries of the South. People suffer from diseases related to hunger.
Literacy	Almost all adults in the North are ***literate*** — able to read and write.	Many people in the South have never had the opportunity to learn how to read and write and so they remain ***illiterate***.
Incomes	Average incomes in the North are more than ten times ***greater*** than average incomes in the South.	Average incomes in the South are ***extremely small***. Many people will earn less in an entire year than we could earn in a month.

These inequalities contribute to the growth of other inequalities.

CHILD MORTALITY in the South is much greater than in the North.
(The term ***child mortality rate*** refers to the number of children under the age of five who die each year, in relation to every 1000 live births during that same year.)

LIFE EXPECTANCY is much shorter in the South than in the North.
(The term ***life expectancy rate*** refers to the average number of years which a newly-born infant can be expected to live.)

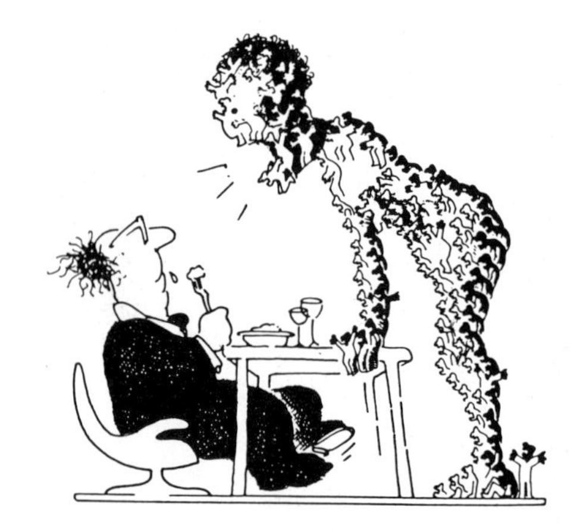

△
Figure 27.3 This cartoon illustrates the inequalities between the North and the South

How is the North represented? How is the South represented? Why?

Child mortality rates throughout the world

Do Activity 1, page 211.

Activity 1 shows that relatively few children die in the richer countries of the North. But child mortality rates are much higher in most countries of the South. (See page 211).

△
Figure 27.4 While many people in the South suffer from hunger, some Northern countries store away unwanted food. In parts of western Europe (EC) this stored food is called 'the food mountain'.

(a) What message do you think is given in this cartoon?

(b) How does the cartoon make you *feel* about the world food situation?

◁
Marasmus is one of the worst hunger-related diseases. It is caused by severe undernourishment and is especially common among babies. Its victims become thin and wasted. They develop large eyes and wrinkled, old-looking skin.

Some causes of child mortality in the South

- ☐ undernourishment
- ☐ malnutrition
- ☐ hunger-related disease
- ☐ poor medical care
- ☐ lack of clean water and sanitation

Why do so many children die in the South?

1. Children in some parts of the South do not have enough to eat. ***Undernourishment*** has caused the deaths of many children during long famines, such as those which have recently struck Ethiopia and the Sudan.
2. Far larger numbers of children in the Third World suffer from an extreme type of hunger called ***malnutrition***. They may eat sufficient ***quantities*** of food, but the food ***lacks nutritional quality***, especially protein. Because of this, they do not have the balanced diets which a healthy body needs.
3. ***Diseases caused by hunger*** are the most common killers in the Third World. These diseases are many and cause terrible suffering. They are common in many areas of the South and are especially serious among young children.
4. Half of the people in the South ***lack clean water and sanitation (toilet) facilities***. This situation causes the spread of many killer diseases. The most common of these is severe diarrhoea which kills an average of ***120 children per minute*** in the Third World.
5. Many Third World countries cannot afford the ***medical care*** which would save many lives. Malaria, for example, kills about 1 million children throughout the South each year. Low cost drugs are available which can prevent malaria, but many poor countries cannot afford even the cheapest drugs.

Women with a baby at a health clinic in Saudi Arabia.

What facilities might this clinic offer the baby and its parents?

Not *all* countries in the South are poor. Find out how Saudi Arabia has become one of the richest countries of the South.

▽

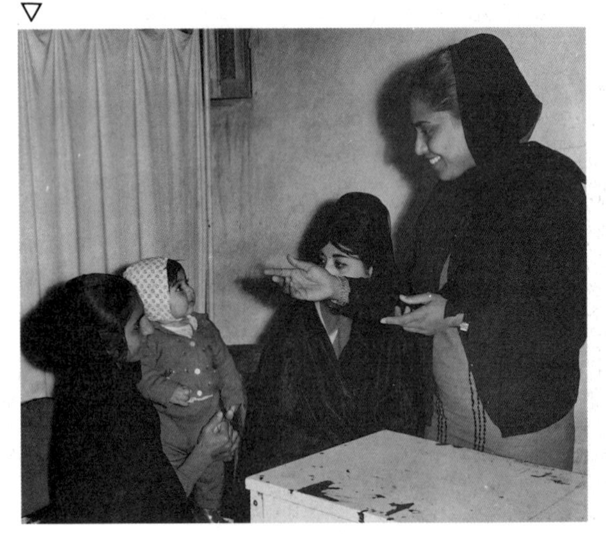

Some ways of reducing child mortality in the South

- **Prevent dehydration:** Each year, millions of children die because of dehydration (loss of body fluids) caused by severe diarrhoea. Dehydration can be prevented by a correct mixture of sugar, salt and water. This cheap remedy could be prepared by parents for their children, although the ingredients would probably have to be provided by relief agencies or governments to begin with.
- **Mass vaccinations:** Killer diseases such as measles and whooping cough can be prevented by mass vaccination programmes. Measles vaccine costs as little as 10 pence per dose.
- **Breast feeding:** Big European and American companies advertise powdered baby foods in Third World countries. Women are encouraged to buy these products because they want to imitate First World women. Third World mothers often use unsafe water and unsanitary containers for mixing and serving this food. Thousands of children die as a result. Such deaths can be avoided if Third World mothers are encouraged to breast feed their children instead. Big multinational companies must also be stopped from selling their products to uneducated Third World women.
- **Health education:** Health education programmes are needed to ensure that these measures are carried out successfully.

△

Figure 27.5 What point is being made by this cartoon?

Life expectancy rates throughout the world

Great differences exist between life expectancy rates in rich Northern countries and those in poor Southern countries.

▽ Figure 27.6a gives the life expectancy rates for a selection of Northern and Southern countries.
Figure 27.6b contains blank life expectancy columns for each country.

1. Use graph paper to copy the columns in figure 27.6b.
2. Use the information in figure 27.6a to shade in the life expectancies in each column. The life expectancy of Ethiopia has already been shaded to help you get started.

Life expectancy rates in selected countries

Country	Life expectancy
Bolivia	53 years
Canada	77
Cuba	76
Ethiopia	44
India	56
Ireland	74
Mali	46
Mozambique	47
Poland	74
United Kingdom	75
USSR	70
Yugoslavia	68
Zambia	52

Figure 27.6a Life expectancy rates in selected countries △

Average life expectancy (years)

Ethiopia | Mali | Mozambique | Zambia | Bolivia | India | Cuba

80
70
60
50
40

Southern countries

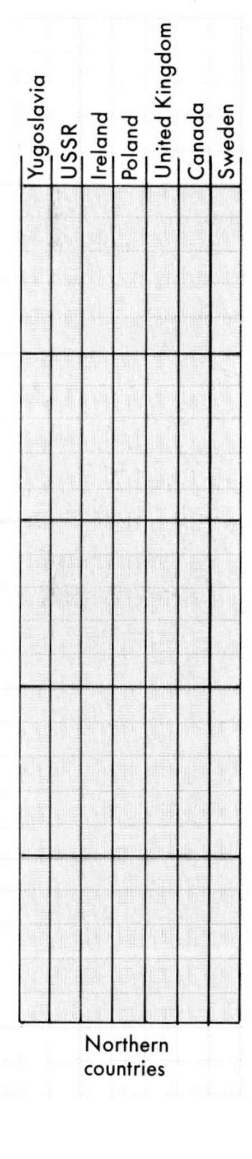

Figure 27.6b Life expectancy columns for the countries named in ▷ figure 27.6a

This activity should help you to realise the following key fact.

People in Northern countries enjoy longer average life expectancy rates than people in most Southern countries.

△
Why is the availability of clean water and sanitation (toilet) facilities so important to this little girl?

Reasons for different average life spans in the North and the South

The difference in life expectancy rates between North and South is caused by North-South differences in the following areas.

- ☐ availability of ***food*** and ***balanced diet***
- ☐ the occurrence of ***hunger-related diseases***
- ☐ the availability of ***clean water*** and ***sanitation***
- ☐ standards of ***medical care***
- ☐ rates of ***child mortality*** (which helps to determine life expectancy rates)
- ☐ world ***injustice*** (which keeps the North rich and the South poor)

Do Activity 2, page 212.

Discuss precisely how each of these factors influences life expectancy rates in the North and South. The information on pages 206 and 207 will also help you.

Things to remember about social inequalities throughout the world

- ☐ ***Some Third World countries have made great advances*** in increasing the life expectancy rates and reducing the child mortality rates of their citizens. Other Third World countries have not made such progress (figure 27.7).

- ☐ ***Social inequalities exist within the North and South*** as well as between the North and South. In some Southern countries such as El Salvador and Chile, a minority of people live in great wealth. In some rich Northern countries such as those in Western Europe, a minority of people live in extreme poverty. In the world's richest country, the United States, 20 million people are hungry because they cannot afford to buy sufficient nutritious food.

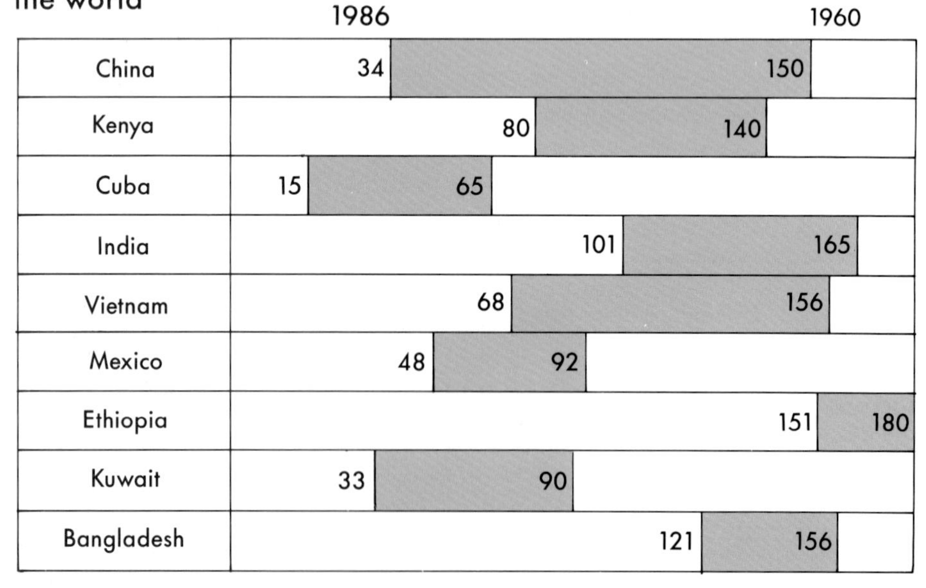

△
Figure 27.7 Selected infant mortality rates

(a) Which country shown achieved the greatest *real reduction* in infant mortality rates between 1960 and 1986?

(b) Which country achieved the greatest *percentage reduction* in child mortality rates during the same period?

(c) Suggest ways by which the countries named in (a) and (b) might have achieved large reductions in their child mortality rates.

Activities

1. The table in figure 27.8 gives child mortality rates (death rates) for a selection of countries. The map in figure 27.9 shows, but does not name, these countries. It also shows a broad general division between North and South.

Figure 27.8 Child mortality rates in selected countries ▷

Child Mortality rates* (per thousand live births)

Country	Rate	Country	Rate
Bangladesh	121	Ireland	12
Brazil	79	Mali	297
China	41	Mexico	50
Cuba	19	Peru	128
Ethiopia	151	Tanzania	179
India	115	USA	13
		USSR	28

*Child death rates refer to the yearly number of deaths of children aged 1-4 years per 1000 population in the same age group.

Figure 27.9 The North and the South △

(a) With the aid of your atlas, locate each of the countries listed in figure 27.8. Use pencil to lightly shade in each of these countries on figure 27.9 following this scheme for shading.

Shading scheme

- very high child death rates (over 100 per 1000 live births)
- high child death rates (50-100 per 1000 live births)
- low child death rates (20-49 per 1000 live births)
- very low child death rates (under 20 per 1000 live births)

(b) Study the work you have done in (a). Then identify the one statement in the boxes below which is correct. Put a tick in the box containing the correct statement.

Tick if correct. ☐	A. All Southern countries have high child mortality rates and all Northern countries have low child mortality rates.
Tick if correct. ☐	B. Southern countries have lower child mortality rates than Northern countries.
Tick if correct. ☐	C. Most Southern countries have high or very high child mortality rates, while most Northern countries have low or very low child mortality rates.
Tick if correct. ☐	D. High and low child mortality rates are equally distributed between Southern and Northern countries.

2. Figure 27.10 shows death records of a sample household in Ireland and for a sample household in the Sudan. Study the table and use the information in it to help you answer the questions which follow.

Death records for a household in Ireland

Person	*Life span*	*Cause of Death*
Mary	91	old age
Margaret	83	heart attack
Patrick	75	stroke
Eliza	73	cancer of lung
John	56	cancer of bowel
Eugene	22	road accident

Death records for a household in the Sudan

Person	*Life span*	*Cause of death*
Mai	82	old age
Mohammed	67	infected wound
Ahmed	59	(cause unknown)
Victoria	42	whooping cough
Ali	3	marasmus/hunger
Yora	1	diarrhoea

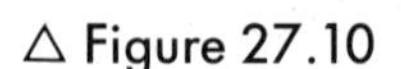
△ Figure 27.10

(a) Calculate the average life spans for each sample household.
(b) How much longer is life expectancy in Ireland than in the Sudan?
(c) In which household is child mortality more common?
(d) What would the life expectancy in the Sudanese household be if the deaths of the two children, Ali and Yora, are not considered? To what extent is the average *life expectancy* level influenced by *child mortality* in this household?
(e) Study the *causes of death* in the case of the three youngest members of the Sudanese household. What do these causes of death suggest about *food, water and medical supplies* in this part of the Sudan?

3. (a) What do the terms 'North' and 'South' mean in describing major world regions?
(b) Define the terms 'child mortality rate' and 'life expectancy rate'.
(c) Explain why child mortality rates are higher in the South than they are in the North.
(d) 'The wealth of our planet is distributed unfairly.' Write a paragraph explaining why you either agree or disagree with this statement.

4. Study the cartoon in figure 27.11.
(a) Who or what do the two figures in the cartoon represent?
(b) In what ways are these figures contrasted in the cartoon?
(c) What message or messages is the cartoonist trying to put across? Be as precise as you can when answering.

△ Figure 27.11

28 PEOPLE ON THE MOVE

Movements of people

Throughout history, people have frequently ***migrated*** or moved from one place to another.

- ☐ Some migrations are ***international.*** People ***emigrate from*** one country and ***immigrate to*** another.
- ☐ Some migrations are ***internal.*** People move from one part of a country to another part of the same country.
  Many internal migrations are ***local.*** These involve only small movements over short distances, such as from a city-centre area in Dublin to one of Dublin's suburbs.
  Other internal migrations are ***regional*** and involve greater movement, as from the West of Ireland to Dublin.
- ☐ Some migrations are ***individual.*** They are planned and are carried out by the migrants themselves.
- ☐ Some migrations are ***organised.*** They are planned by governments or by other powerful groups of people. Organised migrations are often forced upon the migrants.

△ Vietnamese migrants in Southeast Asia

△ Moving house

Discuss some probable differences between the two kinds of migration shown in these two photographs.

Why do people migrate?

A ***combination of reasons*** usually cause people to migrate.

- ☐ People may decide to leave a place if they dislike certain things about it. These unattractive things are called ***push factors*** or ***repellent reasons*** for migration. They push or repel the people from an area.
- ☐ People may move to a place because they think certain things about it are attractive. These are called ***pull factors*** or ***attractive reasons*** for migration.

Figure 28.1 Push factors, pull factors ▷ and barriers to migration

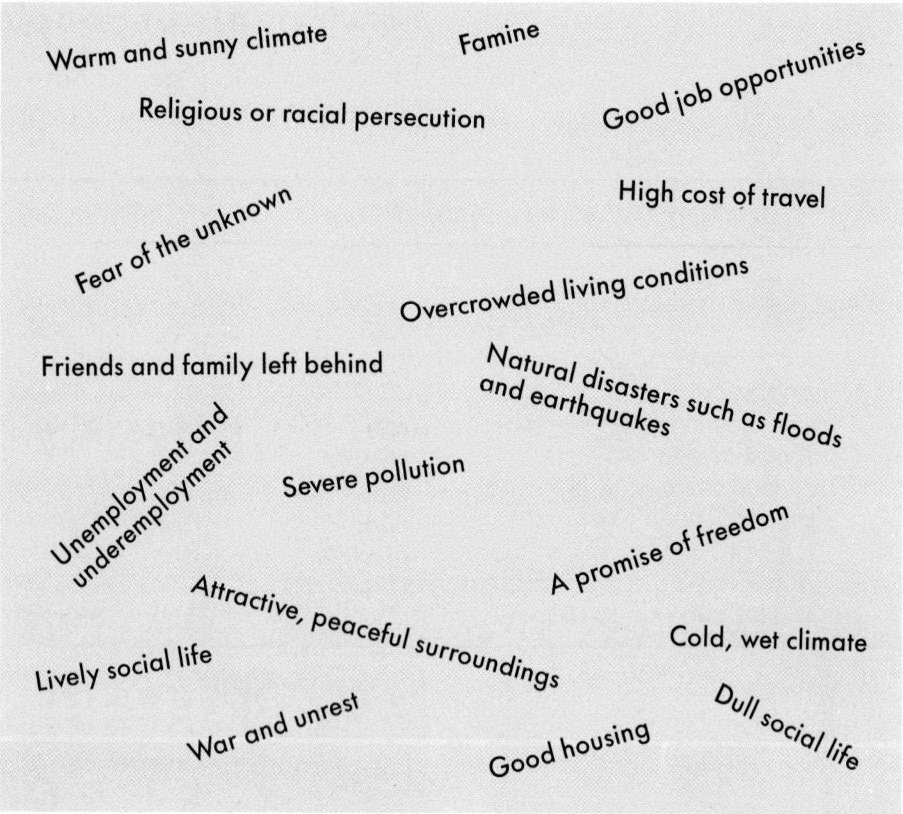

Warm and sunny climate
Famine
Religious or racial persecution
Good job opportunities
High cost of travel
Fear of the unknown
Overcrowded living conditions
Friends and family left behind
Natural disasters such as floods and earthquakes
Unemployment and underemployment
Severe pollution
A promise of freedom
Attractive, peaceful surroundings
Cold, wet climate
Lively social life
Dull social life
War and unrest
Good housing

Copy the diagram in figure 28.2 (you will want to enlarge it). In the appropriate spaces, write the repellent reasons for migration, the attractive reasons for migration and the barriers to migration named in figure 28.1.

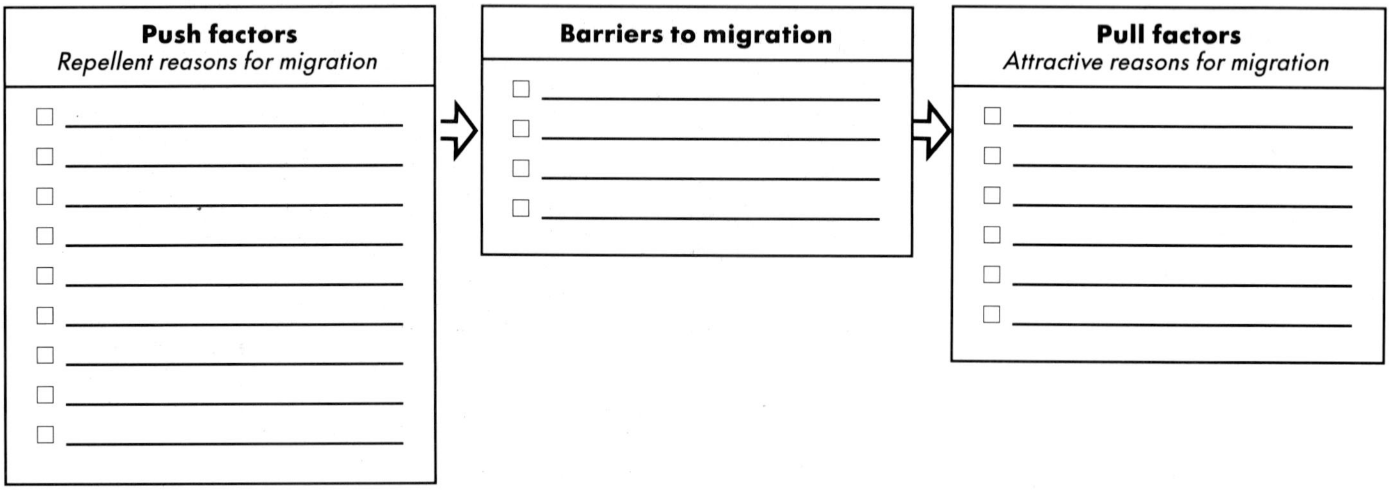

Push factors *Repellent reasons for migration*		**Barriers to migration**		**Pull factors** *Attractive reasons for migration*
☐ ______	⇨	☐ ______	⇨	☐ ______
☐ ______		☐ ______		☐ ______
☐ ______		☐ ______		☐ ______
☐ ______		☐ ______		☐ ______
☐ ______				☐ ______
☐ ______				☐ ______
☐ ______				
☐ ______				
☐ ______				

△
Figure 28.2

Individual migrations

Most migrations involve individual movements of single people or families. They usually leave their homes in search of ***better economic opportunities.***

- ☐ In Ireland, many individuals leave rural areas in the West and migrate to cities, especially Dublin, in search of work.
- ☐ Many Irish people have emigrated, and continue to emigrate, to Britain and the United States in search of employment and improved living standards.

The 'flight' from the West of Ireland

Terri, a typical migrant from a rural area in the West of Ireland, is moving to Dublin. Read Terri's letter to her friend Marie and try to discover three reasons for her migration. Say whether each of these reasons is an attractive (pull) factor or a repellent (push) factor.

What factors make Terri reluctant to migrate? What other attractions of the countryside are mentioned in Terri's letter?

Loughmore
Connemara
Co. Galway
20 March 1989

Dear Marie,

Thank you for agreeing to meet me at the train station in Dublin on Saturday next.

As you say in your letter, it will be a big change for me to leave the familiar sights of Loughmore to go and live in a strange city. I know I shall miss all my old school friends – not to mention my family and home.

But I'm glad I'm going. I'm sure I'll manage to get a job in Dublin and, as you know, there is little or no employment around here for me. In any case, I'm beginning to find life a bit too dull here. Not even a cinema in ten kilometres! I can't wait to go to some of the Dublin discos and cinemas and big stores. We'll have a super time!

I'm delighted you've settled down so well in Bray and that you can commute to work in Dublin so easily. That DART rail system seems to be a great way to get in and out of the city quickly. And wait till you buy that car! I can just see you zipping along on those new, wide suburban roads!

May O'Brien from London is here now on a cycling holiday. She says she's *glad* to be away from the noise and bustle of the city for a while. She's enjoying the beaches and mountains and the quiet roads of Connemara. When her father retires next year, she says her parents may return to live here permanently. I think they miss the unhurried pace of country life!

That's it for now. I look forward to seeing you on Saturday, Marie.

Wish me luck!

Terri

Effects of rural-to-urban migration	
In the countryside	**In the city**
☐ The number of people in the countryside decreases. This is called ***rural depopulation.***	☐ The city ***population increases.***
☐ This may ***reduce unemployment*** in the area.	☐ Young, energetic migrants may help to ***develop the city's economy.***
☐ But it may ***deprive the area*** of many young, energetic people who are vital to economic growth.	☐ But too many incoming migrants may result in ***overpopulation***—too many people for the available jobs, houses and other resources.

Migration from Ireland to Britain and the United States

Since the time of Ireland's Great Famine (1845-49), large numbers of Irish people have migrated to Britain and the United States of America. In the past, most migrants left because of the following factors.

Pull factors *attracted people to destination areas in the UK and the USA*	**Push factors** *repelled people from their source (home) areas.*
☐ the prospect of profitable ***work*** in cities, on the railroads, in mining etc.	☐ ***unemployment*** and poorly paid jobs at home
☐ the prospect of good ***land*** in America, which was offered to European settlers in the last century (see figure 28.3)	☐ a ***shortage*** of good ***land,*** especially in the West of Ireland where farms were small, overcrowded and subject to high rents
☐ the prospect of ***'freedom'***: the USA was seen as 'the land of the free'	☐ a feeling of ***oppression*** at the hands of some landlords or other authorities
☐ the prospect of ***adventure*** and a more lively social life	☐ a feeling of ***boredom,*** especially in areas with limited social opportunities

Many people still migrate from Ireland to the UK and America. Make lists of the 'pull' and 'push' factors shown here which would influence present-day migrants. Rank the factors in each list according to their importance today.

Figure 28.3 During parts of the 19th century, the US government offered free farms to European immigrants. This poster was printed by the Chicago and North-Western Railway Company which made money by transporting immigrants to their farms.

1. Why do you think the US government was prepared to attract immigrants with offers of free land?
2. Can you think of any element of injustice which might have been involved in offering vast tracts of land to immigrants? Explain your answer.

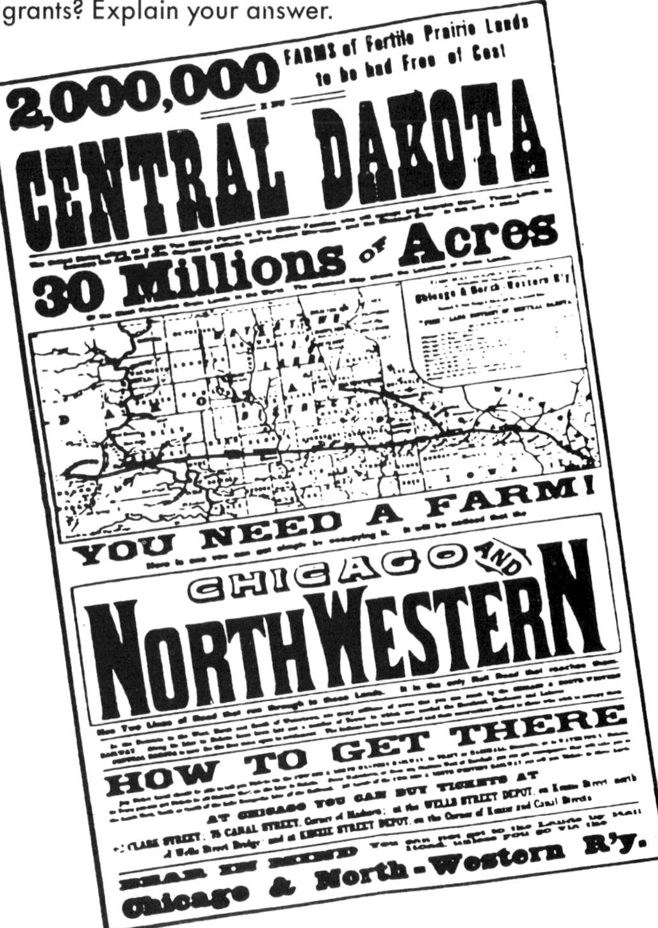

Good-bye, Muirtín Durkin.
I'm sick and tired of workin'.
No more I'll dig the praties,*
No longer I'll be fooled.
For sure as my name is Dorney,
I'll be off to Califor-nee.
And instead of diggin' praties
I'll be diggin' lots of gold.

*praties = potatoes

Figure 28.4 A migrant's idea of America in the last century

Why does the person in the ballad wish to migrate to California? Does the person have a realistic idea of life in California? Explain.

Emigration from Ireland to the USA and the UK continues to this day. With unemployment levels high in Ireland, most emigrants leave the country in search of work. Many go to cities on America's east coast or to the London area of the UK (although other cities also attract migrants). These areas are attractive for two basic reasons.

- ☐ They offer opportunities of well-paid employment.
- ☐ They already contain many Irish people, including relatives and friends of the new migrants.

Now, however, entry into the United States is very ***strictly controlled*** by the American government. Special ***work visas*** are issued to a limited number of people, allowing them to take up permanent residence in the USA. In spite of ***heavy penalties,*** young Irish people are still willing to go to America as illegal immigrants.

Figure 28.5 Nett population losses or population gains in Ireland due to migration, 1980–1987

(a) In the boxes provided, write in the nett population loss or gain for each year. (Use pencil.) (b) Describe the trends shown. (c) What do these trends suggest about the economic situation in Ireland throughout the 1980s?

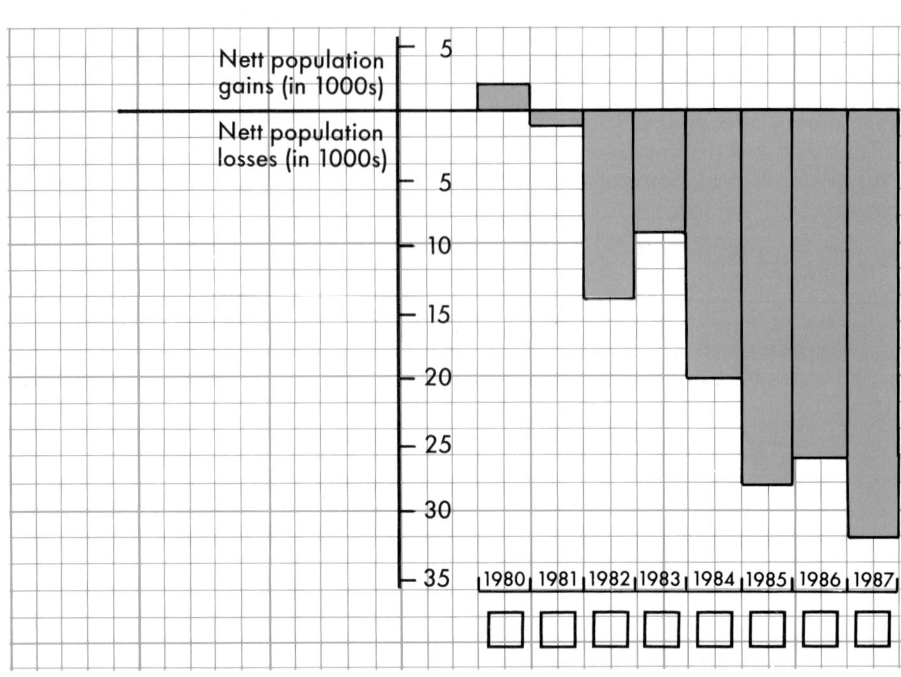

Organised migrations

You may need to consult your atlas.

Use pencil to fill in the blanks.

Large-scale human migrations have sometimes been organised by governments or other powerful groups of people in different parts of the world. Some examples of this include:

- ☐ the Plantation of Ulster
- ☐ the European colonisation of South America
- ☐ forced migrations in present-day South Africa

The Plantation of Ulster

In Irish history, ***plantation*** meant the removal of Irish people from their lands, replacing them with English and Scots settlers who were described as 'planters'.

The Plantation of Ulster began in 1609 when almost 4 million acres in the province of Ulster were confiscated by the king of England. Most of this land was divided into estates which were rented cheaply to English and Scots planters (see figure 28.6).

Figure 28.6 The Ulster Plantation, 1609 ▷

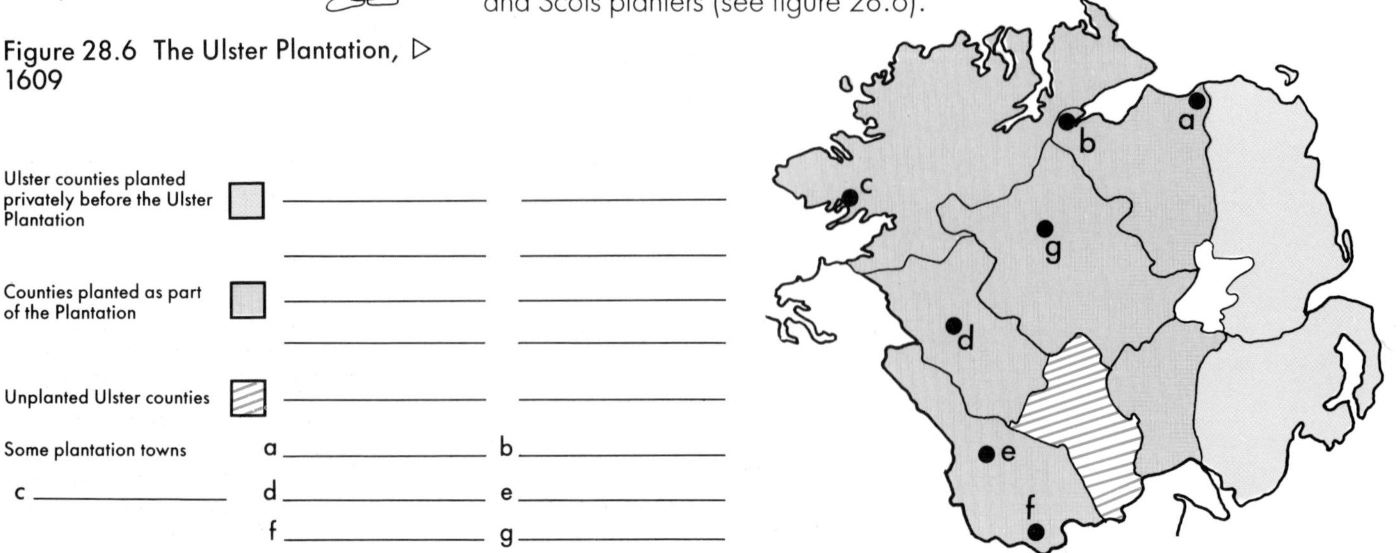

Why people migrated to Ulster

Pull factors: the planters

- ☐ Most planters were ***attracted*** to Ireland by promises of cheap, fertile ***land***. They received estates of between 1000 and 2000 acres, with annual rents of between £5 and £10 per thousand acres.

- ☐ Some were attracted by a sense of ***adventure***. . . or by the ***desire to 'civilise' Ulster*** by introducing the Protestant religion and English and Scots cultures.

Push factors: the Irish

- ☐ Many Irish landowners were ***forced*** to move from the land by the planters. Some moved to smaller estates given to them as part of the Plantation system. Others fled to forests. These 'woodkerns' often attacked the planters.

Some effects of the Ulster Plantation

Large-scale migratons may have lasting and varied effects on an area. The migration of planters to Ulster had three main effects.

- ☐ **Culture:** English language, customs and farming methods took root in Ulster.

- ☐ **Divisions:** Deep and sometimes violent divisions arose between the dispossessed (mostly Catholic) Irish and the newly-arrived (mostly Protestant) planters. Some people feel that these divisions can still be seen today as the cause of the unrest in Northern Ireland.

- ☐ **Settlement patterns:** The planters built ***planned towns*** (figure 28.6) in many parts of Ulster. The major streets of these towns often led to central squares or diamonds. Each square or diamond served as a market place. It also provided the town with an easily defended central area in the event of attack (figure 28.7).

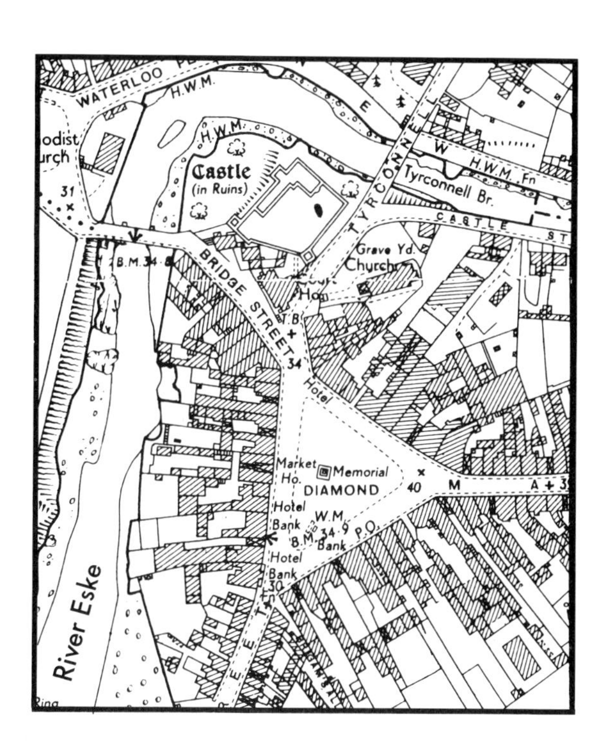

Figure 28.7 The centre of Donegal, an old plantation town

1. Identify the central area around which Donegal was built.
2. What evidence from the map shows that the area was used as a market place?
3. Which streets converge on the central area?
4. *The Ulster planters lived in fear of being attacked*.
   (a) From whom might the planters fear attack, and why?
   (b) Explain how the central area of Donegal might have been easily defended from attack.
   (c) Apart from the central area, what other major defence feature is evident in Donegal? Is this feature well sited for defence? Explain.

Figure 28.8 The Treaty of Tordesillas divided South America between Spain and Portugal.

(a) Today, Portuguese is the official language of one South American country. Which country do you think this is, and why? (b) Name some modern South American countries in the Spanish 'sector'. (c) Was the 1494 treaty a just one? Explain.

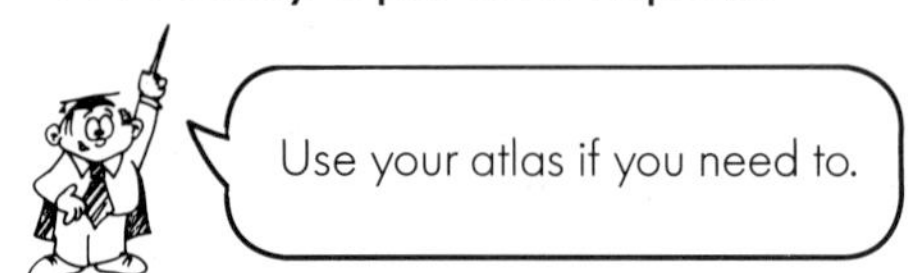

The European colonisation of South America

During the 15th century, ***Spain and Portugal*** were among the most powerful nations in Europe. Each was anxious to capture foreign land in which they could establish colonies and from which they could extract wealth.

In 1494, Spain and Portugal signed the ***Treaty of Tordesillas*** in which they agreed to divide the conquest of South America between them. Spain was to take land to the west of the 49°W meridian. Portugal was to have the land to the east of this meridian.

From the early sixteenth century onwards, Spain and Portugal began to colonise South America. ***Many groups of people began to migrate to South America for a number of reasons***.

- ☐ ***Military men*** and their armies came in search of conquest and loot. Spanish 'conquistadores' (conquerors) overran local peoples, completely destroying such ancient civilisations as the Incas in Peru. The Spaniards then sent vast amounts of gold and other wealth back to Spain.
- ☐ Some ***Catholic missionaries*** came to convert the native South Americans to the Catholic religion.
- ☐ ***Planters*** came in search of land, minerals and other wealth. They took the land from the native people and set up estates on which they grew cotton, sugar cane or other cash crops for profit. They also took the rich silver and gold mines from the local people and forced them to work as slaves on colonial estates and mines.
- ☐ ***Black slaves*** were brought into South America from West Africa. Millions of West Africans were kidnapped by slave traders and brought to South America where they were forced to work for the European colonists. The slaves received no pay and had to live in miserable conditions.

Some effects of colonial migration in South America

Study photographs C and D. Then write a brief account of the effects which colonial migrations may have had on South America. Refer to: population make-up; religious beliefs; style of buildings; likely differences between rich and poor.

◁ Photograph C

Photograph D ▷

Present-day South Africa

Forced migration under Apartheid

The organised migrations which you have just studied took place in the past. But organised, often forced, migrations still happen today.

Today's most notable example of ***forced, organised migration*** has taken place in South Africa. This country operates a system known as Apartheid, which has forced black people to live apart from white people. Apartheid has meant that over 4 million black people have been evicted from their homes and forced to live in poor areas of the country reserved for blacks only (figure 28.9).

Figure 28.9 The effects of Apartheid in South Africa

The ***white government*** of South Africa has set aside 87% of the country for the white minority.

Black people are removed from these white areas and forced to move to poor parts of the country called ***homelands***.

In the homelands, blacks are given special passports which declare that they are ***no longer citizens of South Africa.***

Some young and healthy blacks are then allowed to return to work in the white areas. But they must do so as ***foreign immigrants***.

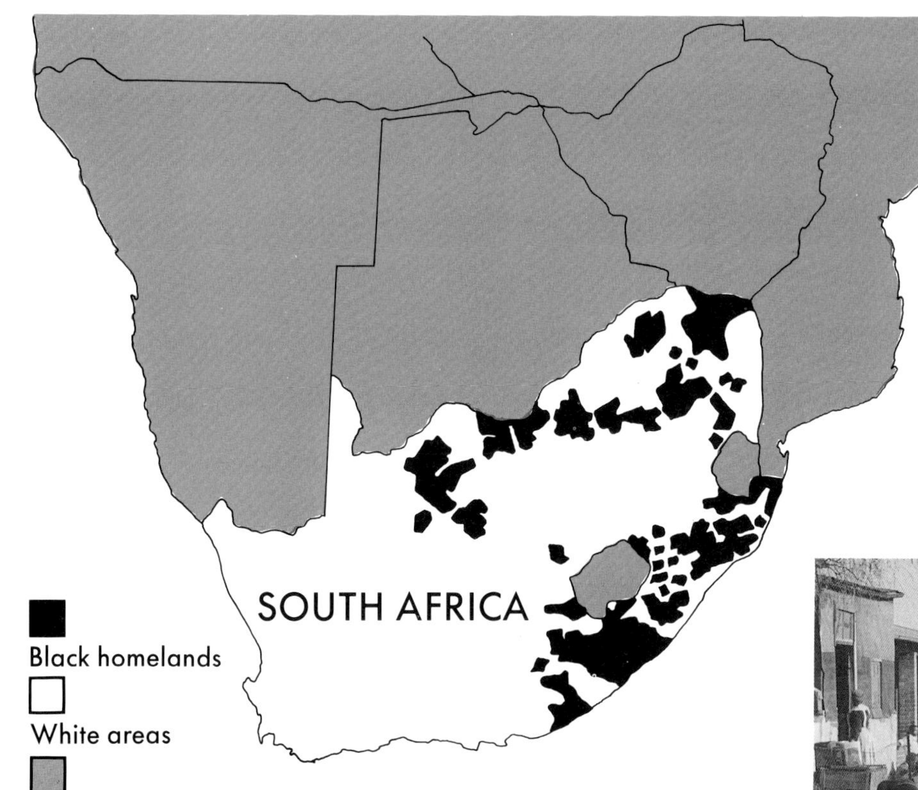

Apartheid in South Africa. Many of these Black people have become forced migrants since this photograph was taken. They have been evicted from their homes and forced to move to poorer parts of South Africa. ▷

Activities

1. (a) Explain the differences between: (i) individual and organised migration; (ii) international and internal migration.
   (b) Name the main 'push' factors and the main 'pull' factors which influence the present migration of Irish people from Ireland to the United Kingdom and to the United States.
   (c) 'Many Irish people who emigrate to the USA are young, energetic and well educated.'

   Name some of the likely effects of the migration of such people: (i) on the population make-up of Ireland and the population make-up of the USA; (ii) on the economy of Ireland and the economy of the USA.

2. *Crossword Puzzle*

   Across
   1. This type of factor may attract migrants to a destination area.
   2. A plantation town in Ulster.
   3. A South American country which experienced Portuguese colonisation.
   4. These places in South Africa have been described as 'human dumping grounds'.
   5. A county in Ulster.
   6. They were forced to migrate from Africa to the Americas.
   7. A reason for migration: opposite to attractive.
   8. To migrate into a country.
   9. Town/county associated with the ulster Plantation.
   10. This country colonised much of South America.

   Down
   11. Ulster had one.

3. People migrate for a variety of reasons. Below are some sets of reasons which might or might not encourage human migration in four sample cases.
   - Rank the reasons given in each sample using numbers 1– 5. Give a rank of 1 to the most important reason for encouraging people to migrate. Give the least important reason a rank of 5. Write the rank numbers in the small boxes.
   - Say whether you think the person in each sample should or should not migrate. Explain your answer in each case.

Sample A — An unskilled Irish worker considers migrating to London

- ☐ Wishes to find work to support family
- ☐ Wishes to find work to save enough money to start a business at home
- ☐ Will leave spouse and family in Ireland
- ☐ Will probably get a monotonous job
- ☐ May live with sister's family in London

Sample B — An 'average' working couple with two children considers moving from a 3-bedroomed terraced house in a city suburb to a 4-bedroomed detached house with a large garden in the same suburb.

- ☐ Much higher mortgage and maintenance costs with bigger house
- ☐ More room for growing family
- ☐ Large garden will be pleasant, but will demand more work
- ☐ No need for children to change schools
- ☐ Bigger house could be seen as a status symbol

Sample C — A doctor working in an inner-city clinic in Dublin considers migrating from the city to a pretty country area of Co. Wicklow

- ☐ Can more easily take part in favourite sport, hillwalking
- ☐ Commuting to and from work will take 2 hours every day
- ☐ Children will have to change schools
- ☐ Family could enjoy open space and quiet surroundings
- ☐ Big shops will be much farther away

Sample D — A young adult suffers from racial or political oppression in his/her own country and considers moving to a neighbouring country.

- ☐ Customs and language are different in neighbouring country
- ☐ Family and friends are left behind
- ☐ Freedom from fear of arrest or oppression
- ☐ May never be able to return home
- ☐ Will find a new life

4. *Imagine* that you are part of a family with two parents and five children. You live in a 3-bedroomed house on a small farm in the West of Ireland. You have just completed post-primary school and have been given the opportunity to leave home and live with your aunt and uncle in Dublin. List the reasons for and against migration to Dublin:

   (a) from your own point of view

   (b) from your family's point of view

29 SECONDARY ACTIVITIES AS SYSTEMS

Review chapter 17.

In chapter 17, you learned that there are three main types of economic activities: primary, secondary and tertiary. ***Secondary activities refer to manufacturing industries.***

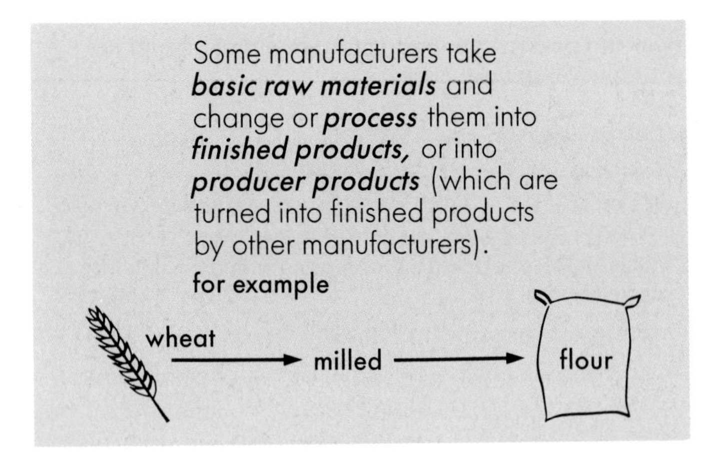

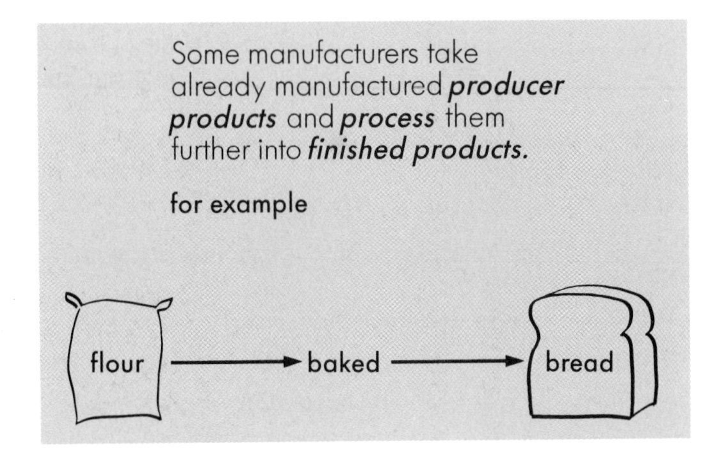

All manufacturing industries are similar in one respect. Each one can be seen as a ***system*** which contains ***inputs, processes*** and ***outputs.***

- ☐ The main task of the manufacturing system is to put together all the ***raw materials*** and anything else needed to make the manufactured product. These things are called the ***inputs*** of the system.
- ☐ The system then ***processes*** or ***changes*** some of the inputs.
- ☐ The processed inputs then become the ***outputs*** of the system. The outputs of industry include ***manufactured products, by-products*** and any ***wastes*** which are produced by the manufacturing process (see the model in figure 29.1).

Figure 29.1 This model or flow chart shows manufacturing activity as a system

(a) What happens to the useful industrial outputs which leave the system? (b) Inputs for the manufacturing system are paid for partly by the money already gained from the sale of outputs. How might a manufacturer obtain capital (money) to purchase inputs for a completely *new* industry? (c) What would happen to the system if any part of it were removed permanently? What does this tell us about systems in general? ▷

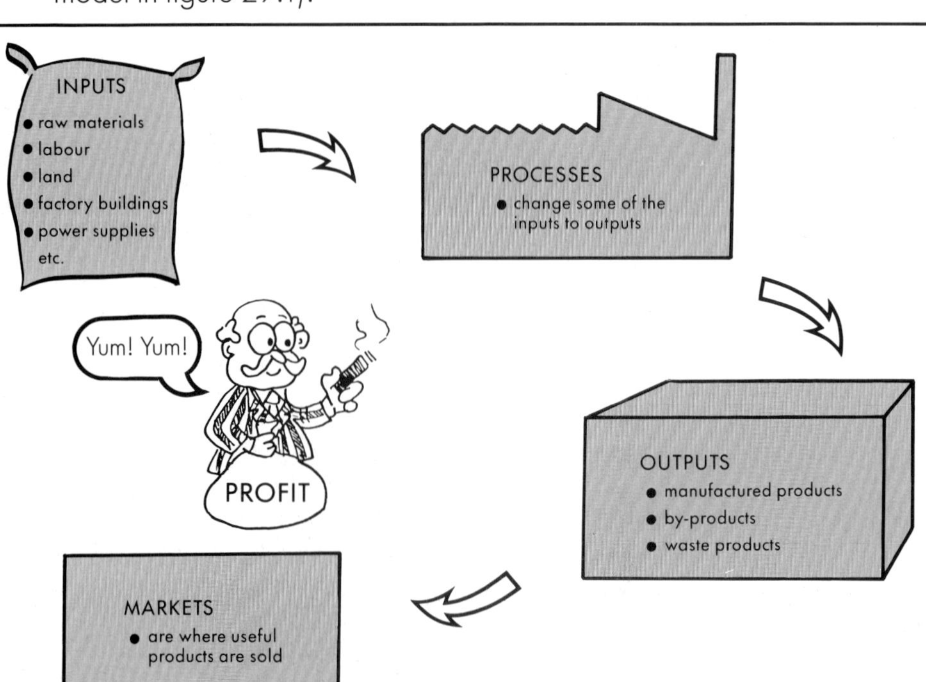

Case Study

Inputs, processes and outputs of Ireland's largest distillery

Read the following newspaper article and then answer the questions which follow.

At the fringe of the busy East Cork town of Midleton stands one of Europe's largest distilleries. It is the chief production centre of Paddy's, Jameson and Power's whiskeys as well as other well-known spirits such as Hussar Vodka, Cork Dry Gin and Mulligan.

Employees at Irish Distillers carry out a wide range of whiskey-making tasks in this famous plant. First there is the grinding of malt and barley into 'flour'. Then there is the brewing process in which the 'flour' is mixed with hot water and left to stand until natural sugars are formed in the mixture.

When brewing is complete, the leftover grain solids are taken out and dried to make a substance called 'dark grain'. This valuable by-product of the distilling process is sold to local farmers' co-operatives which use it as an additive for farm-animals' foodstuffs. Meanwhile, the liquid brew is piped into large fermenting vessels. Yeast is added and the liquid begins to ferment. Alcohol is formed during fermentation.

The fermented alcohol is then distilled. It is heated until it forms a vapour and then cooled until it recondenses into a liquid. This distillation process is repeated until a pure, colourless alcohol is formed. The alcohol is called ethenol or 'new whiskey'. This new whiskey is transferred to timber barrels (the barrels must be made of oak!) and left to mature for between three and twelve years. At this stage the whiskey is deemed ready for bottling.

Spirits have been made at Midleton since 1825. The plant now draws on a variety of sources for the ingredients it needs. Vast amounts of barley are purchased from many parts of Ireland, while small amounts of wheat come from local farms. Malt and maize must be imported, the latter mainly from France. The oak barrels come from Spain and the United States. Huge quantities of high-quality local water are also used, as are large volumes of energy in the form of electricity and natural gas.

The well-known products made at the Midleton plant find their way to every part of Ireland and to many parts of the world.

Great care is taken that the effluent or waste from the plant does not damage the local environment. The waste is piped into settling tanks in which it is filtered. Then, only clean water is finally allowed to flow into the picturesque Dungourney river which flows through the 25ha distillery site.

1. Make full lists of the inputs and outputs of the distillery which have been referred to in this article.
2. Make a flow chart to show the various whiskey-making processes described in the article.
3. Make a model (flow chart) to show the distillery as a system with inputs, processes and outputs.

The pot stills at Irish Distillers' plant at Midleton, Co. Cork. Irish Distillers is known the world over for the quality of its products. ▷

Activities

1. Study photographs A – E below.
   (a) State whether each photograph represents a primary, secondary or tertiary economic activity.
   (b) In the case of each photograph showing a secondary activity: (i) name the inputs, processes and outputs involved in the activity; (ii) say whether the manufactured product is a producer product or a finished product.

This could be carried out as part of the fieldwork activity recommended in chapter 30 on page 237.

2. *Fieldwork – a local factory*
   Investigate the manufacturing system of a factory in your locality.
   (a) Make a list of the inputs needed by the factory. Subdivide this list into *raw material inputs* (actual ingredients from which the finished product is made) and *other inputs.* In the case of each raw material input, state where it comes from and how it is transported to the factory.
   (b) Learn about the processes involved in this factory's manufacturing system. Draw a model to illustrate these processes (figure 29.1 may help you).
   (c) Make a list of the outputs of the system. Subdivide this list into *useful outputs* and *other outputs.* In the case of each of the useful outputs, state the market(s) in which it is sold and the type(s) of transport used to bring it to the market(s).

30 THE LOCATION OF INDUSTRY

△
Describe what is happening in this photograph? Why is a port location suitable for some factories?

Depending on the area in which you live, you may pass a number of different types of factories on your way to and from school each day. None of these factories just happened to locate themselves where they are as if by accident. Very careful planning is involved before a factory decides on its final location.

There are a number of important factors which must be considered by the owners of any factory before a location is chosen. These important ***locational factors*** which must be considered by modern factories include:

- ☐ resource (raw) materials
- ☐ labour force
- ☐ transport facilities
- ☐ markets
- ☐ services
- ☐ capital
- ☐ government/EC policies
- ☐ personal factors
- ☐ local factors

All these ***factors interact*** with each other to determine the location of industry.

Resource (raw) materials

The resources or ***raw materials*** are the ***basic inputs*** of any factory system. A factory owner usually tries to locate near the raw material source or near a port if the raw materials have to be imported. This helps to keep down the factory's transport costs.

Labour force

The availability of a well-educated and skilled ***workforce*** is another important factor in industrial location. The fact that English is spoken in Ireland is an important advantage, particularly for American-owned businesses.

Why is a well-educated workforce so important for the success of some manufacturing industries? ▷

Transport facilities

A good ***transport infrastructure*** is necessary if raw materials and finished products are to be easily transported to and from the factory. Roads, railways, ports and airports are all parts of this transport infrastructure.

△
Does the area shown here appear to have adequate transport routeways? How is this likely to affect local industry?

Markets

Before an industrialist sets up a factory, he or she must be sure that there is a ***market*** for the factory's outputs or ***finished products.*** By setting up in Ireland, an industrialist can sell to over 3 million Irish people, as well as to the European Community market with a population of over 360 million.

Services

Many other ***services*** must be found near any factory. Water and power must be available on the site. Financial services such as banks, accounting firms and insurance companies must also be available nearby.

△
What banking or other financial institutions can be found in your own local area? Why are such facilities important to business people?

△
Even the availability of a golf course may influence industrial location. Name some other *local factors* which can influence a factory's location.

Capital

Capital or ***investment money*** must be available from banks or other financial institutions. This money is particularly vital to local business people who need to borrow the capital required to set up a factory.

Government/EC policies

National governments and the ***European Community*** have specific policies which have set up systems for providing ***grants*** and ***incentives*** to industrialists setting up factories within the EC. There are often special grants and other incentives for industries which set up in poorer parts of the EC such as the West of Ireland, southern Italy and the Central Massif of France. These government and EC policies help to spread industries more evenly throughout Europe—a policy known as ***decentralisation.***

Personal factors

Sometimes ***personal factors*** can influence the location of a factory. Henry Ford, the American industrialist, is said to have located his tractor and car factory in Cork because it had been the birthplace of his grandfather. William Morris set up his car factory near Oxford in England in the old school which his father had attended as a boy.

Local factors

Other ***local factors*** can also influence a factory's location. These include the availability of a factory site, local co-operation and the availability of recreational and educational amenities for the workforce—schools, golf courses, beaches etc.

Heavy and light industries

Industries can be divided into two categories—***heavy industry*** and ***light industry.***

- ☐ A ***heavy industry*** usually processes 'heavy' or 'bulky' raw materials into finished products. Steel manufacture is a good example of a heavy industry.
- ☐ A ***light industry*** usually deals with less bulky raw materials. Light industries may also assemble components (parts) into finished products. The manufacture of computers and computer parts is an example of a light industry.

Case Study 1: The Pfizer Chemical Corporation – A heavy industry

Figure 30.1a The Pfizer factory at Ringaskiddy, Co. Cork

(a) Draw a sketch of this photograph. On it, show and label: the jetty where raw materials are landed; storage tanks; an administrative (office) building; a container truck used to export products; a water tower; a chimney. Also show and label the nearby town of Cobh (with its tall cathedral spire) and a dockyard. (b) Do you think the Pfizer factory found a suitable location for its factory? Explain your answer. (c) What direction is it from the Pfizer factory to Cobh? Approximately how far is the shortest distance between the factory and Cobh? (Use figure 30.1b to help you answer).

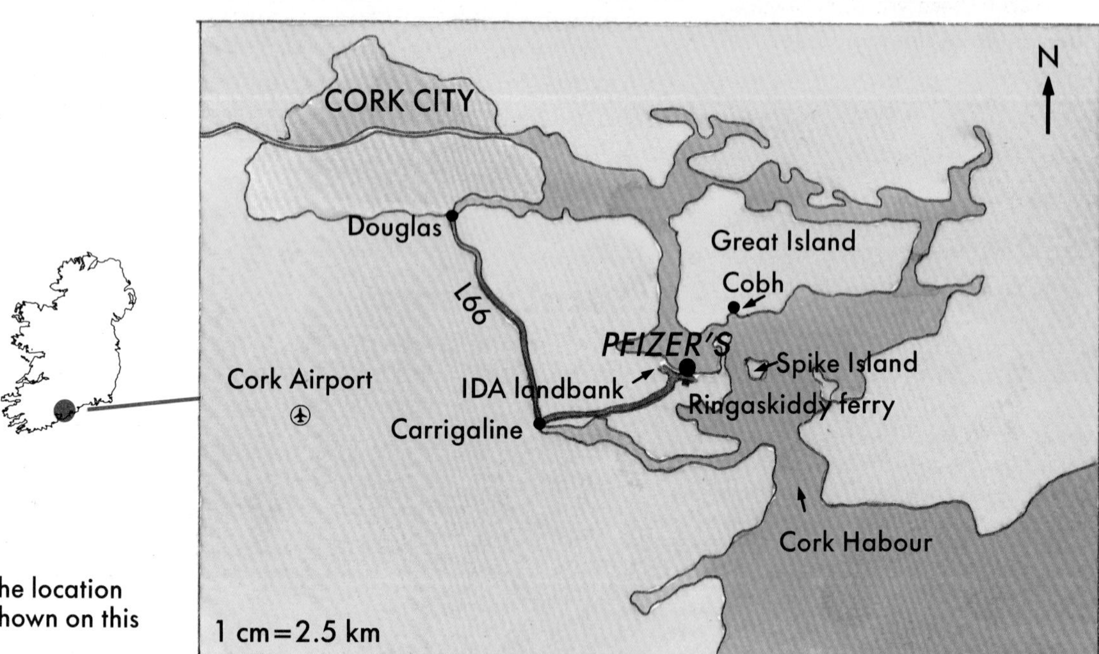

Figure 30.1b Describe the location of the Pfizer factory as shown on this map.

The Pfizer Chemical Corporation

The Pfizer Chemical Corporation has been making industrial ***chemicals*** and ***pharmaceuticals*** (medicines) at Ringaskiddy (figure 30.1) in Cork since 1971. Pfizer's is a ***multinational company (MNC)*** established in the United States in 1849. The company employs over 40 000 people worldwide, with its Cork factory employing over 500 directly and many more indirectly.

Pfizer's export their Cork-made products to fifty countries around the world. Molasses and sulphuric acid are the 'heavy' raw materials which are imported mainly from Europe and processed into products such as citric acid. Citric acid is used in many foodstuffs, including wines and soft drinks.

Some of the wide range of products which contain the citric acid manufactured by Pfizer's. ▷

Why did Pfizer locate at Ringaskiddy in Co. Cork?

To answer this question, read the following extract from Pfizer's own brochure. Then answer the questions which follow.

> Pfizer's story in the Republic of Ireland began in the 1950s when Pfizer first established pharmaceutical and agricultural ***marketing operations*** in Ireland . . . Because of the ***favourable financial terms*** offered by the government of the Republic of Ireland, and because ***labour and transport*** factors were good, Pfizer selected the Ringaskiddy site, near Cork, as the place to create a major manufacturing facility. The decision to acquire the site arose out of the ***increasing international demand*** for high-quality organic acids and the company's desire to improve its capacity to produce these fine chemicals.

1. What type of operations did Pfizer's begin in Ireland in the 1950s?
2. List and describe briefly 3 factors referred to here which caused Pfizer's to locate at Ringaskiddy.
3. Name one other factor, not referred to here, which might have influenced the Pfizer decision to locate at Ringaskiddy.
4. Pfizer's decided to locate its factory in Cork in the late 1960s. Do you think that Ireland's probable entry into the EC may have influenced the company's decision to locate in Ireland? Explain your answer.

The Pfizer chemical factory as a system

INPUTS

Raw materials
- ☐ molasses
- ☐ sulphuric acid

Energy
- ☐ gas
- ☐ oil

Water
- ☐ ¾m gallons stored in tanks

PROCESSES

Fermentation
- ☐ which changes the sugar in the molasses into citric acid

OUTPUTS

Citric acid
- ☐ for the food industry

Medicines
- ☐ antibiotics

Effluent (waste)
- ☐ discharged about 25km south of Cork habour

1. List the inputs and outputs of this factory system.
2. Can you think of any other inputs which are not shown on this flow chart? (hint: L____R, C__I__L)

Case Study 2: A light industry

Analog Devices of Limerick

The Analog company is an ***electronics firm*** which makes computer parts. The Limerick factory, which is located in an ***industrial estate*** (figure 30.2), is part of an American ***multinational*** company—Analog Devices of Boston, Massachusetts.

Where Analog is located

Limerick

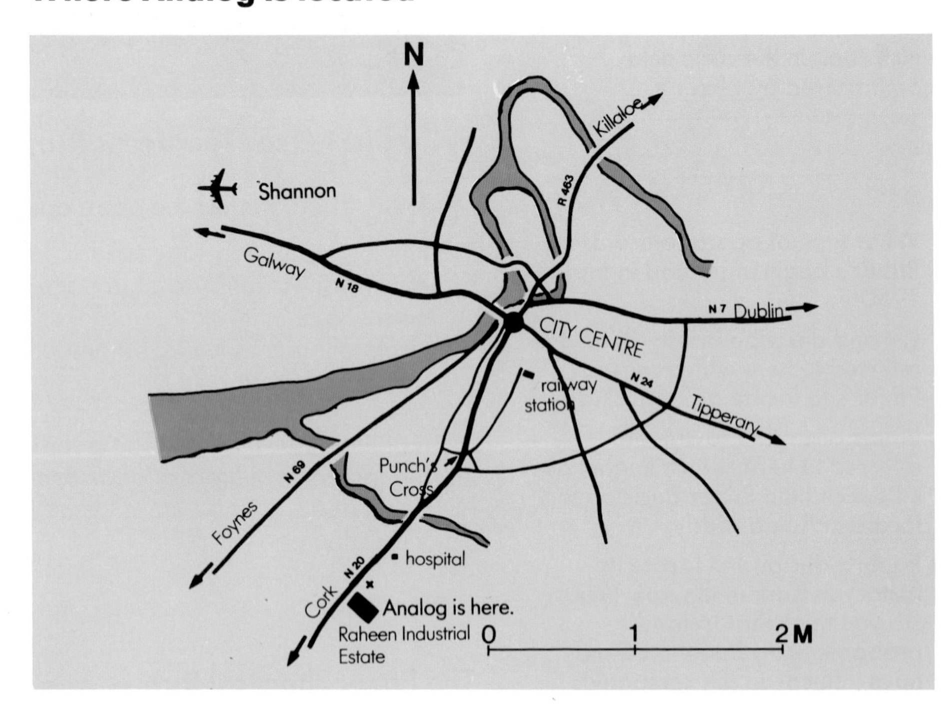

Figure 30.2 Study the map and then fill in the blanks.

The Analog Devices plant is located ____ km (____ miles) ____ (direction) of________city. It is in the________ industrial estate, which is near the ______roadway which links Limerick with ______.

Why Analog came . . .

Clues

1. to Ireland ⟶
   - ☐ Free access to EC markets
   - ☐ Large IDA grants (see page 235)
   - ☐ A skilled workforce (electrical engineers)

2. to Limerick ⟶
   - ☐ Shannon airport

3. to an industrial estate ⟶
   - ☐ Planned site
   - ☐ Infrastructure

What Analog does

Raw materials	processed into	*Finished products*	sold to . . . *Markets*
Its raw material is prepared ***silicon***— a sand-like substance.	→	*Wafers of* ***integrated circuits (microchips)*** which are sealed in tiny plastic cases for easier handling.	→ ***Electronics firms*** in the USA, Japan, and Britain which make computers for spacecraft, hospital equipment etc.

'Footloose' Industries

The traditional industries of the 19th century—steel making, shipbuilding and textiles, for example—were all tied to certain locations which were near their raw material and energy sources.

- ☐ ***Steel*** factories were usually located alongside coalfields and iron ore fields—for example, the Ruhr area in Germany and the Nord area in France.
- ☐ ***Textile*** factories tended to locate on the river estuaries of major ports, as near Manchester in Great Britain.
- ☐ ***Shipbuilding*** yards were (and indeed still are) tied to coastal locations where raw materials can be imported easily and from which newly-built ships can be launched. An example of this is Harland and Wolff in Belfast.

Footloose industries are those which can locate in a wide range of places with ***no single locational factor*** dominating their choice. The electronics industry and the healthcare industry in Ireland are both examples of footloose industries.

Today, most modern industry is no longer tied to specific locations. There are a number of reasons for this modern trend.

- ☐ Coal has been ***replaced by electricity*** as a modern industrial energy source.
- ☐ ***Improvements in transport*** methods and in ***infrastructure*** have meant that raw materials can be more easily transported to and from virtually every area in Europe, and indeed worldwide.

Because of these developments, modern industries tend to locate near their markets. The attraction of a skilled workforce is also important.

Since modern industries are no longer tied to traditional locations, they are now referred to as ***footloose.***

Case Study 3: A footloose industry

Becton Dickinson of Dun Laoghaire, Co. Dublin

Becton Dickinson is an American multinational company. It manufactures ***healthcare products*** and is located at Dun Laoghaire, a suburb of Dublin. The Irish factory manufactures syringes and needles used by doctors both in their private surgeries and in hospitals. The company employs over 21 000 people worldwide; over 600 are employed in its Dun Laoghaire factory.

Why is Becton Dickinson an example of a footloose industry?

Because no one factor dominated Becton Dickinson's decision to locate in Dun Laoghaire. Here are some of the factors which they considered.

- ☐ ***A skilled labour force*** was available in the region.
- ☐ IDA ***grants*** were available.
- ☐ The factory is located ***close to the port of Dun Laoghaire.*** This makes the importation of raw materials and the exportation of finished products much easier for the company.
- ☐ ***Electricity supplies*** were already available.
- ☐ The company could sell to a ***potential EC market*** of over 360 million people.

Some industries tend to be more footloose than others. To what extent would you consider Pfizer's and Analog Devices to be footloose industries?

△
The Becton Dickinson plant at Dun Laoghaire, Co. Dublin. Is this plant an example of a heavy or light industry? Explain your answer.

The IDA: The Industrial Development Authority

Almost every European country has an organisation which is responsible for ***developing its own native Industries*** and ***attracting foreign business***. In Ireland, this organisation is the ***IDA – the Industrial Development Authority***.

The IDA operates in three main areas of business:

- ☐ ***Small Irish businesses*** (employing fewer than 50 people)
- ☐ ***Irish industry***
- ☐ ***overseas-owned companies***

The IDA also operates in the area of ***regional development***, creating many new jobs throughout the country. It provides ***grants*** and ***incentive programmes*** and operates an ***advice service*** for both native and foreign business people.

Study the recent IDA poster and then answer these questions. ▷

1. Describe the scene in the main photograph. Why do you think the IDA chose this area to illustrate its message?
2. The main photograph is called 'The Young Scholars'. Why might this be an important part of the IDA message?
3. What message is the IDA conveying in its slogan 'Knowledge is Power'?
4. Why might the people in the smaller photograph want to call themselves 'The Young Europeans'? Why would this attitude be important to an organisation such as the IDA?

KNOWLEDGE IS POWER.

The Irish never underestimate the importance of learning.
We spend a higher proportion of GDP on education than Britain, France or Germany. More than a quarter of the population is in full-time education.
Today, Ireland's share of US manufacturing investment in the EEC is nine times greater than it was a decade ago.
That's the Power of Knowledge.
Ireland. Home of the Irish. The young Europeans.

IDA Ireland
INDUSTRIAL DEVELOPMENT AUTHORITY

REPUBLIC OF IRELAND

"WE'RE THE YOUNG EUROPEANS."

In 1984, one-third of Ireland's top 300 companies were totally foreign-owned.

Many of the companies which the IDA attracts to Ireland are ***multinational companies*** (MNCs) or ***transnational companies*** (TNCs). The headquarters of these companies are usually located in the United States, Japan, Great Britain or Germany. Their ***branches*** or ***subsidiaries*** locate in many different countries. They are looking for favourable rates of pay, government incentives, transport facilities, good labour relations and other factors as they research their worldwide locations. Many branch factories manufacture components or parts which are then assembled at other branches.

Multinationals in Ireland: For or Against?

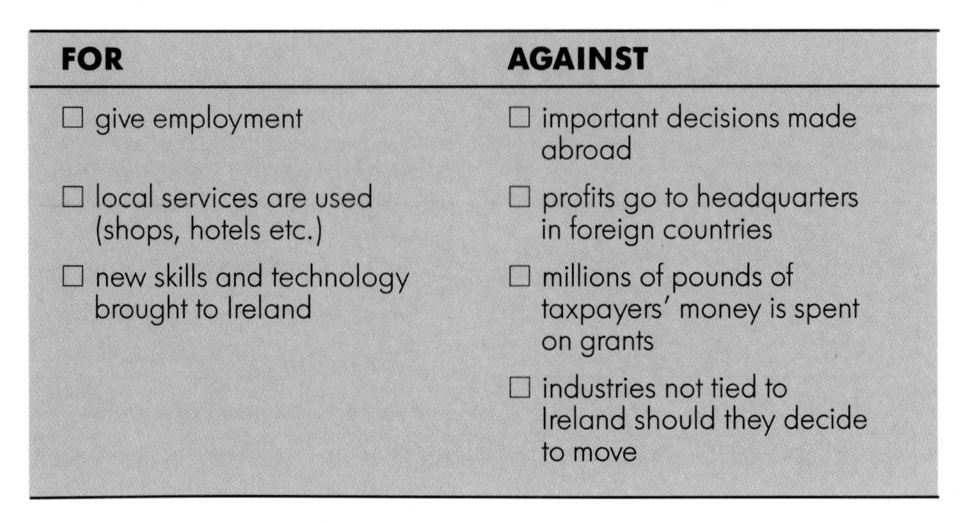

FOR	AGAINST
☐ give employment	☐ important decisions made abroad
☐ local services are used (shops, hotels etc.)	☐ profits go to headquarters in foreign countries
☐ new skills and technology brought to Ireland	☐ millions of pounds of taxpayers' money is spent on grants
	☐ industries not tied to Ireland should they decide to move

Industrial estates

Many new industries in Ireland are located in ***industrial estates.*** Each carefully-planned estate contains ***several factories.*** All of these take advantage of the site location and the services provided on the estate.

Location

The estate is usually located near a major road or other ***good transport facilities.*** The Shannon Industrial Estate, shown here, is beside the ***airport.***

It is usually at the edge of a ***large town:*** close enough to have a labour force nearby and far enough away to avoid traffic congestion. A ***new town*** has been built at Shannon. Limerick City is about 12 km away.

Figure 30.3 All the cities and towns shown here have industrial estates. Complete the name of each place shown.

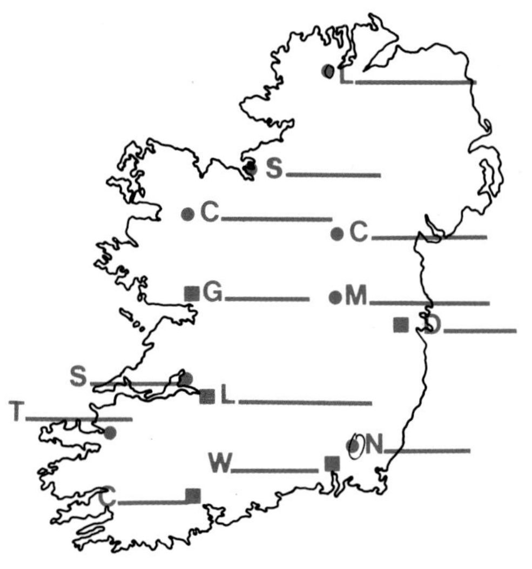

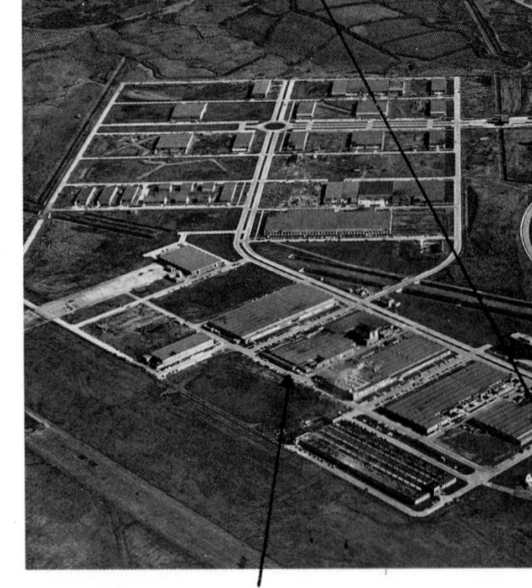

Development

The estate is usually divided into ***industrial plots*** which are rented, leased or sold to industrialists.

The new estate is first supplied with an ***infrastructure—services*** including roads, telephones, electricity, water and sewerage.

Figure 30.4 The location and development of an industrial estate.

Activities

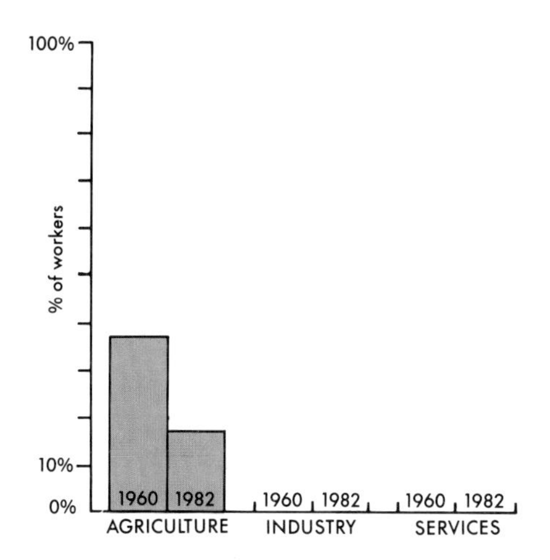

Figure 30.6 Employment in Ireland: 1960 and 1982

1. Examine the following table (figure 30.5) which gives information on the percentage of workers in Ireland and the EC involved in various sectors of employment.

Figure 30.5

	Agriculture		Industry		Services	
	1960	1982	1960	1982	1960	1982
Ireland	37%	17%	24%	31%	39%	52%
EC	19%	8%	41%	36%	40%	56%

   (a) Identify any trends which you see illustrated in this table.
   (b) Why do you think Ireland had over twice the number of workers employed in agriculture, compared with the European average?
   (c) What sector do you think will provide the most jobs in the future? Give reasons for your answer.
   (d) Copy the unfinished bar graph (figure 30.6) into your notebook. Complete the bars for industry and services in Ireland.

2. Imagine that you are an IDA representative and that you are to give a short presentation for foreign business people on the benefits of setting up their factories in your own local area. Outline the main points of your presentation. Make a list of the possible negative factors which the business people might raise and be prepared to counter any objections.

3. *Fieldwork: Visiting a local factory*
   Visit a factory in your area and write a report on why this factory is located in your area. Discuss its impact on the area as well as any other relevant facts. Use sketch maps, graphs, photographs and other illustrations in your report.

   OR

   Fieldwork: Visiting a local industrial estate

   Arrange a visit to a local industrial estate. Then do the following:

   (a) Draw a sketch of the layout of the estate.
   (b) List and then group the factories in the estate according to type, for example, electronics, food components etc. Does any trend emerge in your list and groupings?
   (c) What services does the estate provide for its factories?
   (d) Visit one factory in the estate. Then write a report on that factory as outlined in the first Fieldwork exercise.

31 CHANGING INDUSTRIAL LOCATIONS

In chapter 30 we saw how different locational factors affect the location of an industry.

Locational factors

- ☐ raw materials
- ☐ labour force
- ☐ transport facilities
- ☐ markets
- ☐ services
- ☐ capital
- ☐ government/EC policies
- ☐ personal factors

The relative ***importance*** of each of these factors ***can change*** when new technology is introduced or when different raw materials or alternative markets are found for the industry. These changes can often lead to a new distribution of an industry. Older locations may decline in importance, leading to high unemployment in the area. Newer locations may in turn become more industrialised and therefore more prosperous.

Perhaps the best example of an industry which has shown marked changes in its location over time is the iron and steel industry.

How iron and steel are made

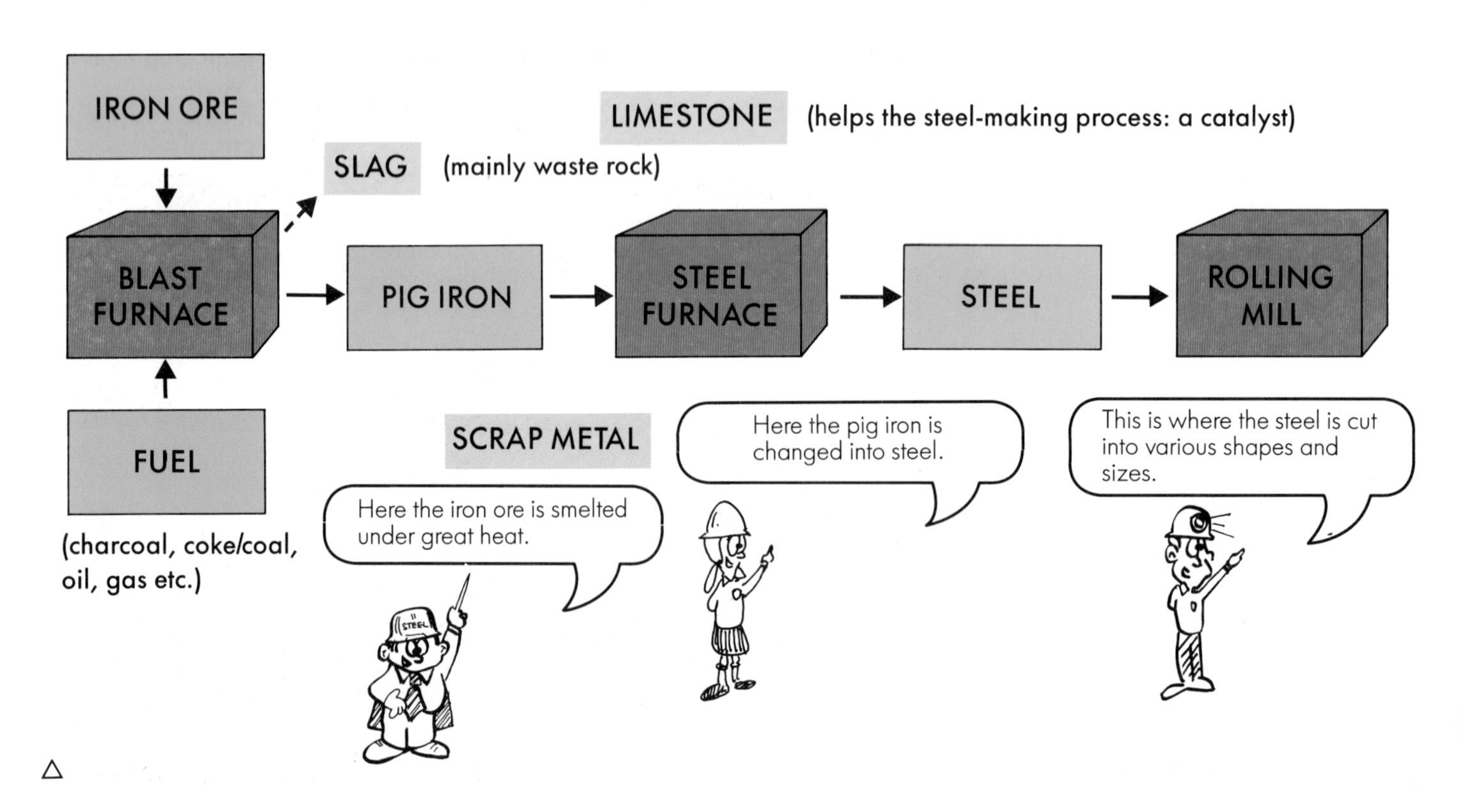

△

Figure 31.1 How iron and steel are made

1. Describe the two principal processes involved in the making of steel.
2. List the inputs and outputs of the first process.
   (a) Which of these inputs is the basic raw material?
   (b) Name a steel-making input *not* referred to in the diagram.
   (c) Identify one environmental difficulty which might arise from the production of slag. Suggest one possible solution to this difficulty.

The British Iron and Steel Industry

For many centuries, iron-making and steel manufacturing industries have been located in Britain. We can identify three stages in the development in Britain's iron and steel industry:

1. The ***pre-industrial*** stage
2. The ***Industrial Revolution***
3. The ***modern*** steel industry

Stage 1 – up to 1750

Near iron ore, timber and water

During this time, ***iron-making*** was the basis of the industry. It used these raw materials.

iron ore + charcoal (wood)

The use of these particular raw materials meant that the iron-making industry was ideally located in those areas where there were:

- ☐ rocks containing ***iron ore***
- ☐ ***forests*** from which charcoal could be made
- ☐ ***rivers and streams*** which could turn the water wheels which powered the forges. The rivers and streams also provided transport routes.

△
Figure 31.2a Iron works were originally located near iron ore deposits, running water and forests

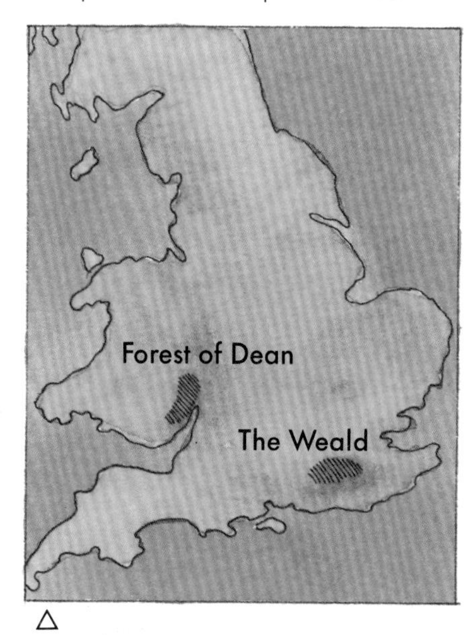

△
Figure 31.2b The locations of The Weald and the Forest of Dean

In Britain, two areas had all these necessary locational requirements (figures 31.2a and 31.2b):

1. The ***Forest of Dean*** near Bristol
2. The ***Weald of Sussex and Kent***

These two locations were also near the populated market areas of the South and Midlands of England. This important factor of nearby markets also helped to make these two locations Britain's major iron-making regions before the Industrial Revolution.

By the beginning of the 18th century, however, the situation was beginning to change. The easily-mined iron ore workings of these regions were becoming exhausted. The nearby forests were being cut down and rapidly used up. Around this time, it was discovered that coke (made from coal) could be used instead of charcoal in the iron-making process. As a result, the ***industry shifted its location from the forests to the coalfields.***

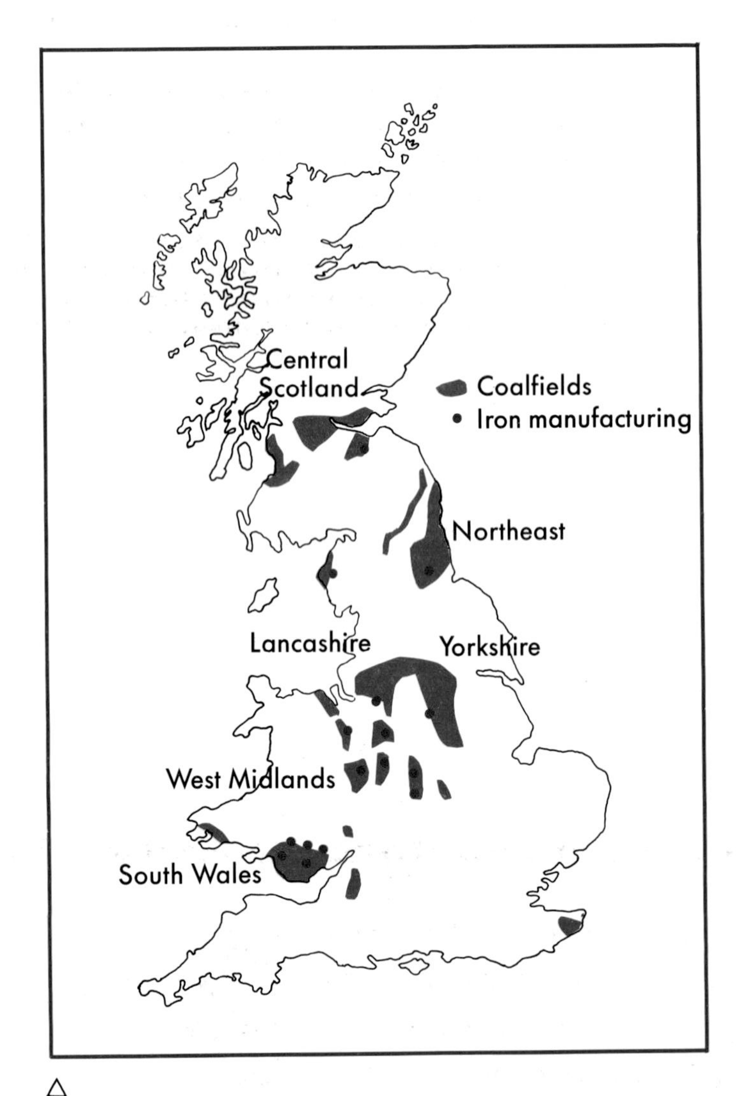

△

Figure 31.3 Centres of the Industrial Revolution in Britain

1. There are six industrial regions show on this map. Using your atlas, identify one industrial city in each region.
2. If for some reason coal was no longer the most important raw material influencing the location of the steel-making factories, where do you think the factories might then decide to locate? (hint: imported iron ore)

Stage 2 – from 1750 to 1900

At the coalfields

Areas which had both ***iron ore and coal deposits*** now became the new centres for the iron and steel industry (figure 31.3). These new locations, such as those in south Wales (figure 31.4) became the focus of Britain's Industrial Revolution.

By the end of the 19th century, the quality and quantity of British iron ore reserves were seriously depleted. At this time, it was possible to import better-quality and cheaper iron ore into Britain by using the newly-developed steamships. Later on, in the 20th century, the discovery and development of new energy sources such as oil, gas and electricity meant that coal was no longer a main locational factor for the industry.

The industry had to adapt to changing trends once again. This time, it moved ***to the ports*** and to other coastal locations.

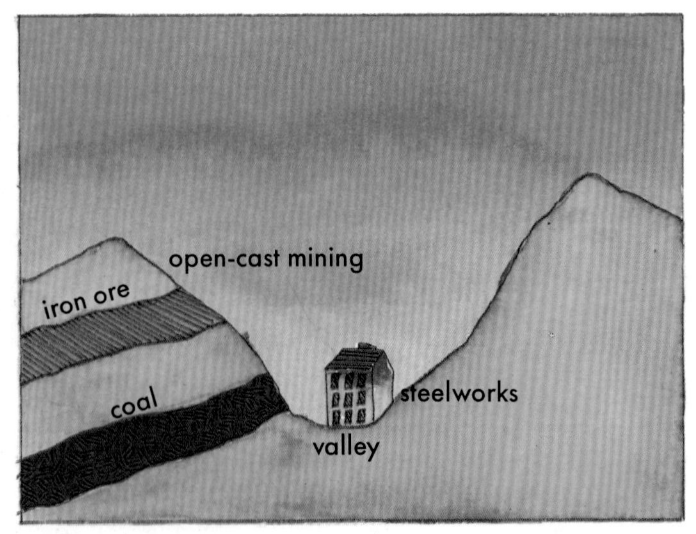

△

Figure 31.4 Steel-making in South Wales

Stage 3 – 1900 to the present

On the coasts

Today, the main locations for Britain's iron and steel industry are ***on the coast. Imported iron ore,*** from Sweden, western Australia and Third World countries, is now used in the ***integrated steelworks*** located at Newport, Scunthorpe, Ravenscraig, Port Talbot and Lackenby/Redcar (figure 31.5).

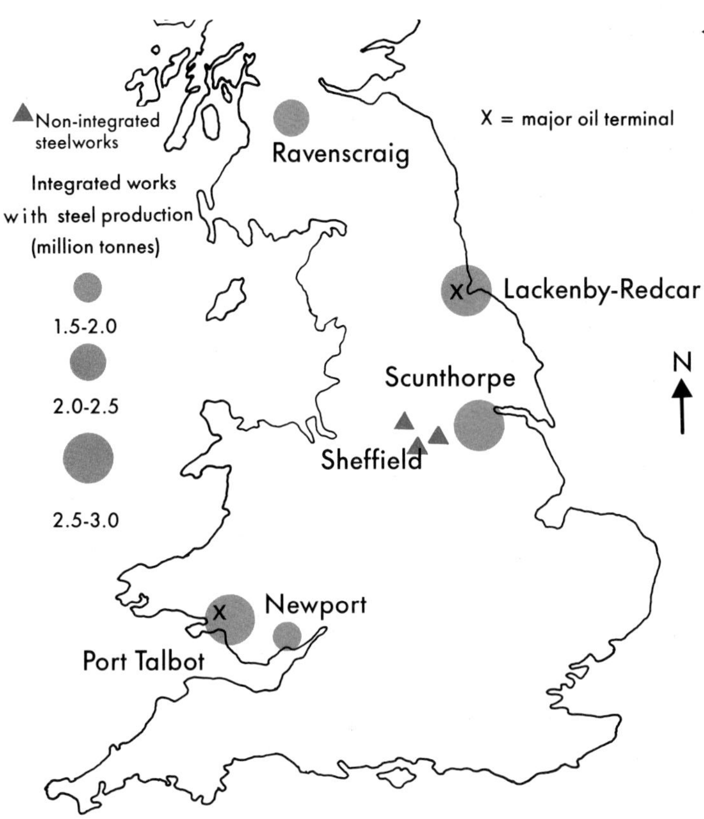

◁ Figure 31.5 British steelworks in 1981

1. Name the five integrated steelworks shown on the map.
2. Which region has steelworks which are not integrated?
3. In terms of its location, why is Sheffield different from the other steelworking regions?
4. Which two steelworks produce the most steel?
5. Compare this map with the map showing the centres of the Industrial Revolution in Britain (figure 31.3, page 240). Which regions have shown the most decline as iron and steel regions? Why might this be so?

Why Britain's iron and steel industry has moved to the coast

- ☐ Iron ore is easily transported through the ports.
- ☐ Transport costs are lower than at inland sites.
- ☐ Plenty of flat land is available for building.
- ☐ Electricity is readily available.
- ☐ Scrap metal can be easily imported.

Integrated steelworks

Up to the 20th century, the different processes in steel making (iron making, steel making and steel rolling) were usually carried out in separate factories which were often far apart from each other. This system has now been replaced largely by the use of ***integrated steelworks*** in which ***all the iron- and steel-making processes are carried out together.***

Industrial inertia

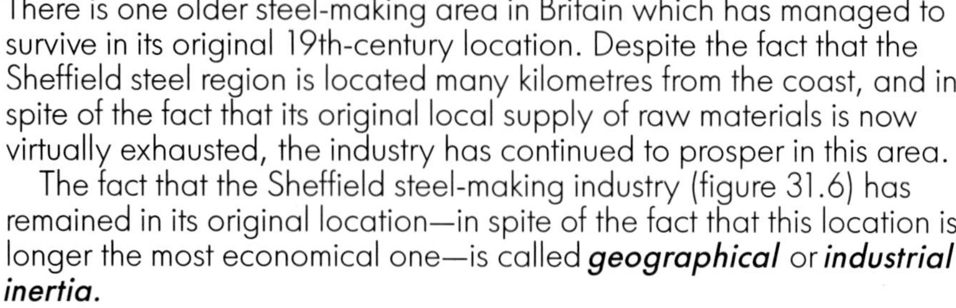

There is one older steel-making area in Britain which has managed to survive in its original 19th-century location. Despite the fact that the Sheffield steel region is located many kilometres from the coast, and in spite of the fact that its original local supply of raw materials is now virtually exhausted, the industry has continued to prosper in this area.

The fact that the Sheffield steel-making industry (figure 31.6) has remained in its original location—in spite of the fact that this location is no longer the most economical one—is called ***geographical*** or ***industrial inertia.***

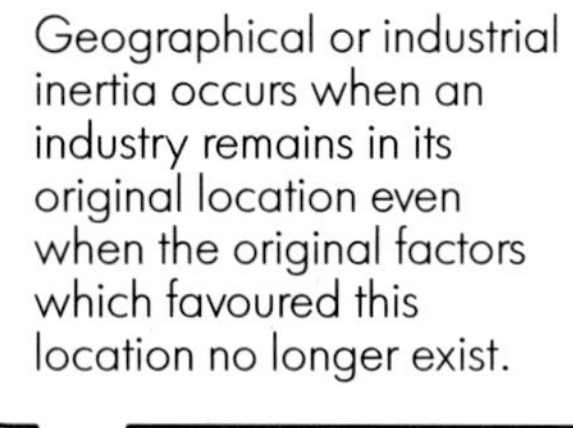

Why has the Sheffield steel industry remained at its original 19th century location?

- ☐ The Sheffield steelworks ***specialise*** in high-quality, expensive steel products. These include cutlery and surgical instruments. This means that Sheffield is not competing directly with those factories which make ordinary steel more cheaply.
- ☐ The region has a ***tradition*** of quality steel-making. This reputation for quality helps to sell Sheffield's products.
- ☐ A ***skilled labour force*** has been built up in the region over the years.
- ☐ ***Capital investment*** in the region has been very high.
- ☐ ***Scrap metal*** is available locally. This can be used as a raw material instead of iron ore.
- ☐ The British ***government supports*** the industry's location around Sheffield because its closure could mean the growth of social and political problems in the area.

△ Figure 31.6 The location of Sheffield

High-quality, expensive steel products have made Sheffield Steel famous. Why might Sheffield steel-makers have specialised in this type of product? ▷

Activities

1. The table in figure 31.7 shows the decline in EC steelmaking between 1974 and 1983. Study the table and then answer the questions.

EC Countries	1974	1977	1980	1983
West Germany	52	39	44	26
France	27	22	23	14
Italy	24	23	26	17
Belgium	16	11	13	9
Luxembourg	6	4	4	2
Netherlands	6	5	5	3
United Kingdom	22	21	11	12
Denmark	1	1	1	1

(a) What is the general trend in steel production in these years?

(b) Which country has suffered the greatest decline in production between 1974 and 1983? Calculate that country's decline: in millions of tonnes; as a percentage of the 1974 production.

(c) State whether the following statements are true or false.

☐ Steel production has declined steadily between 1974 and 1983 in all the countries shown.

☐ The greatest period of production decline was between 1980 and 1983.

(d) Design a graph to illustrate the decline in steel production in the United Kingdom from 1974 to 1983.

2. Imagine that you are a Member of Parliament for the Sheffield area and that a steel factory is about to be closed. Outline three arguments which the British Steel Corporation (the owners of the factory) might put to you in favour of closing the factory. Reply to these points by outlining three arguments in favour of keeping the factory open.

3. With the help of your teacher, try to locate an older local industry which is now experiencing problems because of its location. Present a report of your findings to your class.
4. What is the location of Ireland's only steel-making factory? Why do you think it is located where it is? What problems does the Irish steel industry face at present? (hint: EC competition)
5. Write a paragraph entitled 'The Future of the British Steel Industry'.

6. A word game

L	O	R	R	A	I	N	E	A	X	R
A	F	H	Z	U	S	O	F	A	I	U
X	O	U	T	W	X	R	H	C	R	O
H	R	R	E	M	A	D	R	I	D	B
S	E	F	I	A	T	H	S	E	U	A
A	S	A	Y	R	Y	R	A	F	N	L
T	T	E	D	K	G	T	T	T	K	S
O	S	B	L	E	X	U	U	Y	I	S
P	B	C	C	T	I	A	R	Y	R	E
E	N	O	T	S	E	M	I	L	K	W
S	C	R	A	P	M	E	T	A	L	Z
T	I	N	T	E	G	R	A	T	E	D

Use these clues to find the names hidden in the puzzle. The names may be written top to bottom, left to right, right to left or bottom to top:

L_ _ _ _ _ _ _ A manufacturing area in France
N_ _ _ Another French manufacturing region
M_ _ _ _ _ _ An important factor in industrial location
L_ _ _ _ _ _ _ _ Used to aid the smelting of iron ore
I_ _ _ _ _ _ _ _ _ These steelworks carry out every stage of steelmaking
F_ _ Site of a large steelworks in France
D_ _ _ _ _ _ An important steelmaking port in France
L_ _ _ _ _ _ _ Another important factor in industrial location
S_ _ _ _ M_ _ _ _ A substitute for iron ore
F_ _ _ _ _ _ Early locations for iron-making
R_ _ _ An important industrial region in Germany
E_ _ _ _ A city in the region referred to in the last clue

32 WOMEN AT WORK

The role of women in manufacturing industry has changed in recent decades.

In 'developing' countries of the South

Most Third World women live in rural areas where there are few factories and other large-scale manufacturing industries. These women continue to spend their days caring for their children, fetching water and fuel, preparing food and tilling the land.

In Third World cities, however, more and more women find work in 'light' manufacturing industries such as textiles and clothing. They usually get the worst-paid jobs and are often exploited by the factory owners. Many women work long hours in bad working conditions for very low pay.

In 'developed' countries of the North

In the North, more and more women have found work outside the home in manufacturing and other economic activities.

But, like Third World women, they are usually paid less than men. Many women are also expected to work a 'double day'. Going 'out to work' in the morning, they return home to cook, clean and look after their children. Studies show that in the Soviet Union, for example, almost 50% of the husbands do no housework. In the United States, fathers spend an average of just 12 minutes per day with their children.

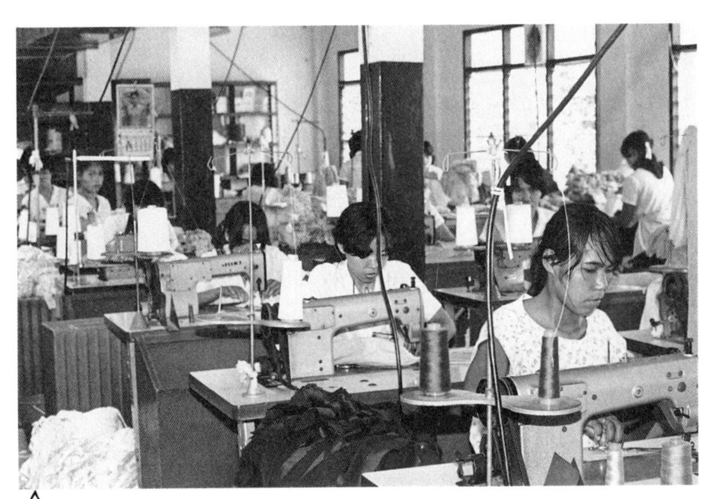

△
Describe the work being done by these poorly-paid women workers in the Philippines. Suggest two reasons why the factory owner has employed women rather than men.

△
The Irish women in this photograph own their own business. Contrast the work they are doing and their working conditions with those of the Filipino women.

Women in Industry

Case Study 1: Ireland

In recent years, some changes have taken place in the roles of women in Irish manufacturing industry. But these changes have not been as great as in some other European countries. In Ireland, women are still far from achieving industrial equality with men.

Changes have taken place

1. There has been a big increase in the number of women in ***paid employment.***

But...

- Many women work only in ***part-time*** jobs. Women make up 70% of all part-time workers.
- Most women continue to be employed in traditionally 'suitable jobs'—office work, catering and other ***service industries.*** The number of women working in manufacturing industry has actually declined in Ireland in recent years (figure 32.1). Eighty per cent of Irish industrial workers are males.

2 Within industry, more women are being employed in ***skilled and semi-skilled work.***

- Within industry, most women still do the ***lower paid jobs*** which have been traditionally seen as suitable female occupations. In engineering, for example, 80% of the bookkeeping but only 4% of the more technical jobs are done by women. The proportion of women who become managers is very small indeed.

3. Women's industrial wages have increased.

- Women's average wages are still ***only 60% those of men.***

Do Activity 1, page 250.

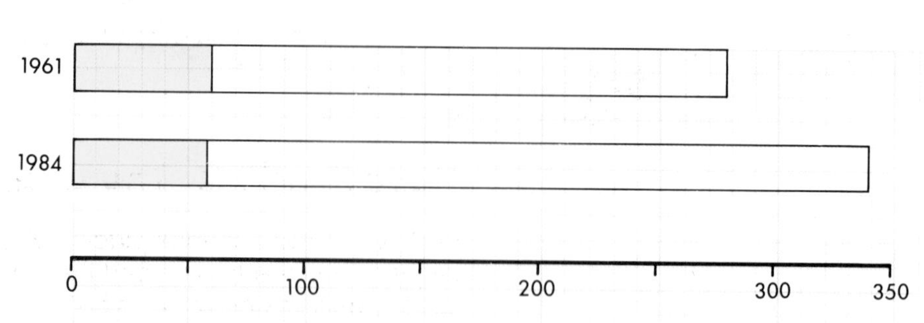

Figure 32.1 Women employed in Irish manufacturing industry: 1961 and 1984.

Describe the changes in the number of women employed between 1961 and 1984: (a) in the entire Irish workforce: (b) in manufacturing industry.

Employed in manufacturing industry

Factors which hinder change

The following factors have hindered change in the roles of women in Irish industry.

☐ Prejudice

Many Irish men—and women—still have set ideas about what they see as 'suitable' working roles for women. A recent EC survey showed that less than 50% of Irish people believe that women are just as capable as men of working as bus drivers, surgeons or TDs.

Now, Miss Murphy. Mary here will bring us some coffee. Then I'll introduce you to our interview board.

CHAIRMAN'S OFFICE

INTERVIEW BOARD

Figure 32.2 How many prejudiced ideas about male and female working roles can you see in this cartoon? Describe how the cartoon should be redrawn in order to eliminate all sexist bias (bias against males *or* females).

☐ Education

Young men and women tend to receive different types of educational training. More girls than boys study domestic sciences and service-training subjects such as typing and home economics. More boys than girls learn industrial subjects such as metalwork, car maintenance and woodwork.

☐ Domestic duties

Women rather than men are expected to perform the childminding and household duties in many homes. This causes many married women to leave the paid labour force at an early age.

What could be done to encourage more positive changes in women's industrial role?

- ☐ A countrywide ***educational campaign*** could be launched to break down people's sexist views on the roles of men and women in society in general and the workplace in particular.
- ☐ A wider ***variety of school subjects*** could be offered to both girls and boys.
- ☐ Industries could appoint ***'equality officers'*** to help ensure that females are not discriminated against in employment.
- ☐ More women could be placed on the ***interview boards*** which appoint people to jobs.
- ☐ . . . . .?

Case Study 2: The USSR

In 1918 a revolution took place in Russia. The leaders of the old Russian Empire were overthrown and the Soviet Union or the USSR (as it is now called) became the world's first communist country.

Since the Russian Revolution, the roles of women in Soviet industry and society have changed greatly.

Changes in Soviet Society

Before the revolution	Since the revolution
☐ Women had very ***few rights.*** They were expected to marry, rear children and obey their husbands in all things.	☐ The Communist government declared that both men and women were to have ***equal rights.***
☐ Most women received little formal ***education.*** Only 16% of women (compared to 40% of men) could read and write.	☐ Women and men received ***equal education.*** Now almost all Soviet women are literate.
☐ Relatively ***few women worked in industry.*** Those who did often worked for 12 hours each day doing boring, unskilled jobs. They were paid much less than those men who did similar work.	☐ Each woman—and man—has ***the right to a job.*** Men and women receive equal pay for equal work. Over 90% of Soviet women of working age are now engaged in fulltime work or study. They are employed in a wider variety of jobs than in any other country in the world (see figure 32.3).

Figure 32.3 Percentages of females and males in a selection of activities in the USSR

(a) State the percentage of women employed in each of the activities shown. (b) Suggest a reason for the employment of a greater percentage of women in catering than in agriculture. (c) Name *two* economic activities, not named here, in which relatively few women are *likely* to be employed.

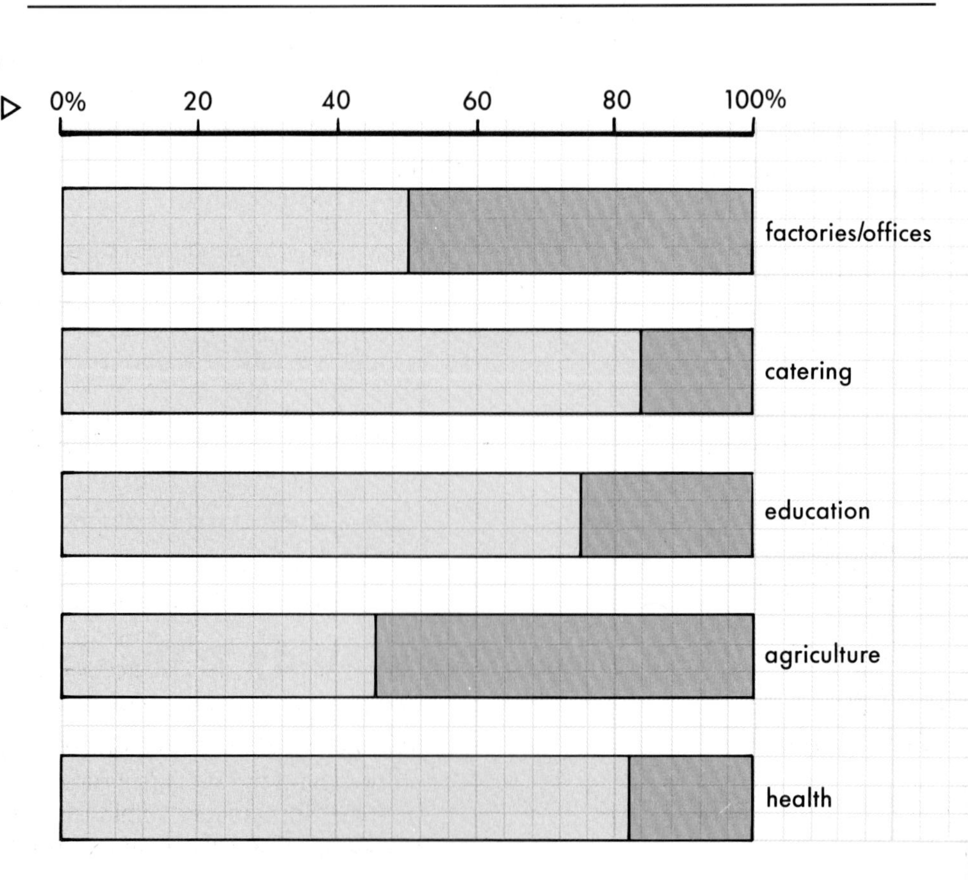

Why increasing numbers of Soviet women work outside the home

Here is a sample of the reasons given by Soviet women for working in industry and other activities outside the home. Rewrite these reasons in your own words, arranging them in three separate groups.

- ☐ those relating to financial ***reward***
- ☐ those relating to the care of ***children***
- ☐ those based on personal ***opinions*** regarding women and work

There are plenty of child-care centres for the children of working mothers. The state pays 80% of the cost of these centres.

When a woman works outside the home, I think her family looks up to her more.

Mothers of young children can work flexible hours to suit their needs.

It must be boring to stay at home when you can get an industrial job!

When a working woman has a child, she can take 18 months' part-paid leave.

My income is very helpful in keeping our family.

It's the patriotic duty of every adult to work for this country's economy.

Women get special pension concessions. A woman who has worked for 20 years can get a retirement pension at the age of 55.

Still not equal...

Women are playing an increasing role in Soviet industry. But in some respects, they have still not achieved equality with men.

1. On average, female industrial workers earn only 70% as much pay as do male industrial workers. This is so because:
   - ☐ Many women are employed in ***light industry*** such as radio engineering or watch-making. Pay is smaller in light industry than it is in male-dominated heavy industry.
   - ☐ A smaller proportion of women obtain high ***management positions.***
2. Women often have to carry out the ***double burden*** of factory work and household duties. On average, Soviet women do three times more housework than Soviet men. Housework often prevents women from taking part in the extra training courses which could lead to promotion to high management positions.
3. Women continue to be regarded as the 'natural' ***childminders*** in the USSR. If a woman finds it difficult to work both in industry and in the home as the main childminder, she (rather than her husband) is encouraged to abandon her industrial work.

Activities

1. Figure 32.4 shows the percentages of Irish female industrial employees in each of 6 job categories in 1972 and 1983. Study the figure and then say whether each of the statements relating to it is true or false.

Figure 32.4

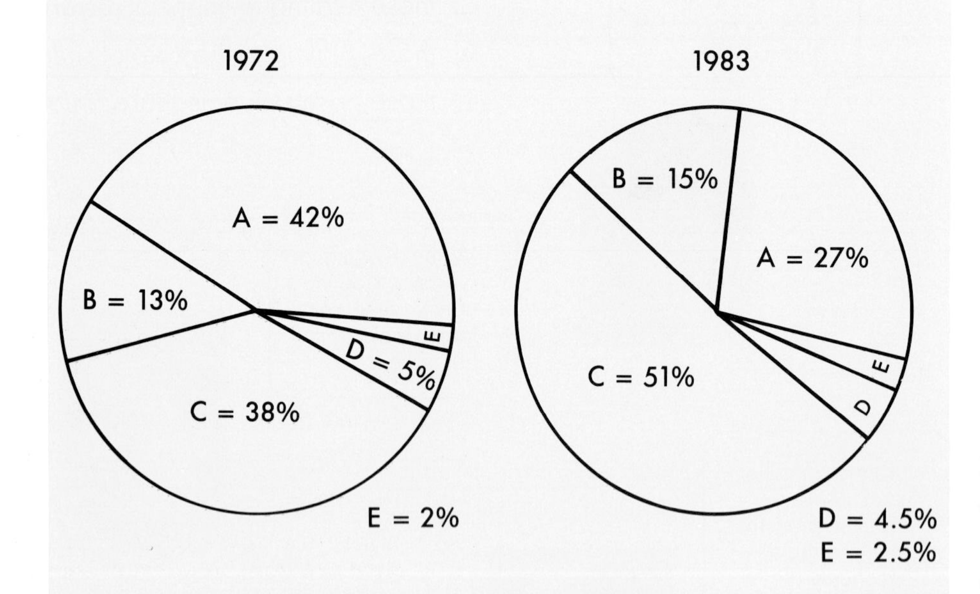

A = unskilled workers

B = semi-skilled and skilled workers

C = clerical (mostly office) workers

D = technicians, supervisors and professional

E = managers

(a) The roles of women have changed greatly in each of the job categories shown.

(b) The percentage of female managers and supervisors remains very small.

(c) Women continue to form the majority of clerical workers.

(d) The percentage of women doing 'traditional' clerical work appears to have increased considerably.

(e) The percentage of women managers has increased only very slightly.

(f) The only big change in the roles of women has been a decline in their roles as unskilled workers and an increase in their roles as clerical workers.

2. Study the statistics in figure 32.5 and then answer the questions which follow.

Figure 32.5

Male/Female composition of the workforce at a Soviet textile factory

Job description	*Male*	*Female*	*Total*
Director (manager)	1	–	1
Other managerial staff	6	3	9
'Factory floor' workers	23	64	87
Catering staff	1	4	5
Cleaning and maintenance staff	2	6	8

(a) What is the total workforce of this factory?

(b) Calculate the percentage of the total staff which is female.

(c) What percentage of the total management staff is female?

(d) Try to account for the contrast between the answers to parts (b) and (c).

3. Study the cartoon and then answer the questions.

(a) What is the message of this cartoon?

(b) Write a suitable caption for the cartoon. This could be in the form of a comment from the woman, a comment from the man, or a general comment on the situation.

(c) Does the cartoon give a fair or a prejudiced view of male and female roles? Explain your answer.

33 A WORLD OF DIFFERENCE: CLASSIFYING REGIONS ACCORDING TO INDUSTRY

Countries or large regions are sometimes classified as industrialised, newly industrialised or industrially emergent (see figure 33.1).

- **Industrialised** — These countries or regions contain many ***long-established industries.*** They have relatively large proportions of their people working in manufacturing industry. Industrialised countries are located mainly in the prosperous North (the First World and Second World).
- **Newly industrialised** — Many of these countries or regions still have relatively small proportions of their people working in manufacturing. But they also have many newly-established industries and their ***percentage rates of industrial growth are rapid.*** Almost all newly industrialising countries are in the South or the Third World.
- **Industrially emergent** — These countries or regions are ***not industrialised.*** Their percentage rates of industrial growth are slow. All of these countries or regions are in the South.

Figure 33.1
▽

Activities

1. Figure 33.1 shows some of the industrialised, newly industrialising and industrially emergent countries of the world. Study figure 33.1 and then do the activities which follow.

   (a) Say whether each of the following *regions* is mainly industrialised, newly industrialised or industrially emergent: (i) Central Africa; (ii) Southeast Asia; (iii) the Middle East (the region around the Iranian Gulf); (iv) Central America (the region between the United States and South America); (v) Western Europe; (vi) Eastern Europe and the USSR.

   (b) Say whether each of the following *countries* is industrialised, newly industrialised or industrially emergent: (i) France; (ii) the Sudan; (iii) Bolivia; (iv) Brazil; (v) Taiwan; (vi) the Philippines; (vii) Mali; (viii) Sweden; (ix) El Salvador

   (c) Describe the general location of 'the industrialised world'.

2. Some countries have deliberately been left blank in figure 33.1. Shade in each of these countries according to whether they are industrialised, newly industrialised or industrially emergent. The information in figure 33.2 will help you to decide the category of each country.

Figure 33.2 ▷

Category	**Signs of category as shown in figure 33.2**
Industrialised	More than 20% of the working population is engaged in industry. Annual industrial growth rate is less than 5%.
Newly industrialised	Annual industrial growth rate is more than 5%.
Industrially emergent	Less than 10% of the working population is engaged in industry. The annual growth rate is less than 5%.

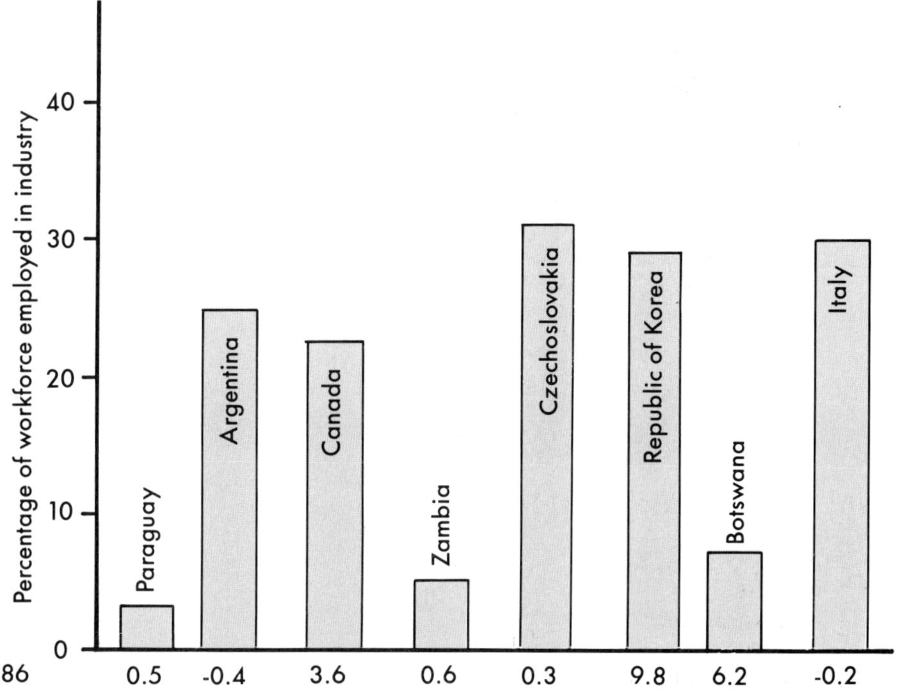

34 ACID RAIN: A BURNING ISSUE

The picture in figure 34.1 shows the new ESB power station at Moneypoint in Co. Clare. Each year, the Moneypoint station will burn 2 million tonnes of coal to produce vast amounts of electricity for Ireland.

Many people, however, fear that large coal-burning plants such as the one at Moneypoint will cause great harm to the environment. They point out that, every ten minutes, the giant chimneys at Moneypoint could emit 1 tonne of sulphur dioxide into the air. This, they say, will greatly increase the amount of acid rain being caused in Ireland.

Figure 34.1 The coal-burning power station at Moneypoint, Co. Clare. Would air pollution from the Moneypoint plant be likely to pass over much of Ireland? Study the map and explain your answer. ▽

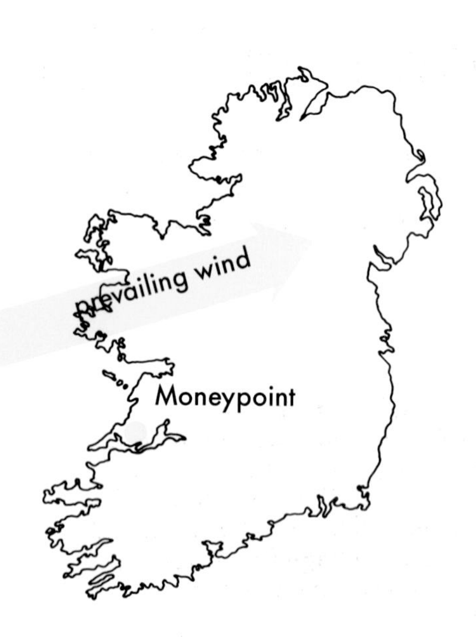

What is acid rain?

Figure 34.2 How acid rain is formed ▽

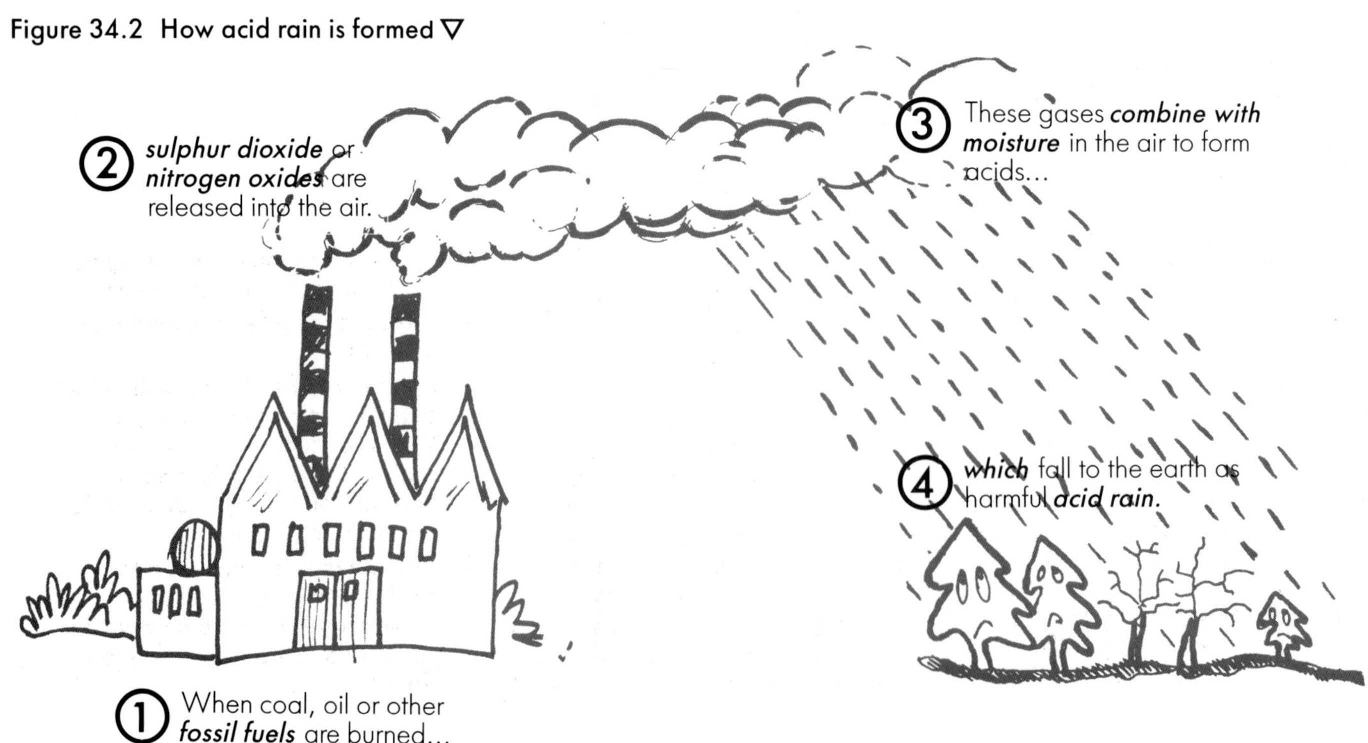

Which of the following are likely to be direct causes of acid rain?

- ☐ turf-burning power stations
- ☐ bicycles
- ☐ electric heaters
- ☐ cars
- ☐ log fires
- ☐ solar-powered electricity stations
- ☐ heavy- oil-burning industries
- ☐ coal fires
- ☐ light industries (those fuelled by electricity)

The effects of acid rain

People are becoming more and more concerned about their environment. One of the greatest concerns is acid rain, especially in high fuel-using industrial regions such as Western Europe. The following newspaper article describes the effects of acid rain.

Acid rain—the silent destroyer

In Sweden they call it *Waldsterben* or 'forest death'—the widespread destruction of trees by acid rain. Conifers suffer especially. Their roots are poisoned; they lose their needle-foliage; their branches become distorted; their twigs drop; their barks split; they stop growing. They die.

Not only in Sweden do forest-owners, foresters and conservationists lament the slow deaths of their woodlands. Almost every country in Europe is affected. Over half of West Germany's trees are said to have been damaged by acid rain. Two-thirds of Britain's conifers are sick. In Ireland, the first signs of acid rain damage to trees have already been spotted.

Acid rain has also caused widespread damage to Europe's lakes—many to the point where they no longer support fish life. In Norway, where 15 000 lakes have been affected, lime is sometimes added to lake waters to reduce acidity. But this is a mere stop-gap measure. It does nothing to prevent the continued production of acid rain.

Acid rain washes valuable nutrients from the soil and so reduces soil fertility. Resulting crop losses can be high. In Scotland, losses are estimated to be £25 million per year. Throughout the whole United Kingdom, annual losses are put at £200 million.

When forests, water supplies and soils suffer, so does the natural balance of the plant and animal kingdom. Woodland animals, deprived of a healthy habitat in which to live, die in their millions. Some plants, such as certain types of mushrooms, are wiped out by excess sulphur and nitrogen in the soil. Others, such as nettles, spread at an alarming rate.

People, too, feel the direct physical effects of acid rain. It is thought to be one of the main causes of bronchitis and lung cancer. It is also being blamed for a new disease known as 'pseudo croup'—a sickness which affects children particularly and which has occasionally proved fatal.

Even great buildings suffer the erosive attacks of our polluted rainwater. The most famous building in Athens—the Parthenon—has decayed more in the past 30 years than it did in the previous 2400. Nearer home, the discoloured faces of buildings such as Trinity College Dublin remind us of the destructive power of acid rain.

Use the information in the newspaper article to answer these questions.

1. Describe the effects of acid rain on forest life.
2. Explain briefly the possible effects of acid rain on tourism: (a) in West Germany's Black Forest region; (b) in the lakelands of Central Ireland; (c) in an historic city such as Athens.
3. How does acid rain affect the *quality* of human life? In your answer, refer to people's health and one other aspect of human life.

A statue at Persepolis, home of the ancient Persian Empire (modern Iran). Describe how acid rain has affected this monument. ▷

Acid rain: an international problem

One of the difficulties with acid rain is that it is an ***international problem***. Sulphur and nitrogen pollutants from one country can travel through the air for hundreds of kilometres and over several other countries before descending elsewhere as acid rain (figure 34.3). ***Countries often pollute each other*** with the acid rain which they produce.

Because of this, only an agreement between the industrialised countries can hope to succeed in greatly reducing the amount of acid rain which is produced. However, such ***international agreements*** on the reduction of pollution are often difficult to achieve. This can be seen by the information in the box entitled 'The 30% Club'.

The 30% Club

The 30% Club consists of ***countries which are pledged to reducing*** the international effects of ***acid rain***.

Each member-country of the 'club' promises to reduce the amounts of sulphur pollution it produces by 30% of its 1980 level.

Fulfilling such a promise can be costly. It may require, for example, that expensive 'scrubbers' be installed in coal-burning industrial plants. The scrubbers filter sulphur out of smoke and so prevent sulphur dioxide from entering the atmosphere.

The 30% Club has been seeking members for several years. So far, however, only 16 countries have fully pledged their support. Ireland is not one of them.

Why do you think Ireland has refused to join The 30% Club? What do you think of this refusal?

Figure 34.3 A recent study carried out at Valentia in Co. Kerry shows that the problem of acid rain is an international one. The study found that sulphur dioxide deposited through the atmosphere at the Valentia weather station came from several different countries. (a) Describe precisely the information shown in the figure. (b) Draw a similar figure to show the following origins of nitrogen oxides deposited at Valentia: United Kingdom—36%; West Germany—22%; France—12%; Ireland—4%; Others or unknown—26%. ▷

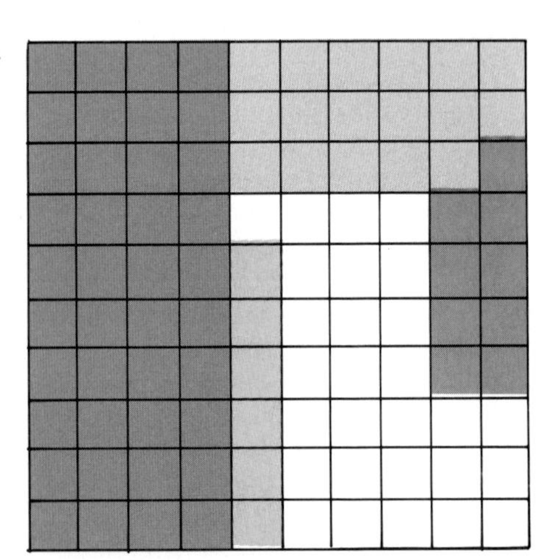

Some solutions to acid rain

Here is a selection of some of the suggestions for government or EC actions to solve the problems of acid rain in Europe.

- ☐ Identify the 3 suggestions which you think are most worthy of support. Explain the reasons for each of your choices.
- ☐ Identify the 3 suggestions which you think are of least value. Give your reasons for rejecting these suggestions.

Activities

1. (a) Explain how acid rain is formed.
   (b) Apart from producing acid rain, manufacturing industry can damage the environment in a number of other ways. Briefly describe 4 such ways. The cartoon in figure 34.4 will help you.

Figure 34.4

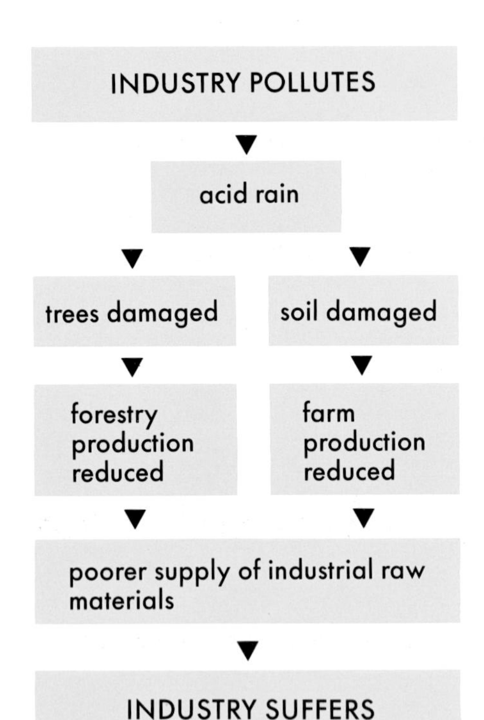

△Figure 34.5

2. *Manufacturing industry suffers because of its own pollution.*
   Study the links or connections between the various parts of the diagram in figure 34.5. Then speak or write briefly on the statement above.

3. The table in figure 34.6 shows sulphur dioxide emissions from a number of European countries in 1980, together with forecasted emissions from these countries in 1995.
   (a) Make three separate *rank order lists* of these countries: (i) according to the amounts of their emissions in 1980; (ii) according to the amounts of their forecasted emissions for 1995; (iii) according to the percentage reductions of their emissions between 1980 and 1995.
   (b) How does the *trend* of Ireland's emissions contrast with those of all the other countries shown in the table? Can you suggest one reason for the trend of emissions in Ireland?
   (c) On squared paper (graph paper) draw two *bar graphs*; one to illustrate Ireland's sulphur dioxide emissions in 1980, the other to illustrate its emissions in 1995. Draw both bar graphs to the same scale.

Annual emissions of sulphur dioxide in thousands of tonnes

Country	*1980*	*1995*	*% Change*
Belgium	799	544	−70
Denmark	438	229	−52
West Germany	3200	1100	−66
Hungary	1633	1140	−30
Ireland	219	251	+15
Norway	141	84	−40
Sweden	483	156	−68

△Figure 34.6

35 THEM AND US: PROBLEMS WITH INDUSTRY

Conflicts of interest and differences of opinion often arise between industrialists and others. In this chapter we will examine:

- ☐ the controversy which arose when the owners of a big American company wanted to set up a ***chemical factory*** on the outskirts of an Irish village.
- ☐ the differences of opinion between employers and workers on whether or not ***working hours*** should be reduced.

Case Study 1: Merrell Dow and the divided village

Merrell Dow is an American pharmaceutical (drugs) company. It is part of the Dow Chemical Corporation—a giant multinational company which has factories in eleven different countries.

In 1988, Merrell Dow applied for permission to open a new IR£30 million factory on a 40ha site, just outside Killeagh village in Co. Cork. There, it intends to manufacture drugs which can be used for the treatment of asthma and heart failure.

The company's plans caused a great controversy which divided the people of Killeagh and the surrounding area. Many local people supported the factory. Many others opposed it. Some of the arguments, both for and against the Merrell Dow factory, are shown here and on the next two pages.

The amount of air pollution will be slight. Waste will enter the atmosphere through a high chimney stack, which will allow the wind to blow any pollution away from the local area.

This river often floods part of my dairy farm. What if the chemical waste spreads out over my grass, which is consumed by my cows, and then on to the people who drink the milk? There's a great danger to human health here!

Cork County Council studied the arguments on both sides. It then gave planning permission for the plant, while adding 27 strict conditions to protect the environment. The decisions of our own County Council experts should be accepted. We need this plant.

Only a minority of people have actually objected to the plant. The objectors do not know what they are talking about. Ignore them!

The 'experts' are sometimes wrong in their predictions about the effects which chemical plants can have on our environment. We can't be certain what the effects of this factory will be. Many Americans didn't want this plant in ***their*** backyard—and this worries me!

If this factory were owned by local people, we could trust the owners to protect the local environment. But big foreign companies are only interested in quick profits. They cannot be trusted!

I work at a big local co-operative creamery. If dairy farming is threatened, so is the future of our creamery and our cheese factory. If these close, hundreds of people will be without work.

I think chemical plants should be set up in industrial zones—not in the middle of rich agricultural areas.

The Dow Chemical Corporation has had a bad record and should not be trusted by anyone! In the past, it produced the deadly 'Agent Orange'—a chemical weapon used by the United States in their war against Vietnam. Agent Orange destroyed much of the Vietnamese countryside, and people who were near it have had many health problems; their children are also suffering from birth defects.

Heavy pollution by the Dow plant in the state of Michigan has been a matter of concern for the US Environmental Protection Agency.

The factory will emit sulphur dioxide into the air. This could cause acid rain.

The Merrell Dow proposals were examined by EOLAS—an official environmental investigation body set up by the Irish government. The proposals were also studied by a group of environmentalists from University College Cork. Both of these groups of experts concluded that the factory should go ahead. We should trust their expert opinions.

Activities

On your own

- ☐ Identify any arguments which you think are based solely on prejudice, and which are therefore valueless.
- ☐ Consider the remaining arguments carefully. Then write a report or prepare a short speech outlining what you think is the best solution to the situation. Explain why you think this is a good solution.

With your class

- ☐ Listen to some of the reports or speeches for and against the establishment of the Merrell Dow factory.
- ☐ Discuss the various views expressed.
- ☐ Take a class vote on whether or not the company should be granted permission to build its factory on the outskirts of Killeagh.

Case Study 2: Should working hours be reduced?

Since 1950, there has been a gradual reduction in industrial working hours throughout much of Europe. Many workers and trades unions argue that the trend towards reduced working hours should continue. Many employers and factory owners disagree.

Some arguments for and against

Here are two 'Letters to the Editor' as they might appear in an Irish newspaper. One favours the reduction of working hours, while the other opposes it. Study the letters and then do the activities.

Dear Sir/Madam:

I support fully the call of our trades unions for shorter working hours in Irish industry.

For over a decade, industry and industrialists have benefited enormously from the use of computers and other advanced technology. This new technology has greatly increased productivity (the amount of goods being produced per hour worked). As productivity has risen, so have the profits of the factory owners. With the help of new technology, we workers are producing more and more goods in less and less time. Our working hours should be reduced, but our pay must stay the same. This is only fair, as we workers should also share in the benefits of the new technology. After all, we are the ones who do the work, not the factory owners.

Many European countries are now developing two-tier societies, with some people working long hours and many others having no work at all. This is creating a dangerously divided society of 'haves' and 'have nots'. If working hours were shortened, existing work could be shared among more people, thus reducing unemployment. This has happened in France, where a mere one-hour reduction of the working week, from 40 to 39 hours, has meant an increase in jobs. According to a French government survey, 70 000 new jobs were created simply by reducing the working week by one hour!

When people work very long hours, they often become bored, exhausted and even unhealthy. This causes their productivity to decline. Shorter working hours are therefore not only desirable from the point of view of the workers. They would also bring about an increase in productivity, and so benefit factory owners as well. Studies have shown that there is always, or nearly always, a link between the amount by which working time is reduced and the amount by which productivity is increased.

Yours sincerely,
Sheila A....
(trade union member)

Dear Editor:

Industrial employers in Europe simply cannot afford to agree to a further reduction in working hours. If working hours are reduced—with wages remaining the same—the cost of manufacturing our products will inevitably increase. With dearer finished products, how can we hope to compete with cheaper goods produced by our rivals in the United States and Japan? In these countries, the average working week is already much longer than it is in most countries in Western Europe. If this happens, our sales will decline; many of our factories will close, leaving many of our workers without jobs.

Even in the short term, a cut in working hours will not greatly increase the number of jobs available. Surveys in Sweden and West Germany have shown that, when weekly working hours were reduced from 45 to 40 hours per week, very few extra jobs were created. In fact, when working hours were reduced, many people continued to work the same number of hours—but with the extra hours being classed as higher-paid overtime.

The real crux of this matter is that workers' trades unions—as always—are demanding something for nothing. If workers agreed to a decrease in wages, shorter lunch breaks or speeding up their pace of work, we employers might find it possible to agree to shorter working hours. But the workers' demand of 'something for nothing' is totally unrealistic. It is a demand which cannot possibly be met.

Yours sincerely,
Jonathan M....
(factory owner)

△
Figure 35.1 An advertisement in an Irish newspaper

(a) On behalf of which organisation was this advertisement published?

(b) Summarise in one sentence the main information contained in this advertisement.

(c) What does the advertisement hope to achieve?

Activities

1. Each of the letters on page 262 contains *three arguments* either for or against a reduction in working hours. In the case of *each argument*, summarise its main point in one sentence.

2. Quote one sentence from each letter which reveals a *prejudice* on the part of the writer against either factory owners or trades unions.

3. It is sometimes said that 'surveys can be used to prove completely contradictory things'. Identify an example in the letters where the results of the surveys appear to contradict each other.

4. *Testing an argument 1*
   In her letter, Sheila A claims that *'there is always, or nearly always, a link between the amount by which working time is reduced and the amount by which productivity is increased'*.
   Test the accuracy of this claim by studying figure 35.2 and doing the activities which follow it.

Figure 35.2 Changes in productivity and working hours ▷

Changes in productivity and working hours in a number of industrialised countries (1960-1986)

Country	Annual % gains in productivity	Annual % reduction in working hours
Belgium	6.3	0.87
Canada	3.3	0.16
France	5.2	0.65
West Germany	4.6	0.46
Netherlands	5.9	0.74
Norway	3.2	0.09
Sweden	4.8	0.73
United States	2.8	−0.09

5. Use the information in Table 35.2 to make two side-by-side *rank order lists* of the countries shown.
   - ☐ Make one rank order list of the countries according to their productivity gains. (Begin with the country which shows the *greatest percentage productivity gain.*).
   - ☐ The second rank order list should present the countries according to the reduction in working hours. (Begin with the country which shows the *greatest percentage reduction*).
   - ☐ Compare the two rank order lists. Do they appear to support or to contradict the worker's claim that *'...there is always, or nearly always, a link between the amount by which working time is reduced and the amount by which productivity is increased'?*
     Explain your answer.

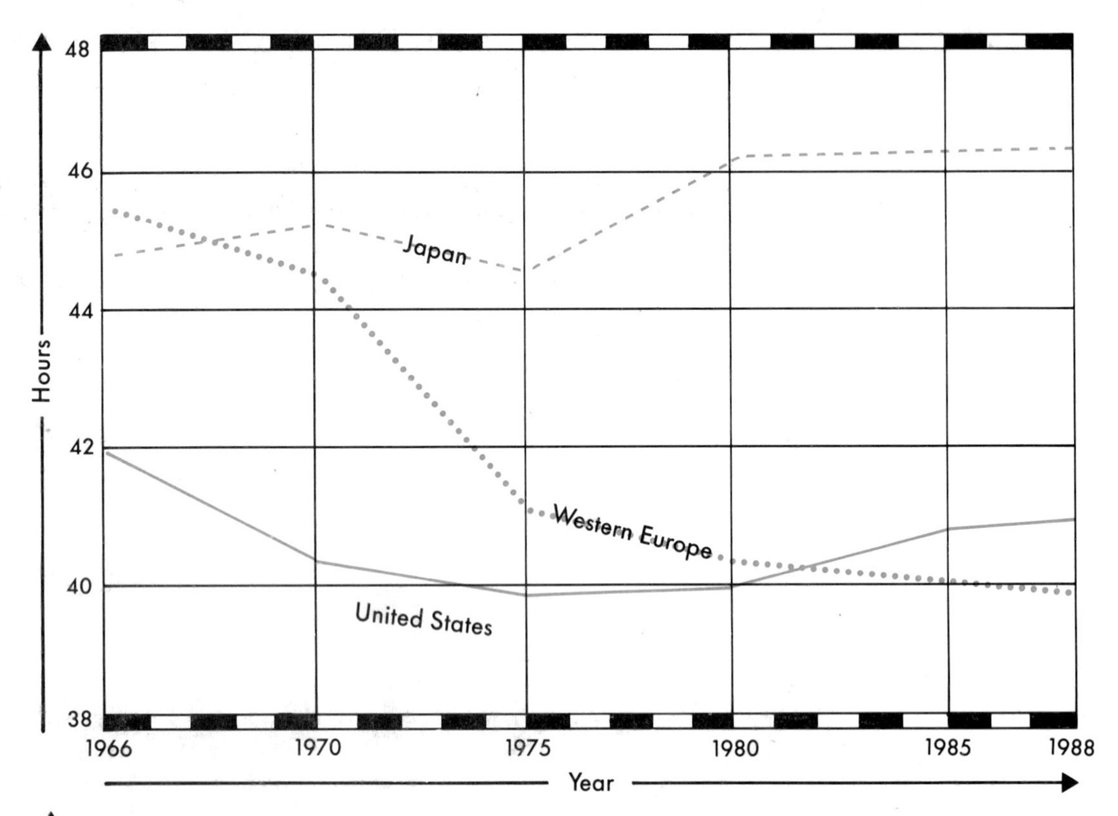

△
Figure 35.3 Trends in weekly industrial working hours in Western Europe, the USA and Japan

6. *Testing an argument 2*
In his letter, Jonathan M claims that average working weeks are already *'much longer'* in the United States and Japan than they are in Western Europe.
Test the accuracy of this claim by studying figure 35.3 and then doing the first activity which follows it.

(a) Calculate the lengths of the industrial working weeks in Western Europe, the USA and Japan for 1988. Do these figures appear to support or contradict the factory owner's claim that average working weeks are *'much longer'* in the USA and Japan than they are in Western Europe? Explain your answer.

(b) Describe how the *overall trend* of working hours in Western Europe differs from the overall trends in both Japan and the USA.

(c) Identify the years during which each of the following occurred:
- ☐ Western Europe's working week became shorter than that of Japan.
- ☐ The working weeks in Western Europe and the USA were of equal length.
- ☐ The working week in the USA became shorter than 40 hours.
- ☐ The greatest difference existed between the working weeks in the USA and in Western Europe.